Can UFOs Advance Science?

Making the Case for a New Electromagnetic Technology

UPDATE 2021

SUNRISE INFORMATION SERVICES

CAN
UFOs
ADVANCE
SCIENCE?

Making the Case for a New Electromagnetic Technology

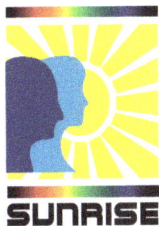

SUNRISE Information Services
Canberra, Australia

ISBN 978 0 6485860 4 3

Printed by Lightning Source
www.lightningsource.com.au

Designer, Illustrator and Typeset by
SUNRISE

To everyone who thought it was impossible

CONTENTS

ACKNOWLEDGMENTS

We would like to thank Linda Brown for her advice and in granting us permission to reproduce some photographs of her father, Thomas Townsend Brown.

The artist impressions of UFOs used in this book were originally published in *The Unexplained – Mysteries of Mind, Space & Time* (1980-84) by Orbis Publishing Ltd in London. As Orbis has been taken over by the Italian publishing firm De Agostini UK Ltd. and with no reply to our e-mails or indications on their website that they own (or are selling) the publication, we believe the publication no longer exists. So, for scientific research purposes, we have decided to republish these artist impression images.

We have also great admiration for the illustrative skills of Sergio Drummond (cartoons and the artist impression picture for the Cash-Landrum UFO case), Edu Torres (for his superb rendition of the Finland UFO case), and Goce Ilievski (artist impression picture of the UFO from Japan used on the book cover). In this edition, SUNRISE has also commissioned 3D artists to accurately recreate environments and UFOs based on where they occurred and the drawings submitted by the witnesses.

And finally we like to thank the handful of brave scientists, and all the UFO investigators and researchers, over the last 60 years for releasing their work for all to see. Without these people, this book may never have been written.

CHAPTER 1

Do UFOs Exist?

[A UFO relates] to any airborne object which by performance, aerodynamic characteristics, or unusual features does not conform to any presently known aircraft or missile type, or which cannot be identified as a familiar object.

—USAF Regulation No. 200-2 released on August 26, 1953.[1]

I F ANY person or organisation knows whether UFOs exist, it has to be the United States Air Force (USAF). As the above quote from 1953 makes it clear, whether we like it or not, there are objects in the skies that are known and unknown in nature. In terms of the unknown objects, scientists categorise them as "UFOs" (an acronym for Unidentified Flying Objects)[2]. Until a better explanation comes along to explain what these enigmatic objects are, this is currently the most scientific way to begin an inquiry into the nature of something unknown, in this case UFOs.

---

1   Good 1997, p.344 and Sachs 1980, p.7.
2   The USAF has created in recent times a new term called Unidentified Aerial Phenomena (UAP) just to add to the knowledge on the subject, but at the end of the day, it should all be called UFOs.

So yes, UFOs exist.

To give examples of two flying objects that have remained UFOs to this day, consider the following cases:

Case 1: Mansfield, Ohio, USA[3]

On October 18, 1973, at 11:05 p.m., near Mansfield, Ohio, USA, four Army Reserve crewmen in a Bell Huey helicopter tried desperately to avoid a mid-air collision with a UFO heading directly toward them.

The crew members quickly put the helicopter into a power descent from 760 meters to 518 meters in an effort to avoid colliding with the mysterious flying object. But somehow, as the crew were distracted by the large UFO—described as "cigar-shaped, metallic grey, with a dome on top" floating just outside and above the helicopter for some ten seconds and filling the cockpit with a greenish haze—the helicopter was lifted to 1,067 meters and continued to climb until the UFO moved away.

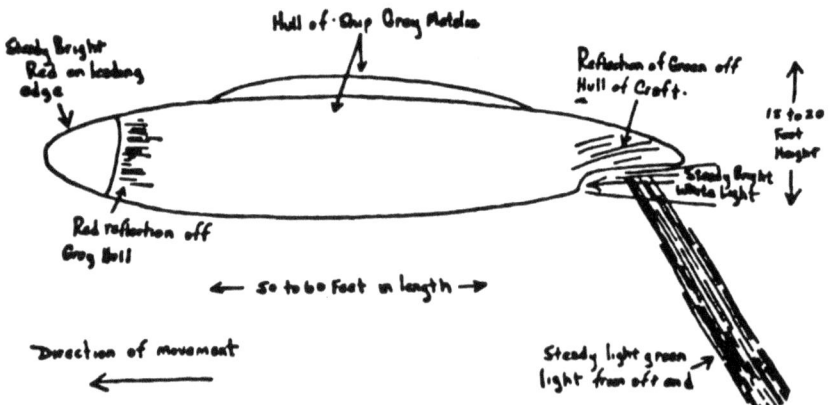

Drawing of UFO by Captain Lawrence Coyne.

---

3   Story 1980, pp.93–95; Randles 1987, pp.102–105; Flammonde 1976, pp.380–382; Stoneley 1976, p.203.

For use of this form, see AR 340-15; the proponent agency is The Adjutant General's Office.

REFERENCE OR OFFICE SYMBOL	SUBJECT
	Near Midair Collision with UFO Report

| TO Commander
83D USARCOM
ATTN: AHRCCG
Columbus Support Facility
Columbus, Ohio 43245 | FROM Flight Operations Off   DATE 23 Nov 73   CMT 1
USAR Flight Facility
Cleveland Hopkins Airport
Cleveland, Ohio 44135 |

1. On 18 October 1973 at 2305 hours in the vicinity of Mansfield, Ohio, Army Helicopter 68-15444 assigned to Cleveland USARFFAC encountered a near midair collision with a unidentified flying object. Four crewmembers assigned to the Cleveland USARFFAC for flying proficiency were on AFTP status when this incident occurred. The flight crew assigned was CPT Lawrence J. Coyne, Pilot in Command, 1LT Arrigo Jezzi, Copilot, SSG Robert Yanacsek, Crew Chief, SSG John Healey, Flight Medic. All the above personnel are members of the 316th MED DET(HEL AMB), a tenant reserve unit of the Cleveland USARFFAC.

2. The reported incident happened as follows: Army Helicopter 68-15444 was returning from Columbus, Ohio to Cleveland, Ohio and at 2305 hours east, south east of Mansfield Airport in the vicinity of Mansfield, Ohio while flying at an altitude of 2500 feet and on a heading of 030 degrees, SSG Yanacsek observed a red light on the east horizon, 90 degrees to the flight path of the helicopter. Approximately 30 seconds later, SSG Yanacsek indicated the object was converging on the helicopter at the same altitude at a airspeed in excess of 600 knots and on a midair collision heading. Cpt Coyne observed the converging object, took over the controls of the aircraft and initiated a power descent from 2500 feet to 1700 feet to avoid impact with the object. A radio call was initiated to Mansfield Tower who acknowledged the helicopter and was asked by CPT Coyne if there were any high performance aircraft flying in the vicinity of Mansfield Airport however there was no response received from the tower. The crew expected impact from the object instead, the object was observed to hesitate momentarily over the helicopter and then slowly continued on a westerly course accelerating at a high rate of speed, clear west of Mansfield Airport then turn 45 degree heading to the Northwest. Cpt Coyne indicated the altimeter read a 1000 fpm climb and read 3500 foot with the collective in the full down position. The aircraft was returned to 2500 feet by CPT Coyne and flown back to Cleveland, Ohio. The Flight plan was closed and the FAA Flight Service Station notified of the incident. The FSS told CPT Coyne to report the incident to the FAA GADO office a Cleveland Hopkins Airport MR. Porter, 83d USARCOM was notified of the incident at 1530 hours on 19 Oct 73.

3. This report has been read and attested to by the crewmembers of the aircraft with signatures acknowledgeing this report.

REPLACES DD FORM 96, EXISTING SUPPLIES OF WHICH WILL BE ISSUED AND USED UNTIL 1 FEB 63 UNLESS SOONER EXHAUSTED. ☆U.S. GPO: 1972—473-063 P.9. 14

Signed report from the witnesses, released under Freedom of Information.

3

"I could hardly believe it," said Captain Lawrence Coyne, the pilot in command. "The altimeter was reading 3,500 feet [1067 meters], climbing to 3,800 [1158 meters]. I had made no attempt to pull up. All the controls were still set for a 20-degree dive. Yet, we had climbed from 1,700 feet [518 meters] to 3,500 with no power in a couple of seconds, with no g-forces or other noticeable strains. There was no noise or turbulence either."

Witnesses on the ground later told investigators that they saw a UFO "as big as a school bus" hovering above the helicopter and shining a green ray over it. Coyne and his colleagues filed a report soon after the incident.

Case 2: Levelland, Texas, USA[4]

At 10:30 p.m. on November 2, 1957, Pedro Saucedo, a 30-year-old Korean War veteran and farm worker, was driving his truck some six kilometers west of Levelland, Texas, with his friend Joe Salaz, when a very bright, yellow-and-white glowing object traveling at high speed slowed as it passed directly overhead, causing the truck's headlights to go out and the engine to stall.

Saucedo said:

> "I saw a big flame, to my right, front, then I thought it was lightning. But when this object had reached to my position, it was different, because it put my truck motor out and lights [too]. Then I stop, got out, and took a look, but, it was so rapid and quite some heat, that I had to hit the ground."[5]

The UFO, shaped like an oblate spheroid, then accelerated and disappeared into the distance. Saucedo's truck soon regained power, and both witnesses, bewildered by their experience, drove on for some distance before telephoning the police headquarters in Levelland.

Interestingly, on the same night, presumably the same oblate spheroid object made its triumphant return to repeat its performance of cutting out, rather inconveniently, all electrical power to motor vehicles and trucks on the outskirts of Levelland before quickly

4   Brookesmith 1984 *The Age of the UFO*, pp.68–69; *Mysteries of the Unknown: The UFO Phenomenon* 1987, p.67; Maney and Hall 1961, pp.73–75; Story 1980, p.210; Peebles 1994, pp.119–121; Flammonde 1976, pp.339–344.
5   Peebles 1994, p.119.

disappearing. At least two dozen drivers complained to the Levelland police.

On one occasion, the oval object's exterior glowing effect began pulsating, and the driver who observed this noticed that his car headlamps went on and off, apparently in synchronisation with the pulsating light.

During another incident on the same night, a truck driver noticed a strange orange oval object sitting on the road ahead. It changed in colour from a fiery orange to a bluish green and then eventually climbed upward. It hovered at an unspecified height above the ground and changed its colour back to orange. The driver's truck then regained its power.

On yet another occasion, Newell Wright, a student from Texas Technical College, driving east of Levelland at 12:05 a.m. on the same night, suddenly saw his ammeter jump to "discharge" then return to normal at the same time his engine cut out. Thinking a major fault had occurred in his car, he got out, raised the hood and inspected the engine, but could find no fault with it. As soon as he slammed the hood down, he noticed, for the first time, a bluish-green oval on the road in front of him. Unable to escape in his car, he sat in the driver's seat for a couple of minutes, watching and waiting to see what the object would do. Then, abruptly the UFO leapt skyward and took off toward the north. As soon as the UFO was gone, Wright's car was able to be restarted.

Apart from the possibility of ball lightning, authorities have remained baffled by this case to this day.

UFOs are a Worldwide Phenomenon

UFOs like the ones you have just read about are a worldwide phenomenon. They have been observed by people from virtually all walks of life, including purportedly rational types such as police officers, military pilots, astronauts, politicians and, perhaps surprisingly, scientists[6].

Major-General John Samford

Even the former Director of Intelligence at the Pentagon, Major-General John Alexander Samford (1905–1968), agrees with this view when he said:

> "Reports have come in from credible observers of relatively incredible things."[7]

Lord Hill-Norton, Chief of Defence Staff at the U.K. Ministry of Defence in 1973 and member of the Military Committee of NATO between 1974 and 1977, also gave unequivocal support to this view when he said:

> "The evidence that there are objects which have been seen in our atmosphere, and even on terra firma, that cannot be accounted for either as man-made objects or as any physical force or effect known to our scientists, seems to me to be overwhelming....A very large number of sightings have been

---

6   In the conclusion, we will learn what it would take for scientists to properly study UFOs even despite the fact that some scientists have already observed the phenomenon.

7   Ruppelt 1956, p.166; Blundell & Boar 1984, p.11; Peebles 1994, p.65.

Dr. Carl Sagan

vouched for by persons whose credentials seem to me unimpeachable. It is striking that so many have been trained observers, such as police officers and airline or military pilots. Their observations have in many instances...been supported either by technical means such as radar or, even more convincingly, by...interference with electrical apparatus of one sort or another...."[8]

Not even the scientific community can deny the existence of UFOs. As the late Dr. Carl Edward Sagan (1934–1996), a distinguished astrophysicist and professor of astronomy at Cornell University, once said:

"I do believe there are objects which have yet to be identified."[9]

This is clearly not a question of whether UFOs exist. We know they exist as the examples we have seen above are testament of this fact. The more fundamental issue we have to deal with here is, what are UFOs?

---

8   Quoted from his foreword to ;*Above Top Secret* by Timothy Good (Morrow & Co's Quill Books) published in 1988.
9   Quote from Associated Press in Los Angeles published in the article "Prof. says Beings from Outer Space have visited Earth". November 26, 1962.

CHAPTER 2

What Are UFOs?

In my opinion, the UFO problem...constitutes an area of extraordinary scientific interest.

—Dr. James E. McDonald, Professor of
Atmospheric Sciences[1]

K NOWING THAT UFOs are a normal part of life, the scientific community has made some efforts to understand the nature of these mysterious denizens of the sky, of which the biggest has to be the study conducted by the University of Colorado in the late 1960s.

The Condon Report

On October 6, 1966, a team of twelve scientists from the University of Colorado headed by a highly respected nuclear physicist, Dr. Edward Uhler Condon (1902–1974), commenced a study to examine a number of UFOs on behalf of the U.S. government.

---

1   Part of his speech presented at the Symposium on UFOs, Boston, December 27, 1969. Quote also published in Sagan and Page 1996.

The 1,465-page Condon report, officially titled *Final Report of the Scientific Study of Unidentified Flying Objects*, was released on January 8, 1969 with the blessing of a special committee of the National Academy of Sciences. The report discussed 91 cases and contained detailed analyses and explanations for 59 of them, showing that these have a natural or man-made explanation. Of the remaining 31 cases, they remained unknown. However, it has been assumed, especially by Dr Condon himself, that these UFOs have insufficient information or had nothing to advance science. As he said:

"Our general conclusion is that nothing has come from the study of UFOs in the past twenty-one years that has added to scientific knowledge. Careful consideration of the record, as it is available to us leads us to conclude that further extensive study of UFOs probably cannot be justified in the expectation that science will be advanced thereby."[2]

Despite the report acknowledging the existence of UFOs and the fact that people throughout the world continued to see UFOs and report them in considerable detail and with clearly defined structures, such as light beams, bright lights, symmetrical metallic objects, and glowing hot regions on the hull, just as we have seen in the previous chapter, Dr. Condon was of the view that UFOs could eventually be classified as Identified Flying Objects (IFOs) based on natural or man-made phenomena if given enough time and information about each unknown case. More importantly, he thought that none of the studied UFO cases were unusual and detailed enough to warrant closer scientific study.

It is a view that still persists to this day when we hear of American space expert at the University of Colorado, Dr. J. W. Warwick, having stated, in no uncertain terms, that:

---

2   Condon 1969, p.2.

"Popular ideas about flying saucers have no basis in scientific fact. I have often spoken to observers of flying saucers and in most cases we can explain the sightings as weather balloons, natural phenomena, or bright planets seen through atmospheric turbulence.

I am convinced all would be explained away if we had enough details of each sighting."[3]

In fact, John Billingham of NASA's Ames Research Center in California and a leader in the Search for Extraterrestrial Intelligence (SETI) movement, put it more bluntly when he said:

"To be quite truthful, we do not pay significant attention to the UFO issue at all. We feel that the whole area is so debatable, uncertain, and unscientific that it is probably not going to help anybody very much. We don't do anything on UFOs, and we do not recommend that anybody else do anything on them."[4]

It is funny that he should say this, as Billingham just so happens to be in the business of searching for the all-elusive alien radio signal in the depths of space as part of his serious SETI work. He believes that this approach is more promising than searching for ETs in UFO reports. If that is true, then how successful has he been so far? Has he found the alien radio signals? Well, we are still waiting…and waiting, and…waiting[5]. Indeed, one has to ask, what makes searching for alien radio signals any more successful than searching in UFO reports for the same kind of evidence?

Indeed, one gets the impression that if anyone talks about the possibility of finding something new (e.g., alien spaceships) in UFO reports will lead to extraordinary fits of uncontrollable laughter among the more skeptical members of the scientific community. In one classic

---

3   Holledge 1965, p.90.
4   Edelson 1980, p.87.
5   On 10 January 2019, Canadian astronomers announced the presence of numerous unexplained repeating fast radio bursts (FRBs) detected by the latest Canadian Hydrogen Intensity Mapping Experiment (CHIME) telescope. As the source of these radio bursts is unknown (other than the fact that the radio bursts are coming from a galaxy 1.5 billion light years away), some scientists are prepared to consider the possibility that aliens generated these bursts if it helps to support the value of SETI in searching for alien radio signals. However, the more likely explanation is that FRBs are from radio pulsars. For further details about these bursts, see https://www.theguardian.com/science/2019/jan/09/repeating-fast-radio-bursts-from-deep-space-could-be-aliens.

example, Professor Edward Ney, an astrophysicist at the University of Michigan, said:

"Respectable scientists don't even discuss UFOs in serious terms."[6]

But not all scientists agree.

The Alien Explanation for UFOs

Professor Hermann Oberth

Professor Hermann Julius Oberth (1894–1989) is considered to be a great pioneer of space travel and the father of modern rocketry. Oberth worked to develop rockets with the U.S. Army Ballistic Missile Agency in the late 1950s and later joined NASA at the George C. Marshall Space Flight Center. He was quite adamant regarding UFOs:

"I have examined all the arguments supporting and denying the existence of flying saucers, and it is my conclusion that UFOs do exist, are very real, and are spaceships from another or more than one other solar system. They are possibly manned by intelligent observers who are members of a race which has been conducting long-range scientific investigations of our Earth for centuries."[7]

Another scientist who argues in favour of this view is Dr. Maurice Anthony Biot (1905–1985), a mathematical physicist who made important contributions to the field of aerodynamics. In a 1952 interview with *Life* magazine, Biot said:

---

6   Flammonde 1976, p.17.
7   Holledge 1965, p.24.

"The least improbable explanation is that these things are artificial and controlled....My opinion for some time has been that they have an extraterrestrial origin."[8]

In other words, the quotes tell us that there is still no definitive and coherent scientific explanation given for those genuine UFOs that remain unexplained to this day, especially among those described as unusual rounded objects revealing clearly defined artificial features as seen in Chapter 1.

Could UFOs Advance Science?

Fortunately, a smaller and more sensible scientific contingent has argued that further study into UFOs is justified on the basis that it could help to advance scientific knowledge no matter how controversial it is. Where this advancement could be made is not entirely clear, but it has been suggested that the field of psychology is a good place to start. For example, UFO cases could provide greater insight into how the brain works, especially in terms of certain chemical imbalances that may occur at certain times and places for the witnesses, leading to *hallucinations*. Or, we could determine whether UFO witnesses have special personality traits that make them more prone to seeing UFOs. As stated by the 1995 edition of the Encyclopaedia *Britannica*, soon after the Condon report was released in 1969, it was learned that:

> "...another group [of scientists] favoured continuing investigations on the grounds that UFO reports are useful in socio-psychological studies."[9]

But why stop there? What about the potential to advance scientific knowledge in the world of physics? Or, perhaps there could be a new technology hidden in the UFO reports? For example, witnesses have regularly experienced electromagnetic side effects in association with UFOs. While we will have more to say about this later in this book, the observations are already indicating the possibility of a new electromagnetic technology waiting to be discovered by the scientists. Also among the UFO observations is the claim by witnesses that UFOs

---

8   Durrach & Ginna 1952, p.96.
9   *Encyclopaedia Britannica* 1995, Volume 12, p.130.

can make themselves invisible and visible in a periodic manner. In that case, what about the potential to understand the mysterious link between gravity and electromagnetism as a natural way to advance scientific knowledge? Even if these UFOs turn out to be perfectly fine examples of the rare ball lightning, would it not be reasonable to see this invisibility effect as potentially able to advance science in certain areas of physics?

Without giving away too much at this early stage, it should suffice to say that there is much we can learn from UFOs. As retired scientist Ken Wear[10] said on August 12, 2007:

> "Let us learn what we can from UFOs. Granting that some of the claimed sightings may have substance, we should undertake to learn what we can from the presence of an unidentified object and from the observer of the object."[11]

All that is needed is for us to find out if there is anything new to be learned from UFOs.

---

10   Ken Wear is a physicist and engineer who worked in space-related research and author of *Love to Live and Live to Love: Making Longevity Worthwhile* (published by Llumina Press on April 28, 2008).

11   http://www.rationallink.org/UFOs.htm.

CHAPTER 3

The Natural and Man-Made Explanations for UFOs

From a scientific and engineering standpoint it is unacceptable to simply ignore substantial numbers of unexplained observations...

—American Institute of Aeronautics and
Astronautics UFO Subcommittee (established in 1967)[1]

BEFORE TAKING a closer look at the more unusual UFO cases, we must acknowledge that most UFOs have a natural or man-explanation. English science writer Ian Ridpath agreed with this view when he said:

---

1   Kuettner, J. P. (Chairman of the UFO Subcommittee of the AIAA) et al. "UFO – An Appraisal of the Problem: A Statement by the UFO Subcommittee of the AIAA": *Journal of Astronautics and Aeronautics.* November 1970, p.49.

Ian Ridpath in the library at the Royal Astronomical Society, London.

"There is no doubt UFOs exist in the sense that people genuinely see things in the sky that they cannot identify. But the real question is: what do these things turn out to be on investigation?

Not all reported UFOs can be extraterrestrial spacecraft, and no one pretends they are. Even UFO believers agree that at least nine reports out of every ten are readily explicable in known terms. The main culprits are bright stars and planets, aircraft, meteors, and satellites—although a complete list of misidentifications would be almost endless. UFO proponents often point to the residue of unexplained cases as proof that there must be something to UFOs after all, but that's not so. An unsolved case is not evidence for any theory. If even the most credible and honest witnesses turn out to be demonstrably wrong nine times out of ten it is not unreasonable to suppose that all cases would be soluble if we had sufficient information."[2]

It is interesting that Ridpath mentions "sufficient information". Does this mean sufficient in the sense that he can find the answer he expects to see (a natural or man-made explanation), or sufficient to find something new in the UFO reports worthy of study? Although care must be taken to define what it means to have sufficient information (if you cannot find what you expect, you would be tempted to say there isn't enough information), there is certainly no shortage of supporters of this view.

Dr. Edward U. Condon was inclined to support the natural and man-made explanation. According to the statistical results in a report published by Condon, most of the 90 cases his team studied were either natural or man-made IFOs. The only slight problem with the report is that he felt comfortable extrapolating the data and concluding

---

2   Ian Ridpath's quote obtained from http://www.ianridpath.com/ufo/ufoindex.htm.

that all UFOs are IFOs. As for those unusual cases revealing electromagnetic effects in association with UFOs, Condon was tempted to explain them as examples of ball lightning. We will have more to say about this later.

The view that most UFOs are IFOs very much mirrors that of every other official study, as can be seen in the following statistical results for Project Sign, Project Grudge, the Robertson Panel and Project Blue Book:

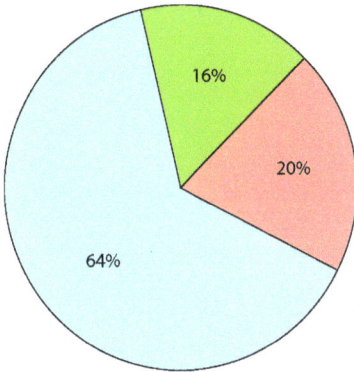

Project Sign (20 percent UFOs, 16 percent "insufficient data", and the rest are IFOs)

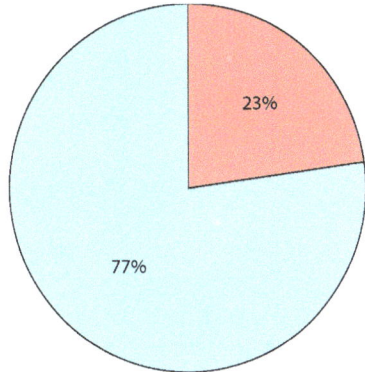

Project Grudge (23 percent UFOs, 77 percent IFOs)

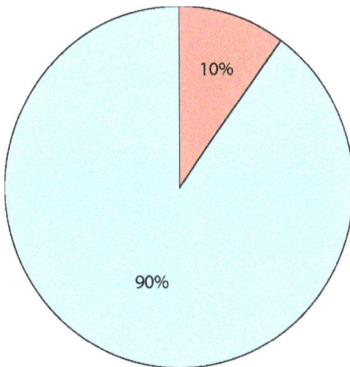

The Robertson Panel (10 percent UFOs, 90 percent IFOs)

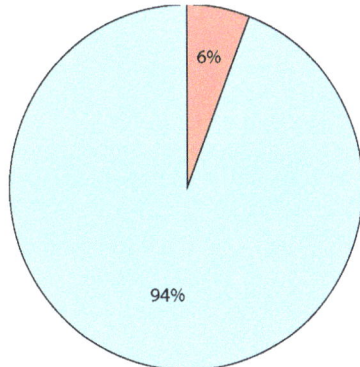

Project Blue Book (6 percent UFOs, 94 percent IFOs)

When faced with these seemingly overwhelming statistical results, one can be forgiven for thinking that if most UFOs are IFOs, the remaining UFOs were probably IFOs if sufficient information was available and sufficient time was devoted to investigating them, as Ridpath suggested. But what kind of IFOs are we talking about here? Just the natural and man-made variety as Ridpath would like to believe? Or is there something new to science if people make the concerted effort to properly study UFOs? This is the fundamental question we should ask.

The planet Venus (shown here in ultraviolet light) is often mistaken for a UFO.

Examples of Physical IFOs

Before answering these questions, we need to discuss IFOs—particularly the ones that are considered familiar but, for some reason, can get interpreted as UFOs. Whether it is because observers may not be fully aware of the wide range of known aerial and astronomical objects, below is a table showing examples of commonly mistaken IFOs.

Common Natural and Man-Made Explanations for UFOs	
Artificial satellites	Balloons
Comets	Swamp gas
Dust and mist	Aircraft
Lenticular clouds	Mirages
Parachutes	Planets
Stars	Auroras
Ball lightning	Birds
Fireworks	Insect swarms
Kites	Meteors
Missile tests	Space rockets
Sun dogs	Searchlights
Light flares	Airships

For a comprehensive look at images of the objects mentioned in the above table, Google search returns plenty of examples of what these objects look like at close range, or may look like when seen at a distance. In Appendix A, we provide a few examples for the reader to look at.

Optical illusions of Physical Phenomena

Things start to get tricky when certain conditions in the environment distort or change the look and/or flight behavior of aerial objects. When this happens, this can fool even the most well-trained and highly rational observers and their "trusty" cameras into thinking that a common object is a UFO. We call these *optical illusions*.

A rather common illusion reported by observers is a bright stationary light that appears to move against a dark background, as stated in the Encyclopaedia *Britannica*:

> "Staring at a single bright light spot in an otherwise darkened room creates the illusion that the stationary light is moving (autokinetic effect)."[3]

Thus, the human eye can be easily fooled into thinking a bright light seen at night, such as the planet Venus, is moving under some kind of "intelligent control". Only a telescope clamped firmly to the ground can help us see that the light is stationary (unless the light is directly heading for us, which usually indicates that it is a man-made aircraft or helicopter, although we can't discount the possibility that it could be a genuine UFO).

Another common optical illusion occurs when a light source, such as a lamp or a fluorescent tube inside a building, is reflected on the surface of a glass window, creating the illusion of a bright circular, oval, or cylindrical light hovering in the sky. For example, the photograph shown on the next page was taken by Shell R. Alpert at 9.35 a.m. on July 16, 1952. It was later revealed that the witness had taken the photograph from inside the control tower at Salem Air Base in Massachusetts, USA, and upon closer examination, investigators identified the glowing objects as light from several lamps that had reflected on the window.

---

3   *Encyclopaedia Britannica* 1995, Volume 25, p.498.

The photograph taken by Shell R. Alpert on July 16, 1952.

Certain defects or imperfections on the surfaces of glass and films can also appear to be UFOs until they are more closely examined. For example, if a glass window has optical defects, it can turn tiny points of light into massive saucer-shaped objects. Similar effects can be observed when there are distortions in the lens of a pair of eyeglasses. When investigators seek the truth in such cases, physically visiting the location of where the witness was standing when observing the UFO and gaining useful information about what he or she was wearing (e.g., spectacles) at the time will often reveal the explanation.

As for the optical illusions produced by imperfections on the surface of a photographic film or the occasional chemical mishap during film processing, these can lead to the formation of some interesting light- or dark-coloured spots in pictures that might look like, or be interpreted as, UFOs. One such example is the photograph on the next page, which was obtained from the USAF's Project Blue Book files. Often, only a close inspection of the film in question using a microscope will reveal the explanation.

Examples of Non-Physical IFOs

In addition to physical IFOs, there are non-physical IFOs created within the human mind. The psychological component of the UFO phenomenon is not as significant as the physical component, but it does exist. There are two types of non-physical IFOs:

• Uncontrolled imaginative episodes called *hallucinations*.

• Controlled imaginative episodes called *hoaxes*.

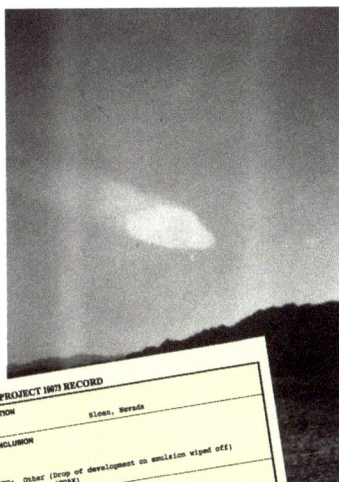

PROJECT 10073 RECORD

(form reproduced in image)

Hallucinations

The process that causes hallucinations is not precisely known. What we do know, and as some psychologists have suggested, is that it may have something to do with the way the brain tries to balance itself when faced with an extreme lack or an extreme overload of information and/or neurochemical transmitters, which may lead to the creation of a strong mental image that is suggestive of a UFO in the mind of the person. Whatever the reason, research has shown that apparitions of UFOs tend to occur for some witnesses during moments of intense excitement, lack of sleep, or extreme mental activity under conditions of stress, desire, or deep relaxation. As *Encyclopedia Britannica* states:

"Hallucinations probably will occur in anyone if wakefulness is sufficiently prolonged; anxiety is likely to hasten or to enhance hallucinatory production."[4]

---

4   *Encyclopaedia Britannica* 1995, Volume 25, p.501.

In other words, people who regularly stay up late after experiencing limited amounts of sleep and long hours of stressful work or anxiety are most likely to experience hallucinations. As for those people who are isolated and given limited or no sensory information input, they may also have a tendency to hallucinate. Could this be an explanation for some UFOs? To this end, *Encyclopaedia Britannica* states the following:

> "When people are kept in isolation (sensory deprivation), information input via the senses (e.g., hearing and sight) is de-patterned or reduced. If such a person remains alert, he is likely to experience vivid fantasies and perhaps hallucinations."[5]

In addition, deep relaxation coupled with a strong mindset or belief in something can induce hallucinations. According to the Encyclopaedia *Britannica*:

> "The mystic achieves hallucinations by gaining control of his own dissociative mechanisms....The hallucinations may take the form of unique visual imagery; for example, the *yantra* is a visual hallucination of a coloured, geometrical image that appears at a level of trance of the sort experienced by practitioners of Yoga. The recurrence of certain designs and patterns in human hallucinatory experience is probably related to structural aspects of the visual system."[6]

It is interesting to see a mention of "a coloured, geometrical image". Does this mean those geometrical glowing UFOs seen at close range (e.g., the oblate spheroid shape in the Levelland UFO case) were hallucinations?

When hallucinations do occur, it may indicate psychiatric disturbances. But generally, hallucinations can occur in anyone. Indeed, age, gender, cultural and racial differences, and other factors in people make no difference. Should hallucinations appear, they will seem so real to the person experiencing them that they will be interpreted as occurring outside the body. Furthermore, the brain can subconsciously control the apparitions, giving them the appearance of "motion" and "intelligent control", which have been reported in some UFO cases.

---

5   *Encyclopaedia Britannica* 1995, Volume 25, p.501.
6   *Encyclopaedia Britannica* 1995, Volume 25, p.501.

Hoaxes

Hoaxes are another technique for creating non-physical UFOs. Practically any healthy and rational person can do it. Like hallucinations, hoaxes require an imaginative mind. However, they differ from hallucinations in that they do not necessarily rely on imbalances in the levels of neurochemicals and information in the brain to *create* UFOs. In other words, the brain can be functioning normally when people create a hoax.

Hoaxes are generally deliberate attempts by people to imagine UFOs for a particular reason. Here is an example of a hoax photograph taken in New Mexico.

A fake photo of a UFO taken by Paul Villa in Albuquerque, New Mexico, on April 18, 1965. It shows two hubcaps joined together and held in the air with a thin wire attached to neighbouring trees.

The reasons for hoaxing a UFO report are many and varied, but they almost always involve some kind of reward or recognition for the hoaxer's efforts by media officials, especially given the media's immense interest in the subject in the early days. The media coverage of reports of UFO sightings is best described in the 1995 edition of Encyclopaedia *Britannica*:

"The publicity given to early sightings in the press undoubtedly helped stimulate further sightings not only in the United States, but also in western Europe, the Soviet Union, Australia, and elsewhere."

People may also perpetuate a UFO hoax simply because they are trying to prove a point. Or perhaps, due to the frustration caused by not finding UFOs, they try to fabricate one for themselves. Whatever the truth, there is almost always a motive for people wanting to hoax a report.

When people do create a hoax report, it is often difficult to initially determine its genuineness unless adequate investigation and research are conducted. Even then, depending on how elaborate and sophisticated the claims are, it may be necessary for the witness to confess to show that it was a hoax. If a confession is not forthcoming, then all we have is the information provided in a report by the hoaxer. But there is another clue. Hoaxers generally do not like to provide a lot of details. If they do, there is a risk that investigators and researchers will "catch them out", so to speak, when the hoaxers say something that cannot be true based on the solid application of appropriate scientific knowledge or a check of the available records. This is the trade-off hoaxers must face: give too much information to investigators means potentially being caught, but give too little means not being taken seriously.

To a scientist, detail is king. If you report a UFO, real or otherwise, to the authorities you will need to provide as much information as possible to the person investigating the case and the scientist tasked with finding an explanation. Every fine detail will likely be important, especially to a scientist. The more information a person can provide in relation to an alleged UFO encounter, the easier it gets for a scientist to find an answer, and the more likely it is that the person is telling the truth (or faking it).

How Reliable Are Photographs and Films as Evidence?

On the issue of providing as much information as possible in support of an alleged UFO, hoaxers are likely to be armed with what they call evidence by way of a photograph or film. From a scientific perspective, the presence of a picture can help to eliminate hallucinations as a possible explanation. However, photographs and films on their own have certain limitations, which is why scientists cannot rely on them to prove that a report is not a hoax. This is not to say that they are not useful as evidence—they certainly can be—but scientists are a conservative bunch. Accordingly, their experiences with photographs and films have led them to take considerable care not to rely solely on this form of evidence when determining an explanation for the object allegedly seen.

A classic example of the difficulties that scientists face in identifying genuine UFO footage and photographs can be observed in an experimental film project funded by the Australian Film Commission, known at the time as the *Australian UFO Wave 2006*. Between June and mid-August 2006, experimental digital video project director Christopher Kenworthy distributed 31 clips of UFOs over the Internet[7]. The purpose of this was to observe the public's reaction to the clips and determine how well UFO researchers could identify which were fakes and which were the two genuine UFO clips thrown in the mix. The results were interesting, to say the least. It took the final two clips to raise researchers' suspicions that several clips were probably fakes. Overall, the project showed, fairly convincingly, that (1) photographs and videos (especially digital ones) can easily recreate many of the UFO pictures and films (genuine or not) distributed on the Internet, and (2) images of anything with tiny moving lights at night or greyish or black blobs hovering or flying during the day seen at a great distance are not sufficient to determine the true nature of the objects.

Considering this and our earlier discussion about how imperfections on the surface of photographic film can lead to strange-looking spots in pictures suggestive of UFOs, it does make one wonder how reliable photographs and films are for recording UFOs. As UFO researcher Fred Beckman writes:

"Scientists and laymen who have an interest in the UFO problem are invariably most strongly interested in the photographic evidence. It is generally believed, and it does seem entirely reasonable, that a good photograph of an unidentified flying object would have 'probative value'. Probative value is a term that was used by the University of

---

7   http://www.australianufowave.com

Colorado's Condon Commission to describe the kind of evidence that would strongly support the hypothesis that they were testing, namely that UFOs are extraterrestrial craft under intelligent control. Their general finding was that no example of UFO photography examined by them had probative value. It was found that most of the photographs examined could have been fabricated by such methods as double exposure, manipulation in the darkroom, or by suspending or throwing a small object and then photographing it against a natural background. In some cases, evidence obtained from a technical analysis of a photograph was inconsistent with the description of the event as recalled by the witness."[8]

Of course, this does not mean that all photographs and films of UFOs have no intrinsic scientific value or are not genuine. It is far better for witnesses to have at least one photograph of a UFO in their possession than no photograph, especially if there is little that scientists can gather from their official statements. Of even greater scientific value is a well-focused film of a UFO (a rarity we hear these days, especially when the object glows intensely at night). But care must be taken to ensure that photographs and films stand up to rigorous scientific analysis before they can be taken seriously.

How do you obtain a good quality UFO photograph? If the object can be recorded on film near the ground, try to show it in reference to some known area or man-made object on the ground. If the UFO has landed, try to get as close to it as possible to show more details. When you do, keep away from metal objects, such as road signs and cars, in case you do encounter a genuine UFO (we will have more to say about genuine UFOs later in this book) as there is a risk you could receive radiation exposure. Other than that, take every opportunity to photograph the object whenever you see it. Use the optical (rather than digital) zoom of the camera to get in close and get the sharpest pictures of the object. It is these landed UFOs or those seen near the ground and witness that will provide the most useful information to scientists compared to those images of UFOs seen at a great distance, which may appear as nothing more than a tiny dot of light or a greyish blob in the sky.

---

8   Hynek & Vallee 1975, p.119.

Here is an example of genuine photographs that have been analysed carefully with no signs of trickery whatsoever. Even the witnesses themselves who took the photographs thought it was a new military aircraft and generally ignored it until a reporter learned about the sighting.

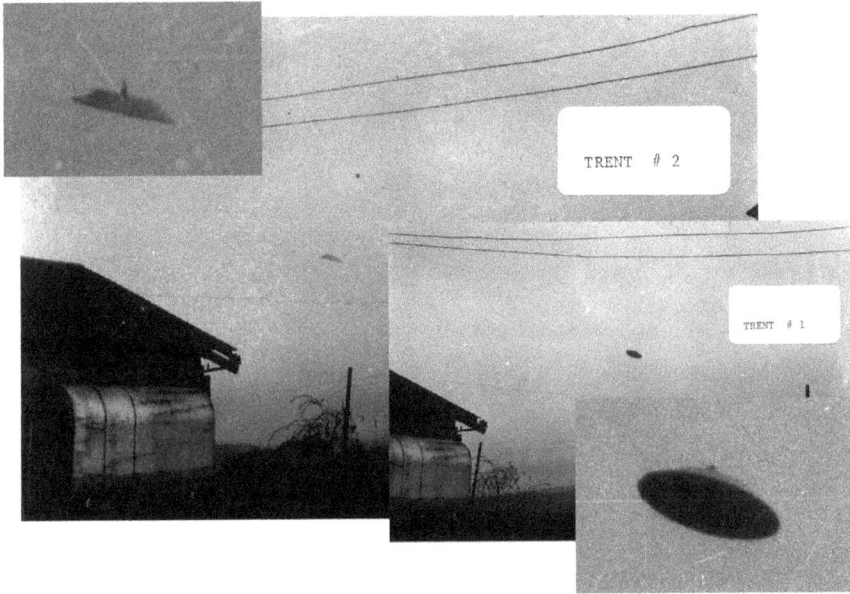

The McMinnville UFO photographs were taken by a farming couple, Paul and Evelyn Trent, near their farmhouse not far from the town of McMinnville, Oregon in May 1950. The photographs were recorded on traditional negative film.

Even better would be if witnesses could somehow record on film the internal structure of a UFO that has landed on the ground (for engineering purposes) and/or anything that might suggest the existence of occupants (for biological purposes). A very difficult feat to achieve we hear given both the considerable effort made by genuine UFO occupants to stop witnesses from gathering this kind of evidence, and the short amount of time they spend on the ground gathering samples. But if witnesses could somehow do this, it would provide superb evidence supporting the existence of UFOs. It would also give scientists plenty of information about the nature of the object and who is flying it (even if the occupants prefer to stay out of the picture at every opportunity).

This neatly brings us to the question of whether alien abduction cases can provide useful scientific information. By their very nature, such cases have the potential to reveal both the internal and external structure of UFOs, as well as superficial biological details of the occupants who allegedly fly them. Even without photographs or films of the internal and external structure of UFOs, excellent testimonials from witnesses and hand drawings can be just as useful to scientists if the details are there and can withstand rigorous scientific analysis.

The only question is, can alien abductees provide sufficiently reliable information in this regard for the scientific community?

Can We Rely on Alien Abduction Reports?

Alien abduction cases feature close and personal contact with both UFOs and their occupants. These cases have the potential to contribute a tremendous amount of useful scientific information, especially when the witnesses are calm and relaxed and can reveal their insights (assuming, of course, that it is not a hoax). Aside from common observations about the UFOs and their occupants, abductees also appear to describe a common trend in the method of abduction. For example, the common approach seems to be to land in some isolated location where at least one human is taken, against his or her will, on board the UFO by the occupants. Then, a series of tests and biological samples are taken. Before the abductee is returned to the site of the abduction, some information might be provided by the UFO occupants, either as warnings to protect the Earth and to do the right thing, or to escort the abductee around the craft to observe various things. Again, such common trends are the kind of thing scientists looks for. Nonetheless, we must still ask, can these common observations of alien abductions give scientists enough information to work with?

The late John Edward Mack (1929–2004), M.D. and Professor of Psychiatry at the University of Harvard, gave some consideration of these controversial claims after studying hundreds of people in the United States and abroad involved in such experiences. His work confirms a common trend in the way the abductions take place with the witnesses:

"The person is in the car or sleeping, and there is a light, there are entities coming to the room. They feel themselves paralysed. They're moved, they are taken. They go into a ship. Something happens there. There may be other people there. They have telepathic communication with aliens and some probing. There is a whole complicated set of events that takes place, which is consistent from one person to another. But there is a basic story here that after several hundred cases in this country and in many other countries it all begins to hold together as something that has a robust kind of truth to it. And, if this is real, then what does it mean?"[9]

However, psychologist Steven Jay Lynn noticed how a number of alien abduction claims had come from people requiring hypnosis to recall the events after suddenly having lost memory of what had happened. To test the reliability of the process in gathering statements, he prepared an experiment[10] in 1994. He asked a random selection of subjects from the wider community to imagine that they had seen bright lights and experienced missing time. Then, he put the subjects under hypnosis and asked to simulate an alien abduction experience. Of those subjects who had heard or read stories about UFOs but never had an alien abduction experience or seen a UFO, 91 per cent stated that they had interacted with aliens. One can imagine, therefore, that if these stories were combined with genuine reports, it would be challenging for scientists to distinguish the real alien abduction reports from the false ones.

Despite the problems of alien abductions involving hypnosis, we must not assume that all alien abductions are fakes based on Lynn's experiment. Nor should we assume all witnesses of alien abduction cases require hypnosis to recall the experience. There are a number of good-quality alien abduction cases on record for which no hypnosis was required.

A classic example of this is the alien abduction case of Antonio Villas Boas (1934–1991) from Brazil in 1957. Details about this case can be found later in this book, but here we have a witness with nothing to gain from his experiences (and more to lose in terms of his health soon after the encounter, and later the ridicule he received from the media). He gave good hand drawings of the UFO he observed and

9   From the 2004 French documentary titled *Experiencers* (a film by Stephane Allix).
10   http://www.psychologytoday.com/articles/%5Byyyy%5D%5Bmm%5D/%5Btitle-raw%5D

descriptions of the aliens and what they did while he was held against his will inside the UFO. He was even given a brief tour of the object before he was permitted to leave. In this situation, scientists found the testimony hard to ignore. It was a classic, top-notch alien abduction report in which absolutely no hypnosis was required and with good scientific information to help scientists analyse the type of object we are dealing with here. In addition, there was excellent evidence of radiation poisoning of the witness immediately after the experience. And if that is not enough, his testimony remained consistent until his death (and so making the hoax theory an unlikely explanation for this UFO case). This is one case that deserves close scientific scrutiny.

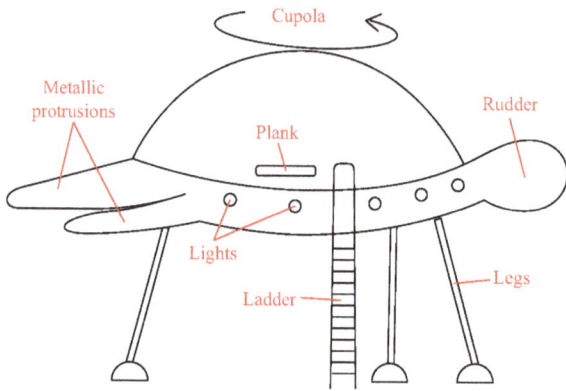

Another classic example requiring no hypnosis is the 1983 case of 77-year-old Alfred John Burtoo (1906–1986), who was abducted from North Town in Aldershot, England. He provided valuable information about the nature of UFOs and their alleged occupants, including the interesting sound produced by the UFO as it started up and moved away from the ground, the excellent internal details of the UFO, and a sketch of the object from the outside. Furthermore, he discussed the aliens' goal in bringing him on board for study, as well as their general appearance (barring their faces, which were covered by large visors and helmets).

When sensible and open-minded scientists see these examples, they reasonably consider the possibility that some UFOs do deserve closer scrutiny. Even if they are not alien in nature (for example, they could be secret military experiments), the possibility of advancing science is potentially still there. It does not matter if this advancement occurs only in the field of psychology because some witnesses are expert hoaxers. Neither would it matter if the brain has a subconscious ability to extrapolate future trends from what we see and later bring them to the witnesses' conscious minds via hallucinations. Why? Studying such cases could still significantly add to our knowledge and understanding of the mind, or why people create hoaxes in the first place. However, given that some witnesses also suffer from radiation poisoning (see the Antonio Villas Boas UFO case), we can safely say the advancements will not be restricted to psychology. These cases are likely to contribute to the field of physics, possibly to the point where we might be dealing with a new technology or an unusual natural phenomenon not yet understood by science.

In summary, alien abduction cases have the potential to advance science even though much of the information is based on testimony. Technically, there is nothing wrong with a good testimony with interesting observations that astonish scientists and have the potential to advance science. However, the value of such testimonies depends on how much useful information they contain. If there isn't much information, interesting reports of unwelcome visits to people's homes by alien-like creatures, even if described emotionally, may be described as possible hoaxes. Even if the alleged testimony contains considerable detail, studies have shown that people who have read books and newspaper articles on alien abductions can produce a similar level of

The hand-drawing of the UFO observed by Alfred Burtoo. The most significant aspects of Burtoo's observation are details of the rounded edges of portholes, the door and various other parts of the object, as well as the sound it made that reminded Burtoo of an electrical generator. Can science afford to ignore such important details when understanding genuine cases of UFOs?

detail when asked to recreate an alien abduction scenario under hypnosis.

The only good alien abduction cases scientists are more likely to look at are those that involve no hypnosis and where the witnesses have experienced radiation poisoning or other symptoms following the alleged UFO encounter. But if there is anything else we can discover from these cases, it could open up the door to something quite sensational if we are willing to open our minds and discover what else is possible.

As good scientists, we have to be ready for the unexpected.

So let us look at those UFO cases where no hypnosis is required and where good details are provided by the witnesses. What can we learn from these cases?

CHAPTER 4

The Alien Explanation for UFOs

We have only had our technology for about 200 years, but supposing that other planets in the galaxy have had theirs for thousands, maybe millions of years, then anything could be possible.[1]

—Brinsley Le Poer Trench

Now we have to consider the opposing view—namely that UFOs could be technological vessels in which one or more alien civilisations are visiting Earth. As scientists are not yet familiar with the kind of technology that could make interstellar travel possible for the UFO occupants, the only other evidence presumed to support this view is primarily found in the numerous and well-documented UFO cases that have been reported by witnesses from around the world, especially those that come with physical effects left behind on witnesses and the environment. Today, there are at least several hundred

---

1   Le Poer Trench 1974, p.175.

highly detailed cases and a further 2,000 cases that are no less scientifically useful but have fewer details. All these genuine UFOs still baffle investigators and researchers alike. Of the more detailed cases, two examples were presented in Chapter 1 (and it should be noted that the most details are contained in the original reports). Still some readers might scoff at these examples as rare or provide insufficient information for any kind of scientific advancement, or determine a reasonable natural or man-made explanation. Thus, let us explore a case in detail—just one of the numerous highly detailed cases we could investigate if we so chose.

The UFO Allegedly Seen by Alfred Burtoo

It began on the night of August 11, 1983. Mr. Alfred Burtoo, a 77-year-old man from the town of Aldershot in the United Kingdom, was fishing in the Basingstoke Canal. He often did so near the old Gasworks Bridge, where Government Road crosses the canal. However, this night would be different.

While fishing, a UFO suddenly appeared. Burtoo saw it land near the canal, not far from the bridge:

> "...[I] saw a vivid light coming towards me from the south, which is over North Town. It wavered over the railway line and then came on again, then settled down. The vivid light went out, though I could see a light through the boughs of the trees."

A few minutes later, Burtoo was approached by two small humanoid beings:

> "...I saw two 'forms' coming towards me, and when they were within five feet [1.5 meters] of me they just stopped and looked at me, and I at them, for a good ten or fifteen seconds. They were about four feet [1.2 meters] high, dressed in pale green coveralls from head to foot, and they had helmets of the same colour with a visor [covering their facial features] that was blacked out.
>
> Then the one on the right beckoned me with his right forearm and turned away, still waving its arm. I took it that he

34

wished me to follow, which I did. He moved off and I fell behind him, and the chap that was on the left fell in behind me."

The manner in which the beings walked seemed normal to Burtoo, but he did notice their movements were rather stiff-jointed. Their one-piece suits were fitted tightly to their unusually thin bodies, and there were no signs of belts, zippers, buttons or fasteners.

A short while later, Burtoo came upon what appeared to be "a large object, about 40 to 45 feet [12 to 14 meters] across". The object had portholes positioned evenly around it, and the object rested on two ski-type runners.

"I thought, Christ, what the hell's that?" Burtoo recalled. "Didn't think about UFOs at the time."

Then Burtoo said:

"…When we got down there this 'form' in front of me went up the steps and I followed….Going in the door, the corners weren't sharp, they were rounded-off. We went into this octagonal room. The 'form' in front of me crossed over the room, and I heard a sound as if a sliding door was being opened and closed. I stood in the room to the right of the door, and the 'form' that had walked behind me stood just inside, between me and the door. I don't know whether it was to stop me going out or not….

I stood there a good ten minutes, taking in everything I could see. The walls, the floor and the ceiling were all black, and looked to me like unfinished metal, whereas the outside looked like burnished aluminium. I did not see any sign of nuts or bolts, nor did I see any seams where the object had been put together. What did interest me most of all was a [central] shaft that rose up from the floor to the ceiling. The shaft was about four feet [1.2 meters] in circumference, and on the right-hand side of it was a Z-shaped handle. On either side of that stood two 'forms' similar to those that walked along…with me."

The ceiling seemed very low to Burtoo. The floor appeared to have a soft covering, as his footsteps did not make a sound. Internal lighting in the room was rather dim and did not appear to emanate from any

definite source. No signs of a control panel, seats, instruments, or other objects could be seen except for a prominent central shaft with a Z-shaped handle.

Burtoo's observations were suddenly interrupted when he heard a voice:

> "All of a sudden a voice said to me, *Come and stand under the amber light.* I could not see any amber light until I took a step to my right, and there it was way up on the wall just under the ceiling. I stood there for about five minutes, then a voice said, *What is your age?* I said 'I shall be seventy-eight next birthday'. And after a while I was asked to turn around, which I did, facing the wall. After about five minutes he said to me, *You can go. You are too old and infirm for our purpose.*
>
> I left the object, and while walking down the steps I used the handrail and found it had two joints in it, so I came to the conclusion it was telescopic. I walked along the towpath to about halfway between the object and the canal bridge, stopped, and looked back and noticed that the dome of the object looked very much like an oversized chimney cowl, and that it was revolving anticlockwise.
>
> I then walked on to the spot where I had left my dog and fishing tackle, and the first thing I did when I got there was to pick up my cold cup of tea and drink it. And then I heard this whining noise, just as if an electric generator was starting up, and this thing lifted off and the bright light came on again. It was so bright that I could see my fishing float in the water 6 feet away from the opposite bank of the canal, and the thin iron bars on the canal bridge. The object took off at a very high speed, out over the military cemetery in the west, and then a little later I saw the light going over the Hog's Back and out of sight. This was around 2:00 a.m."

Burtoo rested until dawn (about 3:30 a.m.) and then stated, "I got into what I had come out for—the fishing!"

He returned home at around 1:00 p.m. that day and mentioned his UFO story to his wife, Marjorie, and a friend of hers. Incredibly, Burtoo refrained from telling the part about being taken on board, thinking, "[T]he wife would say, 'No more fishing for you, old man!'"

For a short period of time, a lack of appetite and general tiredness were the only side effects Burtoo experienced after his remarkable encounter with the UFO. Until his death from cancer on August 31, 1986, Burtoo never retracted his statements concerning the incident. As Marjorie said:

> "What Alf told you [without the use of hypnosis] was the absolute truth. My friend who was with me when Alf came home can verify what he said. He looked absolutely shaken and he told both of us about his experience that he had with the UFO. He was just like a man who had seen a miracle happen and we knew he was telling the truth because no one could believe otherwise if they had heard him and saw him that morning. My husband was not a man who believed in fantasies or had hallucinations. He was down to earth, and you can take it from me that Alf never changed his mind on the story of what he had seen and experienced."

Are Aliens Visiting our Planet?

When faced with this sort of UFO case, the presence of occupants does give a reasonable person food for thought regarding the alien explanation. Well, why not? It is certainly not unscientific. In Chapter 7 we will learn the considerable effort by the USAF to deny the existence of anything alien in the UFO reports, and in Chapter 2 we have seen the skeptical view of one SETI scientist when asked about the UFO situation. However, we know this is not the correct scientific approach. No one should ever think it is preposterous to consider the alien explanation in any serious scientific discussion on UFOs. Unless the U.S. military has something to hide of great importance to the scientific community, we have to re-iterate the words of Giuseppe Cocconi and Philip Morrison of Cornell University in their peer-reviewed article *Searching for Interstellar Communication* in 1959:

> "The probability of success is difficult to estimate, but if we never search the chance of success is zero."[2]

---

2   Cocconi & Morrison 1959, p.846.

So if SETI scientists are prepared to look for alien radio signals as we speak, it should be no different when it comes to looking at UFO reports. Whether it is an alien radio signal or a spaceship, both should be valid scientific pursuits.

But does this mean that all UFOs could be evidence of something alien? Not necessarily.

What makes people think UFOs could have an alien explanation has to do with the UFO occupants. Clearly, they are not exactly the familiar human type other than the fact that they all are humanoid in shape. Beyond that, the UFO itself isn't exactly your standard run-of-the-mill aircraft humans have built.

For example, in the previous UFO case we read, we cannot overlook the highly symmetrical nature of the metallic flying object. Why symmetrical? A helicopter would have a hard time fitting in with the descriptions of the UFO, nor would any drones currently available on the market. All man-made flying objects require propellers to move air and wings to provide lift. Looking at the UFO in question, it seems all it needs to do is glow its metallic surface, create a buzzing sound reminiscent of an electrical generator, and off it goes. Definitely not a common sight for humans in any part of the world. With this in mind, the symmetrical nature of the UFO arriving without wings should be a tell-tale sign that we have something unusual and potentially able to advance science even if we do not want to think that aliens could be visiting the Earth.

The same could be said of the UFO occupants. It is not every day that people observe very thin and small humanoid creatures with large heads and eyes emerging from flying objects. But some skeptics might argue that there are a number of secretive, anorexic individuals in our society that love to wear large motorcycle helmets and dress up in alien-like outfits in order to scare the living daylights out of others for a bit of amusement. The only problem is, the witnesses find the occupants looking a bit too real to be anything other than a real biological entity. Furthermore, we have seen UFO cases in which the witnesses have nothing to gain from the experience, and a number of them have already died without signs of a confession of any kind. Are we to assume that these are hoaxes?

The Electromagnetic Observations

One of the things that perplexes scientists is the electromagnetic nature of the observations reported by the witnesses. Such observations are not likely to be a coincidence given the number of cases in which they appear. Mr. Burtoo's case featured a few highly notable electromagnetic observations. For example, the sound of the UFO when it started up had reminded Mr. Burtoo of the sound of an electrical generator. Also, he commented on the rounded corners as he entered the front door of the UFO; the electrically insulating, rubber-like black floor; the symmetrical nature of the UFO; the occupants' skin-tight metallic suits and the lack of sharp points, such as fasteners or belts. In addition, there was a glowing effect on the exterior metal of the UFO as it arrived and started to take off and move away, suggesting something had to be electrically charged to generate this glow. But what was the purpose of this glowing effect? Was it a giant warning light for everyone to keep away, or does it serve another function that we don't understand?

To boil it all down, what is so special about the association between electromagnetism and UFOs?

Dr. James E. McDonald (1920–1971) was one such scientist who became aware of the electromagnetic nature of UFOs. As a former senior physicist at the Institute of Atmospheric Physics and a professor in the Department of Meteorology at the University of Arizona, Tucson, Dr. McDonald said to the Office for Outer Space Affairs of the United Nations on June 7, 1967:

> "...A wide range of electromagnetic disturbances accompanying close passage or hovering of the UFOs is now on record throughout the world....
>
> ...Disturbance of internal-combustion engines coincident with close passage of disk-like or cylindrical unconventional objects is on record in at least several hundred instances....Often the disturbances are accompanied by broad-spectrum electromagnetic noise picked up on radio devices.
>
> In many instances compasses, both on ships and in aircraft, have been disturbed. Magnetometers and even watches have been affected. All these reports, far too numerous to cite in

detail, point to some kind of electromagnetic noise or electromagnetic side effects…"[3]

Do not think for a moment that we are talking about only one or a handful of UFO cases; there are plenty of cases on record that show a remarkable link to electromagnetism. If you want more examples, it may be worth your while to read the next chapter. In it, you will discover a remarkably strong electromagnetic streak running through a number of UFO reports.

What is the reason for this link? Either the UFO reports indicate the existence of a new electromagnetic technology, or there is another type of electromagnetic phenomenon we have yet to understand after all this time.

Could There be a Natural or Man-Made Explanation for these Electromagnetic Observations?

Referring back to Mr. Burtoo's UFO case, it is interesting to find the witness had not changed his story, even on his deathbed. Considering that he died a few years after his alleged encounter—plenty of time to re-think about his story, clear his conscience and consider telling the truth—there is no evidence from his family or a written confession to indicate that he saw anything other than what he claimed to UFO investigators. Something deeply moved him at the time. When faced with these kinds of impressive cases, it is highly unlikely we are dealing with hoaxes.

As there are enough detailed UFO cases on record with witnesses willing to testify to their grave of what they saw is genuine, we can safely rule out the hoaxes explanation. We have to be dealing with something else. Of course, we should not assume all UFOs are alien. In most cases for which enough details are provided and enough time is spent investigating the claims, it is possible to find a logical natural or man-made explanation.

One possible explanation is that we could be dealing with a rare but particularly persistent form of ball lightning that doesn't want to go away. The other explanation is that we could be dealing with a secret military experiment on a new electromagnetic flying object not yet unveiled to this day.

---

3   Deyo 1992, p.13–14.

Yet at the same time, we need to be careful not to eliminate the "alien explanation". We should not be surprised to find in the literature a number of "down-to-earth" proponents (affectionately known as *skeptics*) doing all they can to find their own natural or man-made explanations for all the UFOs they hear about. This would be fine, but it can create some problems.

For instance, many skeptics are quick to brush aside any possibility of an alien explanation. A personal favourite of the USAF no doubt as we shall see in Chapter 7. However, scientists know very well that taking on this position is unscientific. Furthermore, no one can be God to know exactly what all UFOs are. By their very nature, UFOs are unknown. There will have to be some objects that we simply do not know precisely what they are. And it would be unscientific to assume all UFOs are IFOs. Well, at least not until every case has been properly investigated. But then to later say there is "insufficient information" to determine the nature of the flying objects is only ignoring the reality that there are UFO cases with plenty of details and are choosing not to do the work to find out.

As physicist Dr Erwin Schrödinger (1887–1961) said:

Bertrand Russell

"The first requirement of a scientist is that he be curious; he must be capable of being astonished and eager to find out."[4]

And as British philosopher, mathematician and historian Bertrand Russell (1872–1970) said in 1928:

"What is wanted is not the will to believe, but the will to find out, which is the exact opposite."[5]

Any skeptic that is not capable of being astonished and chooses not to find out cannot be described as a scientist. A true scientist must be open-minded and curious to understand a genuine mystery. Otherwise, as Albert Einstein once said:

---

4   From his book, *Nature and the Greeks*, p.55; Hynek 1972, p.6.
5   Russell 1928 in *Sceptical Essays* (W. W. Norton: New York).

"The most beautiful experience we can have is the mysterious. It is the fundamental emotion which stands at the cradle of true art and true science. Whoever does not know it and can no longer wonder, no longer marvel, is as good as dead, and his eyes are dimmed."[6]

Remember, while a mystery remains, UFOs deserve our utmost attention. And that means we, as genuine scientists, must do the proper investigations to find out exactly what we are dealing with here in case there might be something to advance science no matter how astonishing it might look on the surface. Sure, UFOs and UFO occupants look odd, but this should not frighten a true scientist. Anything different just means that there is an opportunity to learn something new.

As for the possibility that UFOs could be a secret military experiment, this is reasonable. We will take a closer look at how likely this is in Chapter 7.

As for ball lightning, Chapter 6 looks at what we know regarding these enigmatic and ephemeral glowing objects.

Nonetheless, could there still be an alien explanation to consider as well? In Chapter 8, we will find out by delving more closely at the most common UFO observations, starting with the ubiquitous glowing effect noted by many witnesses. We will take this glowing effect to its logical conclusion and see if there is anything new we can learn from this humble observation.

But we need to emphasise again that we are not dealing with hoaxes here. Remember, a hoaxer would need to apply a certain amount of creativity to formulate his story, resulting in more diversified reports of the flying objects and their occupants. In Chapter 5, we will see how the UFO observations are not so different as to confuse a scientist. There are a lot of observations in common across many of these genuine UFO reports.

This is the thing. The observations are not exactly different when we get done to the details. Sure, UFOs are symmetrical glowing or metallic objects. Not a common sight at our local airports. But if we get to the nitty-gritty of what is actually part of the object, for example, the glowing regions and even right down to the protrusions appearing out

6   "The World As I See It—An Essay by Einstein" originally published in Forum and Century, 1931. This quote and the full essay can be read at http://www.aip.org/history/einstein/essay.htm as of February 2010.

of the central symmetrical body for different UFO cases, they are not exactly unheard of or so different that no scientist can ever understand the pattern.

The same is true of the UFO occupants. Sure, there are superficial differences. But again, deep down the occupants are not exactly horrendous creatures of such diversity in shape, size and other physical attributes to really have scientists in a state of apoplexy in trying to understand the sheer differences. There is actually something in common, including the humanoid shape of these enigmatic individuals flying these UFOs

So why do UFOs have a symmetrical, geometric and metallic shape (what is wrong with a flying object shaped like the ones in the science fiction films *Star Wars* or *Battlestar Galactica*?). And why are occupants described as universally very thin and usually short humanoid creatures with large heads? Indeed, why do we not see obese-looking octopus-shaped creatures with tentacles for arms and legs, such as the Daleks in *Doctor Who*?

In addition, many UFO cases feature the same curious link to certain electromagnetic side effects associated with the flying objects. Forget any form of propellant like we see in many modern science fiction films or man-made spacecraft that would suggest the possibility of a nuclear or some kind of chemical propulsion system. Somehow, the energy coming off these objects to make them move is invisible, with no signs of an exhaust pipe to explain how it can recoil in the opposite direction. When faced with this kind of electromagnetic UFO, the world has hundreds of such cases providing great details of a consistently symmetrical metallic flying object emitting an intense level of electromagnetic fields into the environment for an apparently unknown purpose. Not even the occupants themselves are entirely immune to this electromagnetic aspect of the UFOs. These odd-looking individuals are prepared to wear skin-tight metallic suits with no sharp points, such as zippers or buttons, as if this is considered critical to the operation of the flying objects. Why should this be so? What is so special about these flying objects that require this kind of attire?

As Dr. Josef Allen Hynek said:

"They are not all imagination, because if they were, we would be having reports of all sorts of things. Instead, we have consistent reports of disk-shaped lights and of craft-like objects."[7]

"Oh, I know," pipes up the skeptic as he searches for any reasonable answer. "The witnesses must be copying from one another. This might explain why we have common observations."

A clever answer, except we again have a problem. For people to copy the descriptions from others, every witness must have read about UFOs.

In the case of Mr. Burtoo, he didn't care about UFOs and claimed to have never read a book on the subject. When he heard about UFOs on the news, he took them with a grain of salt until he had his own experience with a UFO. Yet he managed to report rather detailed electromagnetic observations, mostly consistent with those noted by other witnesses around the world, as well as some new observations, such as the dimming of the UFO's glow as it moved over metal railway lines. Now how did he manage to figure that last one out if no one else mentioned it?

Still not sure? Well, there is another point we need to make that should put the final nail in the skeptics' argument: there are enough UFO cases on record in which the witnesses suffered symptoms of radiation poisoning, as confirmed by a number of doctors. It would be rather extreme for one individual to not only copy another's observations, but also find a substantial amount of radioactive material and make himself sick just to prove a point or show his belief in the subject.

Here are a couple of UFO cases to support this radiation poisoning claim.

The Cash-Landrum UFO Case

In 1980, three witnesses observed a slow-moving and glowing diamond-shaped UFO as it was being closely followed by over a dozen black military helicopters in the state of Texas.

It began in the late evening of December 29, 1980. A 51-year-old businesswoman, Betty Cash, was driving home in her car with two other occupants: 57-year-old Vickie Landrum, an employee of a restaurant owned by Betty Cash, and her 7-year-old grandson, Colby

7   Hynek & Vallee 1975, p.17.

Landrum. They drove along Highway FM148, which is used only by a few people who live in this sparsely populated area because of its isolation, to get to their home in Dayton, Texas.

At around 9:00 p.m., Colby Landrum was the first to notice a brightly lit UFO moving over the trees that bounded the highway on both sides. The distance between the witnesses and the UFO at that time was about 5 kilometres. Then, the object changed direction and approached the witnesses. Frightened by this latest action from the object and its unusual appearance, Betty thought of turning the car around and speeding off, but the road was too narrow and the dirt on either side of the road was too wet from recent rains for her Toyota Corolla to make the turn. Realising there was nothing she could do, Betty quickly slowed the car and pulled it off to the side of the road. By the time she stopped, the UFO had slowed down and straddled the road ahead of them.

The object was described as diamond-shaped with blue lights centred around its outermost rim. The entire object glowed intensely. A large, intermittent flame could also be seen underneath the object, keeping it aloft. Despite the bright glow, it looked as if the object was made of dull aluminium, and it seemed to be devoid of sharp points and edges.

Betty told Vickie and her son to stay inside as she got out of the car and closed the door to protect them. However, curiosity got to the better of Vickie, who also got out. Colby followed, staying close behind Vickie for protection.

All three witnesses were in view of the object, which was hovering only 50 meters away. Vickie was standing just behind the open door on the right-hand side of the car with her left hand resting on the car roof, Colby was standing next to Vickie for protection, and Betty walked around to the front of the car. For the next three minutes or so, Vickie and Betty stared intently at the brightly lit object ahead of them, while Colby pleaded with Vickie to get back inside the car. Upon seeing the object rising in the air, Vickie called out to Betty. She responded but when she got back to the door of her car, the handle was painfully hot to touch. Betty successfully opened the door with her leather jacket. As they entered the car, all three witnesses felt the intense heat of the interior and were forced to turn on the air conditioner.

As they watched the UFO depart, a group of black helicopters suddenly arrived on the scene. "They seemed to rush in from all directions," Betty recalls. "[I]t seemed like they were trying to encircle the thing."

The last they saw of the UFO was when they drove off and joined a larger highway, where they could just make out what looked like a small cylindrical object lighting up the surrounding area and the helicopters following it closely. From this new vantage point, they counted 23 helicopters, some of which were double-rotor (CH-47 Chinooks) and some of which were single-rotor.

By the time the witnesses arrived home, they all noticed that their skin had turned red as if they were badly sunburned (especially Vickie's left hand), and a collection of blisters had appeared (especially on Betty's face). Other medical symptoms included headaches, diarrhoea, swellings on the neck and eyes, and feelings of tiredness. All experienced hair loss of varying degrees of severity, with Colby being the least affected and Betty the worst.

Betty was the sickest, as she had stepped outside closest to the UFO and in front of her car to receive the full brunt of what appeared to be high-speed electrons streaming off the glowing surface of the UFO to hit the car's stainless steel body, resulting in a dose of intense ionising X-rays to emerge from the car itself. Furthermore, the high-speed electrons were likely to have caused the facial burns. She was taken to the hospital on January 3, 1981, and later she purchased a wig to cover her head due to the amount of hair loss she experienced, worse than the other two witnesses. After medical examination, it was concluded that the witnesses had probably been exposed to radiation of some type.

"This is a very important case providing physical evidence of the existence of UFOs," said John F. Schuessler, a NASA aerospace engineer who investigated the case. "A radiologist who examined the women's records said they were apparently suffering from the symptoms of radiation poisoning."[8]

As for proving whether the United States government was directly involved in the incident, it is extremely difficult, if not impossible. For, as the *New York Times* reported on January 20, 1981:

"Finding out what goes on in the CIA [and other clandestine organisations] is like performing acupuncture on a rock."

---

8   Blundell & Boar 1984, pp.146–148; Brookesmith 1984, pp.153–164.

While Vickie and Colby eventually recovered from their symptoms, this wasn't the case for Betty. She was beset with numerous health problems and was never able to return to work. She died at the age of 71 on December 29, 1998, exactly 18 years after the UFO encounter.

So, how did the witnesses manage to fake this UFO incident? Total geniuses if they did. And why would these witnesses be prepared to sacrifice their own health by receiving dangerous doses of radiation from an unexplained source?

Need a more extreme example? Try this one from Brazil.

The Inàcio de Souza UFO Case

On August 13, 1967, 41-year-old Brazilian Inàcio de Souza and his wife were returning on foot from a shopping trip to the ranch where they worked. As they approached the first building, they were surprised to see three slender individuals, wearing yellow skin-tight garments, apparently playing on the ranch's landing strip nearby. The intruders seemed to notice the couple and started to walk toward them. When Inàcio glanced elsewhere and noticed a strange craft that looked like an overturned washbasin at the end of the runway, fear and panic set in. He unslung his .44 carbine rifle, took aim, and fired at the nearest intruder.

Almost immediately, a green ray shot out of the craft and hit Inàcio on his head and shoulders. He fell to the ground. As his wife ran to his aid, she noticed that all three individuals were walking back to the craft. They went inside and started up the machine, which sounded like the humming of bees, and took off vertically at high speed.

Over the next few days, Inàcio complained of severe headaches, nausea, numbness, and tingling throughout his body. On the third day after the UFO encounter, his head and hands began to tremble uncontrollably. The owner of the ranch took his employee to a doctor for examination. The doctor found circular burns about 15 centimetres in diameter on Inàcio's head and torso. Worst of all, it was learned that Inàcio had contracted leukaemia and was not expected to live no more than 60 days.

Inàcio died on October 11, 1967.[9]

If this witness was a hoaxer, dying so soon was clearly not the ideal outcome he had hoped to achieve. Why waste one's life over UFOs by copying others right down to receiving lethal doses of radiation? Seems like a rather pointless act.

What About Hallucinations?

Still, this has not stopped the skeptics from coming up with other possible natural or man-made explanations. Perfectly fine by us. So how about the possibility that the minds of these witnesses could be playing tricks in what scientists call a *hallucination?* An interesting argument, if one may say so.

If hallucinations are meant to be the panacea for all UFO reports, it would have to be one hell of a hallucination for cases in which multiple witnesses were present to observe a UFO. A classic example of this is a UFO case from France.

The Oloron-Sainte-Marie UFO Case

On the sunny Friday afternoon of October 17, 1952, hundreds of witnesses from Oloron-Sainte-Marie and nearby villages in the southwest of France saw as many as 30 small objects in the sky

9   Brookesmith 1980-84 *The UFO Casebook*, pp.77–78.

followed by a larger white cylindrical object. The smaller objects looked like balls of mist at a distance. All the UFOs were traveling at the same speed in a southwesterly direction. Puffs of white smoke could be seen churning at the upper end of the cylinder while it was tilted at a 45° angle, which seemed responsible for the formation of a solitary cloud in an otherwise clear blue sky some 2,000 to 3,000 meters above the ground, just below the smaller UFOs.

The general superintendent of Oloron High School said:

> "…To the naked eye, [the small objects] appeared as featureless balls resembling puffs of smoke. But with the help of opera glasses it was possible to make out a central red sphere, surrounded by a sort of yellowish ring [or halo] inclined at an angle. The angle was such as to conceal almost entirely the lower part of the central sphere, while revealing its upper surface. These 'saucers' moved in pairs, following a broken path characterised in general by rapid and short zig-zags. When two saucers drew away from one another, a whitish streak, like an electric arc, was produced between them. All these strange objects left an abundant trail [of angel hair] behind them [possibly produced from the water-like vapour emitted by the cylinder], which slowly fell to the ground as it dispersed. For several hours, clumps of it hung in the trees, on the telephone wires, and on the roofs of the houses."

These "threads of the Virgin", as the villagers called them, were like strands of nylon, except they felt gelatinous and vaporises upon contact with the skin. Most of the remaining untouched material soon vanished after a few minutes.[10]

Just when the people of these peaceful French villages were starting to settle down, the UFOs repeated their performance on October 27, 1952. This time, the sighting and angel-hair fallout occurred over the town of Gaillac and neighbouring villages at around 5:00 p.m.[11]

---

10   Angel hair is essentially water molecules in the clouds that are aligned electrically by a strong electric field between the UFOs (presumed to have a negatively-charged underside with a positively-charged top). Water molecules get into an orderly fashion and held temporarily in place by the weak electrical charge dipoles of the molecules after the electric field is removed, forming a clear gelatinous material. However, the structure can easily be broken apart by touching it, forcing the water molecules to suddenly revert to a random state of either a liquid or vapour state.

11   Brookesmith 1980–1984 *The UFO Casebook*, pp.37–38.

How do we explain this UFO case? Are we to assume that something in the waters in this part of France affected so many people in the towns at these precise times and places, because, afterward, the people never hallucinated the vision of UFOs again? Or maybe magic mushrooms were growing prolifically in the countryside and people couldn't differentiate them from ordinary mushrooms? Afterward, they must have realised what they had been eating? As the skeptics say, there must surely be a simple and logical explanation for the UFO sighting, right?

Even if we are dealing with a form of mass hallucination not yet understood by scientists, how is it that there are UFO cases in which physical marks were left on the ground, trees were burned and branches were ripped off, tarmac was melted and ripped up, and radioactive symptoms affected the witnesses, as later supported by scientists and doctors? No hallucination on its own can account for this kind of physical evidence.

Need another example?

The Viljo and Heinonen UFO Case

On Wednesday January 7, 1970, 36-year-old forester Aarno Heinonen and 38-year-old farmer Esko Viljo were out orienteering on the ski slopes of southern Finland, not far from the village of Imjarvi, approximately 16 kilometers from the town of Heinola. It was a cold evening, nearly −17°C and windless. The sun was setting, and it would not be long before it would be too dark for the men to continue with their competition skiing.

After an exhilarating run through the snow, at about 4:45 p.m., Heinonen and Viljo decided to stop on a clear patch of snow just below a small hill to look at the stars. Roughly five minutes later, they heard a buzzing noise. Glancing over their shoulders, the two men observed a strong bright light surrounded by a cloud scurrying across the sky from the north. It made a wide sweep of the area and approached the witnesses from the south, descending as it came closer. Soon, the bright light was over the same clear patch of snow on which Viljo and Heinonen were standing.

The object reduced its intensity and descended to a height of about 15 meters above the ground. It was at this height that the witnesses noticed, appearing out of the gradually dispersing cloud, a round,

turtle-shaped metallic object with a diameter of 2.7 meters and three large metal spheres set into its base. The cloud itself was a glowing reddish-grey mist that was swirling and pulsating around the edge of the object.

Then without warning, as the cloud cleared even more and the object hovered between three and four meters above the ground, a strong beam of light fired from a centrally positioned, 25-centimetre tube underneath and touched the ground, creating a brightly illuminated circle about one meter in diameter with a black edge around it in the snow.

"I was standing completely still," recalled Heinonen. "Suddenly I felt as if somebody had seized my waist from behind and pulled me backwards. I think I took a step backwards, and in the same second I caught sight of the creature."

Standing within throwing distance, Heinonen described the creature he observed over a period of twenty seconds as:

"It was standing in the middle of the light beam [on the ground] with a black box in its hands. From around the opening in the box there came a yellow light, pulsating. The creature was about 35 inches (90 centimetres) tall, with very thin arms and legs. Its face was pale like wax. I didn't notice the eyes, but the nose was very strange. It was a hook rather than a nose. The ears were very small and narrow towards the head. The creature wore some kind of overall in a light green material. On its feet were boots of a darker green colour, which stretched above the knee. There were also white gauntlets going up to the elbows, and the fingers were bent like claws around the black box."

Viljo noticed the same features on the creature, but he also added that he saw the creature wearing a conical metallic helmet.

While the skiers watched, the creature made a half turn and pointed the opening in the black box toward Heinonen. Then, quite abruptly, a dazzlingly bright, white pulsating beam of light shone out of the black box, hitting Heinonen and forcing him to the ground. Immediately afterwards, the forest was very quiet.

The red-grey mist reappeared, increased in thickness and began to expand while moving around the craft. The mist descended and surrounded both witnesses until they could not see the creature or each other. Just before touching the ground, the red-grey mist suddenly disappeared in a spectacular shower of red, green and violet sparks, each about ten centimetres long and emanating from the mist to the snow. Some of the sparks hit Heinonen's feet, but he felt nothing.

Heinonen recounted this moment:

"Suddenly a red-grey mist came flowing down from the object and large sparks started to fly from the illuminated circle of snow. The sparks were like tapers, about 10 cm long, red, green and violet. They floated out in long curves, rather slowly; many of them hit me, but though I expected them to burn me, I did not feel anything."

Viljo's version of events was very similar:

"The sparks were shining in several colours. It was very beautiful. At the same time the red mist became thicker and hid the creature. Suddenly it was so dense that I could not see Aarno even though I knew he was standing only a few meters away from me."

Soon thereafter, the light beam and entity returned to the object. Then, the craft left with haste.

Both men stayed where they stood for a further three minutes as they looked at each other. "We were laughing and talking about this light," Viljo said. "But at the same time we felt a little uneasy."

Eventually it was time to move on. However, as soon as Heinonen stepped forward on his skis, his right leg would not support him and he fell in the snow. Viljo thought his friend was playing a game, but he quickly realised there was something wrong. He discovered that Heinonen had been paralysed on his right side after exposure to the beam of light. As Heinonen said:

"My right leg had been nearest the light. The whole leg was stiff and aching. My foot was as if anaesthetised."

It was getting dark, so Heinonen had to be dragged across three kilometres of snow back to their home.

Over the next few days, Heinonen started to complain of aching joints and headaches.

"I felt ill," said Heinonen. "My back was aching and all my joints were painful. My head ached and after a while I had to vomit. When I went to pee the urine was nearly black; it was like pouring black coffee onto the snow. This continued for a couple of months."

Other symptoms included severe amnesia and a red, swollen face. Heinonen's paralysis subsided five months after the incident, enabling him to walk. However, he still had trouble balancing properly for some time thereafter.

Viljo was not entirely unaffected by the incident. His face was red and swollen, and for a while he could not speak coherently and was often absent-minded.

Realising the seriousness of his friend's symptoms, Viljo became concerned about his own health as well. So he hurriedly went over to his neighbour's house to use the telephone and contact a local doctor:

> "I hurried to the nearest neighbour, who lived some 600 m away; he has a telephone. The first two doctors I called couldn't come, but Dr. Kajanoja said he would meet us at Heinola clinic in an hour's time. The neighbour drove us there."

Dr. Pauli Kajanoja examined and interviewed the witnesses. He said:

> "I could not find anything actually wrong with them, but Heinonen was complaining of nausea and pain in the limbs. I thought it might be the beginning of Hong Kong flu which was going round the Heinola area at the time. But when I took Heinonen's blood pressure I began to wonder. It was far below normal. It looked as if the man had had a real shock. Esko Viljo was very red in the face as if excited. At the same time both seemed absent-minded. They were talking excitedly and rather disconnectedly. In violent shock the blood pressure drops appreciably. Both seemed confused in their account of what had happened. I could not understand anything but that they had encountered some sort of electrical phenomenon. It was impossible to make a definite diagnosis. I was therefore

unable to prescribe any particular medicine. I gave them two sleeping tablets each.

The only symptom I was able to observe was Viljo's red and swollen face. Heinonen's sickness might be stomach trouble and fear. The symptoms he described were very similar to those of radioactive radiation. Unfortunately I had no radiation meter, although one can seldom observe damage from radioactive radiation after only a few hours. The colour of the urine was not mentioned on their visit. It might be the result of blood in the urine. But that cannot go on for several months. I did not think any blood tests were needed.

They had definitely encountered something. Both gave the impression of reliable men. I do not think that they had made all this up. I am convinced that they were shocked when they came to me. Something must have frightened them."

Further credence was given to the witnesses' claims after unusual "distant lights in the sky" were reported by two other witnesses in the same area at the same time that Viljo and Heinonen had their UFO encounter.

Matti Haapaniemi, a farmer and a neighbour who knew the witnesses well, said:

"Many people in this neighbourhood have laughed at this story. But I don't think it's anything to joke about. I have known both Aarno and Esko since they were little boys. Both are quiet, rational fellows and moreover they are abstainers. I am sure their story is true!"

The incredible story of Viljo and Heinonen's encounter with a UFO remains unexplained to this day.[12]

Can the Hallucination Theory Explain All UFOs?

So here we have two lucid, apparently rational witnesses from Finland with nothing to gain from the experience (certainly no money or fame) other than damage to their health, while allegedly observing the same

---

12   Spencer 1991, pp.98–99; Le Poer Trench 1975, pp.42–43; Rehn 1974, pp.105–111; Sachs 1980, p.158.

object and its occupant. Both described what they saw in the same way and in quite remarkable detail, which neither witness could have guessed unless one of them was a scientific genius and/or had read extensive books on UFOs. A rather amazing feat in its own right, for two separate minds to experience what skeptics fervently claim are hallucinations while reporting electromagnetic observations that is perfectly reasonable from a scientific standpoint. Also, it would take one powerful hallucination to explain the witnesses' medical conditions. Continuing to believe in the hallucination theory in the face of this and the previous UFO case would be especially foolish. Too many witnesses for UFOs seen at the same time and place in the French case, and too serious the medical symptoms involving radiation exposure in the Finland case are not exactly lending themselves well to the hallucination theory.

It should be clear by now that relying on hallucinations or hoaxes as the sole explanation for this case and every other UFO case with radiation effects and multiple witnesses (especially when we are talking about hundreds) is far too easy, a kind of cop-out for non-curious skeptics who are unwilling to get to the heart of the matter. The consistent nature of the electromagnetic observations, together with the physical traces left behind on witnesses (and premature deaths of some) and the environment tell us that hallucinations and hoaxes cannot be the underlying cause of all UFOs reported. It must be something real.

To determine how real these UFOs are, we shall now look more closely at the most common features observed in UFO cases.

CHAPTER 5

The Most Common Observations of UFOs

The mind does not perceive just detailed bits and pieces but is constantly weaving a large pattern from our experiences.

—Ivan Barzakov and Pamela Rand, the directors of
Optima Learning Institute in San Francisco, California,
USA

THERE IS a strong common thread passing through the observations of many genuine UFO reports. A closer examination of these observations reveals a remarkable link with the laws of electromagnetism, which seems important and central to understanding the true nature of these enigmatic flying objects. Whatever the explanation for this link, this chapter is devoted to highlighting and explaining those essential common observations reported by witnesses.

Summary List of
Common UFO Observations

#	*Observations*	*Page*
1	Daytime UFOs are often described as grey or silvery metallic objects.	62
2	Daytime UFOs have exceptionally smooth outer surfaces.	63
3	UFOs seen at night tend to be glowing.	66
4	The position of the glowing region indicates the UFOs' direction.	69
5	The glowing region can diminish its luminosity when hovering.	73
6	The glowing region can increase its luminosity when accelerating.	74
7	The main central body of UFOs is symmetrical.	74
8	UFOs can come with protrusions (or fuselages).	76
9	Protrusions can have a glowing region at the end.	79
10	Symmetric and asymmetric configurations of UFO protrusions.	80
11	Three protrusions forming triangular or boomerang UFOs.	82
12	UFOs can replace protrusions with glowing domes underneath.	89
13	A multitude of small bright lights around the rim of UFOs.	90
14	The glowing region can blink on-and-off.	93
15	A flash of the glowing region can initiate movement of the UFOs.	94
16	The rate of flashing of glowing regions reveals the speed of UFOs.	95
17	Zig-zag / sinusoidal flight paths of UFOs.	96
18	Luminosity of glowing region changes near certain objects/substances.	98
19	UFOs can mimic behaviour of electrons in a magnetic field.	99
20	UFOs can cause metal objects to vibrate.	102
21	UFOs can heat up nearby metal objects.	103
22	UFOs leave behind coloured glows in the air.	104
23	UFOs can masquerade as cloud-like structures.	105
24	UFOs can cause disruption to man-made electrical circuits.	107
25	UFOs can bend light beams.	108
26	UFOs can render themselves invisible.	110
27	UFOs exhibit unusual acceleration behaviour.	113
28	UFOs emit a buzzing/humming sound at close range.	114
29	UFOs may come with antennae sticking out of the central body.	115

UFO seen during the daytime are often described as silver or grey metallic objects.

Evidence

The artist's impression below of a UFO was observed by 41-year-old Maurice Masse, a lavender farmer from Valensole, France, on July 1, 1965:

Notice the definite use of a metal (or alloy) in the construction of the object's exterior parts, including the legs to stabilise the object when on the ground, and a central cylinder that comes down to touch the ground. The latter is an unusual feature if the aim is to stabilise the object, as the legs should already be performing this task. This suggests there is something special about this central cylinder and the need for it to touch the ground. It is almost as if it is designed to discharge something into the ground.

The occupants were also observed to have worn bluish, skin-tight metallic suits covering almost all parts of their bodies (except the heads and hands). If the suit is made of a metal, then why is it necessary to use that metal on the body? What is wrong with normal insulating materials, like cotton, as a form of clothing?

2 UFOs have exceptionally smooth outer surfaces.

At close range, UFOs have been noted by witnesses for their remarkably smooth surfaces. There are virtually no signs of bolts, welding, rivets or other means to construct the objects on the outside.

Evidence 1

Dr. James M. McCampbell, a nuclear engineer who worked for NASA and Westinghouse Electric Company, commented on this "smooth surface" observation during his study:

> "Usually, the exterior surface of UFOs is reported to be extremely smooth. Many witnesses have commented upon this aspect, expressing surprise that they were unable to detect any line of adjoining plates on the surface or any rivets…"

A possible reason for this is to reduce air friction as the device moves through Earth's atmosphere, as American journalist turned UFO researcher John A. Keel has suggested:

> "…[A]ny kind of lumpy or irregular surface would present considerable drag and greatly reduce the potential speed of the object. Modern airplanes are made with as smooth a surface as possible. Even exposed rivet heads can cut down speed appreciably."[1]

---

1   Keel 1970, p.125.

Then again, it is a well-known fact in electromagnetism that a smooth surface can minimise the loss of energy around sharp edges and points, or regions of higher curvature where an electric charge is present on the surface.

Evidence 2

The previous French UFO case of the lavender farmer is typical of how a daylight UFO on the ground looks at close range with its smooth metal surface. The suits of the occupants were also described as smooth (i.e., a fine-textured and flexible material designed to follow the contours of the body), without zippers or buttons of any kind. Hard to imagine the need for any occupant to have a smooth suit for reducing air friction when the UFO does all the moving around through the air.

Evidence 3

It has been claimed by a scientific source working for the U.S. government that up to three UFOs were allegedly recovered by the U.S. military between 1947 and 1949. Details of the disks as known by the mysterious source at the time were published in 1950 in Frank Scully's *Behind the Flying Saucers*.

According to Scully's book, a UFO allegedly crashed to Earth approximately 19 kilometres northeast of Aztec, New Mexico, on a plateau (still fenced off today by the U.S. military) on March 25, 1948. In a matter of hours, the disc was found by a special team called the Interplanetary Phenomenon Unit (IPU), whose main function was to deliver crashed disks to specific secret locations.

The IPU brought a group of scientists to the site to investigate the crash. They were all sworn to secrecy. However, it is believed that a magnetic science expert named Dr. Carl Heiland broke the oath and passed on information to one of Scully's sources by the name of Leo GeBauer.

The magnetic science expert[2] claimed he saw no rivets, bolts, screws or signs of welding to suggest how the disc was put together. The metal surface was described as extremely smooth. Entry to the disc could be achieved only after a tiny crack was found in one of six portholes spread evenly around the disc's perimeter. It was enlarged, and a long pole was used to press a small knob inside the cabin, revealing a hidden door. The portholes looked metallic and appeared translucent only on close inspection.

Strange hieroglyphic symbols and numerous buttons and levers covered the instrument panel. Small display screens on the panel illuminated more of these strange symbols.

The disc was taken apart only after several interlocking key devices were uncovered, which opened seams on the inside at specific positions around the base. Scientists examining the craft realised the disc was made of a number of segments that could be slotted into grooves and pinned down internally without the need for nuts and bolts.

A number of extremely slender and lightweight occupants with unusually large heads, all of which had died, were allegedly

---

2   If this were true, it would show that the U.S. military was aware of the electromagnetic nature of UFOs by 1948 according to this crashed disc case. Does this mean UFOs are secret military experiments conducted by the USAF on a new electromagnetic technology?

recovered from inside the disc.

3   UFOs seen at night tend to be glowing.

UFOs seen at night at close range are usually described as highly luminous objects. When UFOs glow, their colors are highly reminiscent of a metal being heated to high temperature—namely red, orange, yellow, white and bluish-white.

Evidence 1

Shown below is an artist's impression of the UFO observed by Ronald Wildman on the intersection of Tringford and Ivinghoe roads in Bedfordshire, England, at 3.00 a.m. on February 9, 1962.

Wildman was a delivery driver for the Vauxhall Motor Company. He was driving a new estate car from the factory to Swansea when he noticed a strange glowing flying object. The object was hovering 6 to 10 metres above the Ivinghoe Road, about 50 metres from the intersection at Tringford. Wildman was heading away from the town of Ivinghoe and was heading toward Aston Clinton. But with the object ahead of him, high above the road, Wildman thought he could drive underneath it.

Wildman estimated the object to be about 12 metres wide. A series of black markings reminiscent of portholes could be seen around the circumferential edge. The entire object glowed as it hovered.

When Wildman was within 18 metres of the object, he noticed the engine power of the car began to fade. The car then slowed down until he was moving at 30 kilometres per hour. Putting the accelerator pedal down to the floor did not help. He tried to change to a lower gear, but it still wouldn't work. His headlights, on the other hand, stayed on throughout the encounter. All that Wildman could do was patiently let the

66

car move along the road as the object decided to move with the car, staying roughly 6 metres ahead of and above the car for a distance of 180 metres.

During this time, he noticed the headlights reflected from the object when it was closest to the road, suggesting to him that the object was solid.

Suddenly, a white haze, described by Wildman as "like a halo around the moon", appeared around the perimeter of the object (as if the air pressure near the surface of the object had gone down), and then it veered off to the right at a high speed. As it moved away, the object brushed fine particles of frost from the trees lining the side of the road and onto the windscreen of the car.

Power was restored to the car. But Wildman was sufficiently and visibly shaken enough by the experience to drive down to Aylesbury to report his encounter to police. Later, Wildman was approached by investigators from *Flying Saucer Review* to check on his story. In the end, everyone who spoke to the witness was convinced the man definitely had seen something real and not a hallucination.[3]

3   Brookesmith 1980–84, *The UFO Casebook*, p.21.

Evidence 2

Here is another artist's impression of the UFO observed near Levelland, Texas, on the night of November 2, 1957. This impressive UFO case revealed other fine details, such as a change in color of the glow as it moved away from the ground. For further details on this case, see Chapter 1.

Evidence 3

In July 1973, 20-year-old university student, Masaaki Kudou, while sitting inside his car overlooking the ocean (located near an industrial coastal town), saw several smooth, white, glowing spheres flying over the southern coast of the island of Hokkaido, Japan, approximately 800 kilometres north of Tokyo.

The incident began when Kudou observed what he thought was a "shooting star". But then the light descended until it was only 20 metres above the ocean and close enough for him to recognise the object for what it was. Still not believing what he was seeing, Kudou stared at it and observed a tube emerging from the object. The tube glowed as its tip finally touched the water.

A short while later, the spherical object retracted the tube into its structure. It then moved in the direction of Kudou's car and hovered noiselessly some 50 metres above him. At that distance, he could see a row of circular windows positioned about mid-way up the object and noticed two human-shaped figures in one window and another individual in another, adjacent window, presumably looking at him.

Afterward, a group of three or four objects of the same design, accompanied by a large, dark brown cylinder, descended on the scene.

Then all the glowing spheres manoeuvred into position around the cylinder, and each entered one end of it, one at a time, until they all disappeared inside. With the spheres now hidden, the dark cylinder scampered out of sight in a northerly direction

The whole incident lasted 12 minutes.[4]

4 The position of the glowing region indicates the UFOs' direction.

Witnesses have reported glows around the circumferential edge of UFOs. There is also another region where this glow is often observed: when the object is accelerating forward, its

Evidence 1

Campers at Wattamolla Beach in the state of New South Wales, Australia, noticed a brightly glowing light "dancing" around on the night of January 13, 1968. The light suddenly became

---

4   Brookesmith 1980–84, *The UFO Casebook*, pp.62–63.

rear surface glows.

larger, and the observers on the ground soon realized it was approaching them because it was already casting light over some nearby rocks. The silent object "switched off" (or dimmed its lights) and hovered 60 metres above the ground. Although it was situated over 400 metres away, the campers could clearly see a disc-shaped object with a dim red band of light along its rim. Witnesses claimed the UFO had responded to their torch signals before moving slowly away.[5]

Evidence 2

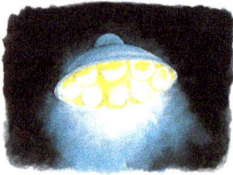

Here is an artist's impression of a UFO observed by a 29-year-old engineer named Paul Green and an anonymous independent witness, early on Sunday September 14, 1965. It was observed near a road going to West Mersea just outside Langenhoe, Essex, England.

Green was riding his motorcycle along the B1025 road in a southerly direction toward West Mersea heading home after his visit with his fiancée. It was around 1.00 a.m. when he passed the town of Langenhoe and overtook a rider on a scooter near the Pete Tye Common. Approximately a minute later, he was approaching the road turnoff to Langenhoe Hall when he suddenly heard a high-pitched humming noise emanating from his left (the east). Green tried to ignore it, but the noise persisted and grew louder.

Green looked up to observe what it was. He saw a blue light in the sky getting brighter and flashing. At first, he thought it was a light aircraft of some sort, but there was something odd about this light. It rapidly got larger, and the

5   *Australian UFO Review*, Number 10, p.25. UFOSH, p.57.

humming sound became more noticeable. He realised the object was approaching him. He tried to escape. Unfortunately, when the object reached the Langenhoe Marsh, about 1.5 to 2 kilometres east of the road, Green's motorcycle lost power. The engine tried to stay alive as it spluttered and coughed, but eventually went dead, including the headlights.

Green looked more closely at the blue flashing light in the sky. Careful observation helped him finally make out the object's shape and size. He noticed how big the object was, describing it as "about as big as a gasometer". Its shape resembled the upper half of a large spinning top with a dome attached on top.

Apparently the top of the object was tilted toward him as it moved. The flashes of light came from beneath the object. He couldn't see underneath until it began to slow down and slowly descend as if aware of his presence. This was when the object tilted its underside towards Green in an attempt to slow down.

The underside was glowing an intense yellow colour, with the brightest glow emanating from spherical or dome-shaped protrusions. The protrusions were spaced equidistantly and gave the impression to Green that they were doing a "luminous ball race". This probably meant one of the protrusions increased slightly in brightness before returning to its original brightness, followed by another protrusion glowing slightly more and then back to normal brightness, as if the object was trying to stay stationary in the air.

By now the object had stopped moving. Green dismounted from his motorcycle and took a few involuntary steps toward the object before he stopped and felt unable to move or speak. As the bright light flashed, he felt pressure hit against his chest in rhythm with the flashing light and with his heartbeat. As Green said:

> "I felt spellbound and unable to move or speak, just as if I had become paralysed. The flashing blue light became so intense that it was painful, and it appeared to fluctuate in rhythm with my heartbeat and hit against my chest. I felt myself tingling all over, rather like the electric shock one gets when handling an electrified cattle fence."

The humming sound suddenly went quiet as the object descended toward some farmhouses in Wick. A young lad in a leather jacket riding his scooter finally caught up with Green. His scooter lost power and he had to stop. He dismounted and stood looking at the object.

Green did not want to stay around much longer. As he said:

> "My head began to throb, and felt as though there were a band tightening around it. With a great effort I made myself move, and I grasped the bike and tried to start it."

Green pushed his motorcycle along the road, eventually achieving a bump start. He mounted his bike and rode off as fast as he could. A line of tall hedges eventually hid the object from view, but he could still see the blue glow in the sky left behind by the object.

He arrived home at 2.00 a.m. to tell his invalid mother what had happened. The next day, Green noticed his hair and clothes were so heavily charged with static electricity that they crackled continually.

Green was interviewed by Dr. Bernard Finch for *Flying Saucer Review*. Dr. Finch listened and felt convinced of the story after noting the symptoms Green had described. They reminded Dr. Finch of the effects of "a very powerful magnetic field on the human body".

Although the second witness on his scooter was never tracked down to help corroborate Green's claims, Green spoke of his experience to a friend who lived 8 kilometres north-west of Langenhoe near Shrub End. At the time of the incident, his friend said his dog started to bark. He opened the door to let the dog out, and it was at this time that he saw a blue light moving rapidly across the sky directly overhead. It was traveling toward the north-west.[6]

5   The glowing region can diminish its luminosity when hovering.

Glowing UFOs hovering in the air seem to require less energy resulting in a change in color to orange or red with reduced luminosity.

German rocket expert Professor Hermann Oberth has studied UFO observations in the 1950s. In 1961, Oberth wrote in the German journal *Mitteilungen Der Gesellschaft Fur Interplanetarik*:

> "They [UFOs] are dark orange and cherry red at night, if there is not much power necessary for the particular movement, for instance, when they are suspended calm. Then, they also do not shine very much."

---

6   Brookesmith 1980–84, *The UFO Casebook*, pp.52–52.

6   The glowing region can increase its luminosity when accelerating.

Glowing UFOs accelerating in the air seem to require more energy resulting in a change in color to yellow, white or bluish-white with increased luminosity.

Professor Hermann Oberth also noted another common UFO observation when he wrote in *Mitteilungen Der Gesellschaft Fur Interplanetarik*:

> "If more driving power is necessary, the shining [on UFOs] increases (brightens) and they appear yellow, yellow-green, green like a copper flame and in a state of highest speed or acceleration extremely white."

7   The main central body is symmetrical.

UFOs are universally symmetrical for the main central body. Even when there are external fuselages attached, there is always a central symmetrical body to be observed.

Below are artists' impressions revealing the symmetrical design of UFOs.

A. Artist's impression of a UFO observed by several witnesses off the coast of Brazil on Sunday June 27, 1970, not far from Rio de Janeiro, Brazil.[7]

B. Artist's impression of the turtle-shaped UFO observed by security watchman Almiro Martins de Freitas at 9.30 p.m. on August 30, 1970 at Funil Dam, not far from Rio de Janeiro, Brazil.[8]

C. Sketch from Carlos Mercado of the triangular UFO observed by the entire

---

7   Brookesmith 1980–84 *The UFO Casebook*, p.70.
8   Brookesmith 1980–84 *The UFO Casebook*, p.78.

communities of four towns on December 28, 1988, over the southwest corner of Puerto Rico.[9]

D. Artist's impression of the spherical UFO observed by Masaaki Kudou near the industrial town of Tomakomai on the southern island coast of Hokkaido, Japan, in July 1973.[10]

E. Artist's impression of the UFO observed near Levelland, Texas, USA, on the night of November 2-3, 1957.

F. Artist's impression of the UFOs observed by Kenneth Arnold in June 24, 1947 over the Cascade Mountains in Washington State, USA.[11]

A. DISC-SHAPED

B. TURTLE-SHAPED

C. TRIANGLE--SHAPED

D. SPHERE-SHAPED

E. OBLATE SPHEROIDAL-SHAPED

F. MISCELLANEOUS

9   Good 1990, p.195.
10   Brookesmith 1980–84 *The UFO Casebook*, p.63.
11   Arnold & Palmer 1952.

8   UFOs can come with protrusions (or fuselages).

UFOs can have three or six protrusions in fixed positions, or four movable protrusions around the main central body.

The metallic protrusions of UFOs around their symmetrical central bodies may move around to certain positions to help control the UFO's speed and direction.

Evidence

On 26 October 1967, a retired flight administrator for the British Overseas Airways Corporation and a former Royal Air Force photographic interpreter, J. B. W. Angus Brooks, was taking his usual morning stroll with his two dogs on Moigne Downs in Dorset, England. He finally rested and took cover in a shallow area because of the blistering wind that prevailed that morning.

Then suddenly, he was distracted by something in the air. He looked up and noticed a cloud of water vapour dispersing to reveal an object that was best described as a central disc with a diameter of approximately 7.5 metres, a thickness of some 3.5 metres and four "fuselages" sticking out of it, each estimated to be about 23 metres long. He could make these estimates because the object approached and descended to a height of about 90 metres above the ground. It was apparently decelerating by moving two of the three parallel rear fuselages forward until it hovered with a fuselage configuration of a cross with the disc at the centre. The entire object rotated through 90 degrees and just stayed there in the air for the next 22 minutes. During this time, one of the dogs, obviously nervous, tried to pester Brooks into leaving the area, but Brooks refused and decided to remain absolutely still. Finally, the object accelerated in an east-northeast direction by physically moving the front and rear fuselages to one side.[12]

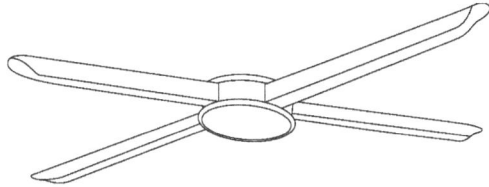

Hovering position

Acceleration to the left

Despite alternative explanations of his sighting suggested by investigators, Brooks remained firmly convinced that what he had seen was a controlled flying object of unique design and performance.

"Before the Moigne Downs sighting," said Brooks, "I was only mildly interested in unidentified flying objects, but now I am convinced there is something to be investigated and the sooner we find what is going on the better it will be."

----

12   Good 1988

In other sightings, the fuselages can be fixed into specific positions around the central body.

Evidence

In the early hours of 16 October 1957, Brazilian farmer Antonio Villas Boas was plowing his fields, alone in the light of his tractor's headlamps, when he suddenly noticed a "large red star" descend from the sky and approach him. The light soon became a large luminous egg-shaped object about 10 metres long and 7 metres wide. It landed on the field just in front of him.

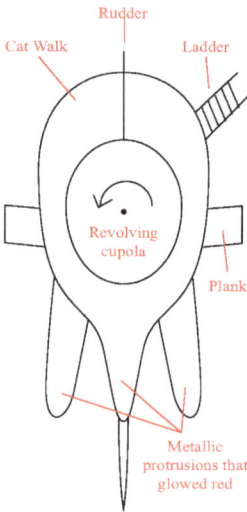

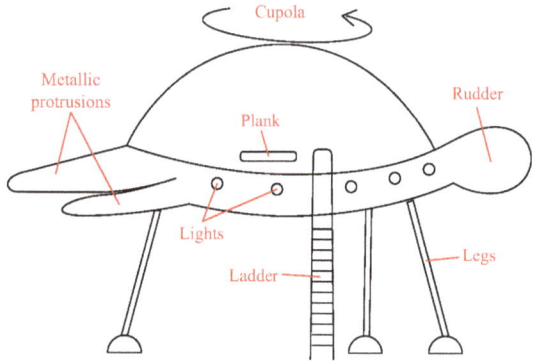

Rudder

Cat Walk

Ladder

Metallic protrusions

Cupola

Rudder

Plank

Revolving cupola

Plank

Lights

Ladder

Legs

Metallic protrusions that glowed red

As soon as the object landed, Antonio promptly decided to escape on foot (since the tractor could not be restarted) but was quickly grabbed by three humanoids in tightly fitting grey suits with large, broad, metallic helmets who dragged him to the strange machine.

The first room Antonio found himself in was square and metallic, brightly lit from above by small lamps. Later, he was escorted to an adjoining room described as large and oval-shaped with a metal column running vertically from floor to ceiling.

After more than four hours of exhaustive examination by the humanoids—including

78

having a blood sample and stomach contents taken, as well as having a sexual encounter with a female alien—he was then set free.

The craft, with its lights turned on and flashing, took off at high speed towards the south.[13]

9 Protrusions can have a glowing region at the end.

Protrusions that are designed to accelerate UFOs tend to glow and have the brightest glows.

Evidence 1

Artist impression of UFO seen by Antonio Villas Boas. When accelerating, the three rear protrusions glowed a red colour.

In the UFO abduction case of Antonio Villas Boas, the witness observed the tail and adjacent rear side protrusions glowing red when moving forward.

----

13   Brookesmith 1980–84, *The UFO Casebook*, pp.66–69.

The UFO allegedly seen by American couple Betty and Barney Hill when they drew it from memory, the UFO moved with two protrusions going out to the side and partially pushed back toward the rear. At the end of each protrusion was a bright light.

An artist impression of the UFO observed by Betty and Barney Hill.

10 Symmetric and asymmetric configurations of UFO protrusions

UFOs tend to position the protrusions in a symmetric configuration around the main central body when hovering, and in the asymmetric configuration when accelerating or moving forward.

Evidence 1

In the Angus Brooks UFO case mentioned in observation 8, the UFO hovered in the air by moving two of the four protrusions to form a cross, and hence a symmetric configuration.

80

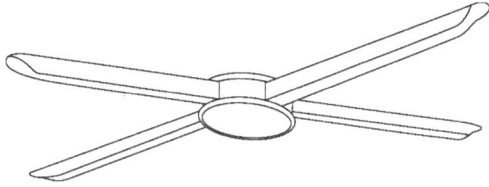

Hovering position

When moving forward, two protrusions are moved around to meet with a third protrusion.

Acceleration to the left

Evidence 2

In the Betty and Barney Hill UFO case, the witnesses draw a picture of the UFO with two side protrusions sticking out and slightly toward the rear when moving forward. As one of the witnesses wrote on the official hand drawing, "This [the protrusions] slip out from side with red lights".

However, the protrusions would get withdrawn into the central body and the lights align to the centre of gravity of the UFO when hovering.

81

In the Antonio Villas Boas UFO case, the protrusions need not have to move. The symmetric and asymmetric configurations of the protrusions can probably be switched on and off as needed.

11   Three protrusions forming triangular or boomerang UFOs.

UFOs can come with three protrusions. With added structural support, triangular shaped objects are common.

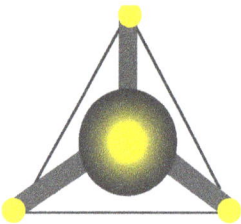

The minimum number of protrusions one would need to achieve a symmetric configuration is three. Depending on the choice of design and how much materials are added, the UFOs may be described as either triangular- or boomerang-shaped objects. At night, if the body is dark, bright lights or glows will appear underneath will appear at the apex points, and often with a main central bright light right in the middle. As such, witnesses will commonly report the UFO as triangular-shaped.

Evidence 1

In support of this observation, a UFO wave hit Belgium on November 29, 1989. A large triangular-shaped object descended over the city of Brussels and flew at low altitude across the country, heading towards the Netherlands and Germany.

The object made no sound as it traveled through the air. As it did, the object revealed its underside consisting of three bright lights at the apex of an equilateral triangle with a central light positioned at the precise centre. Either a dome on top or a central sphere with three wings attached around the perimeter was observed.

The triangular-shaped object made numerous visits to the Belgium region over a 7-month period.

The Belgian military was forced to get involved on the night of 30-31 March 1990, when a Belgian national police captain, Alain Renkin, called military headquarters at 11.00 p.m. stating that the object was flying in the air over Ramillies. Confirmation that the object was definitely in the air came from a ground radar station at Glons, where it was reported that the object was of unknown origin and with no transponder signal to identify itself. As Renkin said to investigators:

> "I was playing cards with my wife and a couple of neighbours when our attention was drawn to a light outside the house. The sky was clear and full of stars. The light made irregular lateral movements and changed colours. Having convinced myself it wasn't a plane, I contacted the base at Beauvechain and asked for information about this phenomenon. Because their radar system wasn't operational, the military contacted the radar centre at Glons...At about 23.05, the point of light became red, and it moved off in the direction of Gembloux, gaining height."

At 11.08 p.m., at the base, Captain Jacques Pinson received the report of the sighting from Renkin. He departed for Ramillies, together with a colleague. Both were able to confirm a visual sighting of the object, appearing as three points of light forming a perfect triangle. Pinson said,

"It was neither an aircraft nor a star."

Further support for the sighting would come from several police patrols in Jauchelette at 11.25 p.m.

A decision was eventually made by the military to send two pilots in F-16 fighter jets from the air base shortly before midnight to investigate and identify the mysterious intruder in the Brabant airspace. With guidance from police on the ground, the pilots made nine separate interceptions of the object and managed on three occasions to lock on to it with radar. The object appeared on the pilots' screens as a clear and solid target. However, a few seconds after locking on the target, the object accelerated out of range of the pilots' radar, and in another situation the object simply changed direction and speed so fast that the radar could not lock on. Police on the ground reported how the points of light on the object became brighter as if in response to the jets flying in its direction.

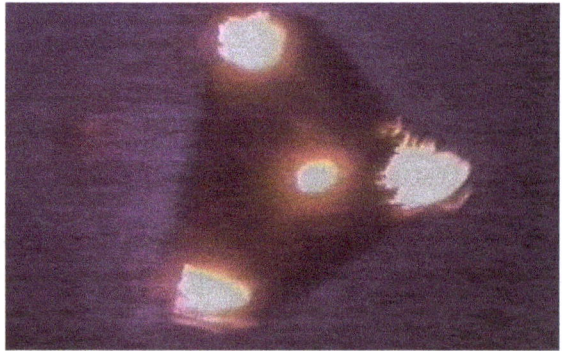

An anonymous witness (and his girlfriend) took this photograph of the triangular UFO observed over Belgium in April 1990.

An hour-long chase ensued, as if the object wanted to play a cat-and-mouse game and perhaps test the capabilities of the pilots' aircraft. But the pilots failed to capture or identify what it was.

Pilots reported the object made unbelievable manoeuvres at speeds beyond the capabilities of their own fighter-jet technology. In the first radar lock, the pilots observed the object accelerating from 240km/h to over 1,770km/h while descending in altitude from 2,700 metres to 1,500 metres in less than two seconds without creating a sonic boom, then ascending to 3,350 metres at high speeds before descending once more to almost ground level. Radar specialists on the ground were also astonished by the object's remarkable speed and manoeuvrability.

The Belgian Air Force could not give an explanation other than to eliminate the following possibilities from its investigations:

1. Balloons
2. Ultra-light aircraft (ULA)
3. Unmanned aerial vehicles (UAV)
4. Aircraft (including stealth)
5. Laser projections or holograms
6. Mirages or other meteorological phenomena.

By night's end, the triangular object was sighted, photographed and tracked on radar by approximately 13,500 people on the ground, of which approximately 2,600 filed written statements detailing what they had seen to the authorities.

Over the next three months, the cheeky aerial rascal would make its triumphant return more than 1,000 times, both day and night, as if whatever was flying this UFO was trying to tell us something, or give us a lesson. People tried to take photographs with their trusty cameras. But nearly all were blurry and too far away to make out sufficient details. However, one good image was obtained on videotape. It was taken in April 1990.

Another photograph surrendered for analysis by a witness, J. S. Henrardi, was possibly the clearest available on still film. The witness alleged the picture was taken in the evening of June 15, 1990.

A massive skywatch was organised by SOBEPS during the Easter long weekend of April 14–16, 1990. The world's television crews camped out on the Ardennes airfield hoping to catch a glimpse of the mystery object. However, as if the object knew a large crowd was present, it kept flying low and just below rooftops. Occasionally, some witnesses reported seeing the three points of light in the sky, but the object managed to disappear before television crews could move into place and take pictures.

Scientists did get involved. According to an article in London's *Financial Times* by Lucy Kellaway on April 18, 1990:

> "Scientists on the ground appear in the past few days to have produced a clear image of the object, which is said to correspond to the reports of eyewitnesses. It is a triangle 30m-50m in diameter, with red, green and white lights at the corners, ten times brighter than any star. It has a convex underbelly and makes a sharp whistling noise..."

By the time the object decided to leave Belgium for good, never to be seen again, more than 15,000 witnesses could testify to having seen the object.

Skeptics, on hearing the reports, have had a hard time trying to find an explanation using prosaic natural or man-made explanations. So

86

far, the best they can suggest is that perhaps it was a helicopter and that the lack of noise could be explained by a strong natural wind blowing in the direction of the witnesses, or the wind noise generated inside a vehicle while the witnesses were driving. The Belgian Air Force had to disagree.[14]

Major-General Wilfred de Brouwer, Deputy Chief of the Royal Belgian Air Force, said:

> "...the Air Force has arrived to the conclusion that a certain number of anomalous phenomena has been produced within Belgian airspace. The numerous testimonies of ground observations compiled in this [SOBEPS] book, reinforced by the reports of the night of March 30-31 [1990], have led us to face the hypothesis that a certain number of unauthorised aerial activities have taken place. Until now, not a single trace of aggressiveness has been signalled; military or civilian air traffic has not been perturbed nor threatened. We can therefore advance that the presumed activities do not constitute a direct menace. The day will come undoubtedly when the phenomenon will be observed with technological means of detection and collection that won't leave a single doubt about its origin. This should lift a part of the veil that has covered the mystery for a long time. A mystery that continues to the present. But it exists, it is real, and that in itself is an important conclusion."

---

14 http://en.wikipedia.org/wiki/Belgian_UFO_wave. More details available in Good 1990 *The UFO Report 1991*, pp.57–63.

Evidence 2

At approximately 7.00 p.m. on 24 January 2015, a retired police officer and U.S. air force veteran reported seeing a low-flying UFO with three bright lights, two at the ends and a central light. It was moving parallel to the Daubenberger Road in Turlock, California, in partially foggy conditions. The configuration of the lights and direction of motion of the UFO is interesting in the sense that the central bright light was behind the other two lights and so was acting as the rear end of the UFO as it moved forward. As the witness stated to investigators, "Three white lights in a backward triangle possibly a boomerang shape."

Passing over some buildings, the UFO sped up and moved up into the sky and vanished in the foggy night. MUFON Case No.62855.

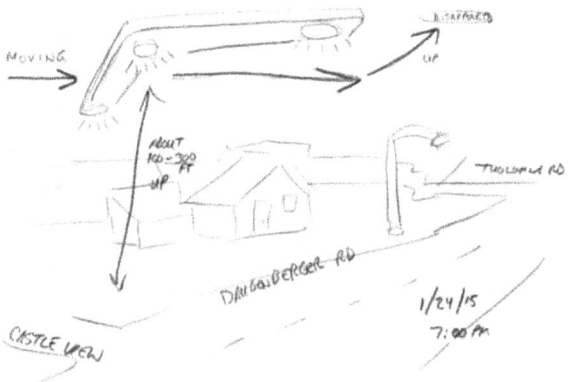

A hand drawing made by the witness, giving an indication of direction, height above ground, relative size to a couple of buildings on the ground, and his interpretation on the shape of the UFO.

UFOs can replace protrusions with glowing domes underneath.

Instead of protrusions, UFOs can have a minimum of three and as many as eight domes underneath. The domes will glow at night, or appear metallic in the daytime together with the usual changes in color and intensity in the luminosity during acceleration and hovering.

Most UFOs will have a large glow or bright light directly underneath the main central body, as probably needed provide lift and keep them in the air. Other UFOs can be harder to detect this glowing region, especially in the daytime. At any rate, the glow can be large and include metallic domes underneath.

Evidence 1

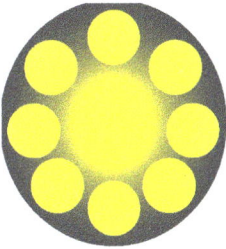

We have seen a good example of this in the Langenhoe UFO case as observed by Paul Green (see Observation 4).

Evidence 2

The Finland UFO case discussed in Chapter 4 reveals another fine example of three metal domes underneath the object together with a central opening to permit a beam of light to emerge.

*Artist impression of the UFO observed by two Finnish skiers
36-year-old forester Aarno Heinonen and 38-year-old farmer
Esko Viljo on January 7, 1970.*

13   A multitude of small bright lights around the rim of UFOs.

Instead of protrusions that would move or lock into fixed positions, extra lights may appear at specific positions around the circumferential edge.

Evidence 1

At 9.30 p.m. on August 30, 1970, security guard Almiro Martins de Freitas was patrolling the Funil Dam at Itatiania in the state of Rio de Janeiro. Not long after heavy rains had fallen, with the ground still wet, de Freitas had just finished his round and was returning when he suddenly noticed a turtle-shaped object, with multi-coloured lights around the rim, sitting on a mound. The lights were flashing orange, red and blue at a high rate.

His initial instinct was to retreat as fast as he could. But he overcame his fear and decided he would investigate more closely. He carefully approached the object, trying to figure out what it was. He came within 15 metres of the object. Unfortunately, the darkness made it hard for him to gather more details.

Suddenly, a deafening noise emanated from the object like the sound of a jet engine. The security guard was startled. He drew his revolver and started firing at the object. The object appeared to take another course of action to protect itself. After the second shot, a bright flash of light hit his face. He was blinded. He fired a third shot, this time wildly. The object emitted a wave of heat that engulfed the man, causing his muscles to stiffen up and become paralysed.

Another watchman and a passing motorist came to his aid. The object had left, leaving behind a dry mound of earth where it had landed, compared to the wet ground that surrounded the spot. De Freitas was standing stiffly next to the mound and holding his revolver as he warned them:

"Don't look! Beware the flash! It has blinded me!"

He was taken to the car, and after a while, he recovered the use of his limbs. However, his sight did not return to normal. After he was examined by doctors at a local hospital, they could find nothing wrong with de Freitas. They assumed the blindness was probably caused by the shock of whatever it was he had seen or heard.[15]

Evidence 2

The witness hand sketch of the UFO.

In Chapter 4, we learned of the case of Alfred Burtoo, in which he claimed to have been taken onboard a UFO, not far from the town of Aldershot in the United Kingdom. In his hand drawings, we can observe large bright lights around the rim of the object as it took off, as well as a bright light underneath when lifting off the ground.

Artist impression of UFO during take off. Scene rendered as in the original location with Basingstoke Canal shown on the right, viewed from the road.

Alternatively, some UFOs may reveal no obvious signs of positioning the glow other than the common observation that the intensity of the glow tends to be greater below the object than on top, especially when lifting off the ground. However, it is altogether conceivable that a similar mechanism of changing the position of the glow to create a symmetric or asymmetric

---

15   Brookesmith 1980–84 *The UFO Casebook*, p.78.

configuration is probably being deployed but done internally.

We see an example of this in the Levelland UFO case. Further details about this case is revealed in Chapter 1.

14   The glowing region can blink on-and-off.

Glowing UFOs or UFOs with bright lights on the ends of protrusions or around the circumferential rim have been observed to blink on-and-off in a periodic manner.

Evidence

On 1 October 1948, between 9.00 p.m. and 9.27 p.m., 25-year-old Lieutenant George F. Gorman of the North Dakota Air National Guard observed and chased a UFO while flying in his F-51 plane over Fargo, North Dakota.

When he was within about a kilometre of the object, he described it as:

> "…about six to eight inches in diameter, clear white, and completely round with a sort of fuzz at the edges. It was blinking on and off. As I approached, however, the light suddenly became steady [and intensified] and [it] pulled into a sharp left bank…"

In a sworn statement to Project Sign investigators taken on 23 October 1948, Gorman said:

> "I am convinced that there was definite thought behind its manoeuvres.
>
> I am further convinced that the object was governed by the laws of inertia because its acceleration was rapid but not immediate and although it was

able to turn fairly tight at considerable speed, it still followed a natural curve. When I attempted to turn with the object I blacked out temporarily due to excessive speed. I am in fairly good physical condition and I do not believe there are many, if any, pilots who could withstand the turn and speed effected by the object, and remain conscious.

The object was not only able to out turn and out speed my aircraft...but was able to attain a far steeper climb and was able to maintain a constant rate of climb far in excess of my aircraft."

This sighting was backed up by two other independent witnesses on the ground—Lloyd D. Jensen and H. E. Johnson. Both were on air traffic control duty on the night in question.[16]

15 A flash of the glowing region can initiate movement of the UFOs.

Witnesses have reported seeing UFOs accelerate after the appearance of a flashing light or glowing effect.

Evidence

A New York physiotherapist and his family were having a picnic on April 11, 1964. At around 6.30 p.m., the witnesses sighted a dark, oblong object in the sky. Through binoculars, the family could see the object was in a vertical position and emitting smoke around its main body. Then, in the words of the physiotherapist:

"As I was observing it with my binoculars, there was a flash of white

---

16   Project Blue Book files, National Archives, Case No.234. Also mentioned in Vallee, Jacques & Janine 1966, pp.138–139; Flammonde 1976, p.259; Story 1980, pp.151–152.

light from the rear of it and it shot forward with incredible speed for a distance of about five times its length."

The object performed other extraordinary feats in the air. The whole sighting lasted 45 minutes.[17]

16 The rate of flashing of glowing regions reveals the speed of UFOs.

The frequency of the on-and-off blinking effect of glowing UFOs controls their speed. Generally the faster the rate of flashing, the faster the UFOs move.

Evidence

There is an interesting case from Russia of a UFO revealing the direct relationship between its speed and the rate at which its bright lights flashed. Generally, the higher the frequency of the flashing, the faster the UFO moves.

On March 21, 1990, over 100 reports of a highly manoeuvrable UFO with two bright flashing lights on its sides were made by ground and aerial observers. The UFO appeared near Pereslavl Zalesskiy, approximately 128 kilometres northeast of Moscow, and the incident was officially reported by General Igor Maltsev.

Captain V. Birin, a Soviet Air Force pilot who witnessed the object, reported what he saw:

"The object looked like a flying saucer with two very bright lights along the edges. Its diameter was approximately 100-200 metres (judging by the shining lights). A less intense light, which looked like a porthole, could be seen between the two bright lights....The

---
17   Vallee, Jacques & Janine 1966, p.43; Vallee 1965, p.214.

trajectory depended on the flashing of the bright side lights: the more often they flashed, the faster the speed of the UFO and vice versa. While hovering, the object extinguished its lights almost completely. At 22.30 hours the object headed off in the direction of Moscow.

The movement of the UFO was not accompanied by sound of any kind and was distinguished by its startling manoeuvrability. It seemed that the UFO was completely devoid of inertia. In other words, they had somehow "come to terms" with gravity. At the present time, terrestrial machines hardly have any such capability."[18]

Slow movement

Faster movement

17 Zig-zag / Sinusoidal flight paths of UFOs.

Flight paths of UFOs can take a zig-zag or sinusoidal pattern.

At low frequencies of the rate of their flashing lights, UFOs have exhibited flight behaviors

18   Good 1991, pp.7–10.

described by some witnesses as zig-zag or like "stones skipping on water". Sinusoidal flight paths are also not uncommon.

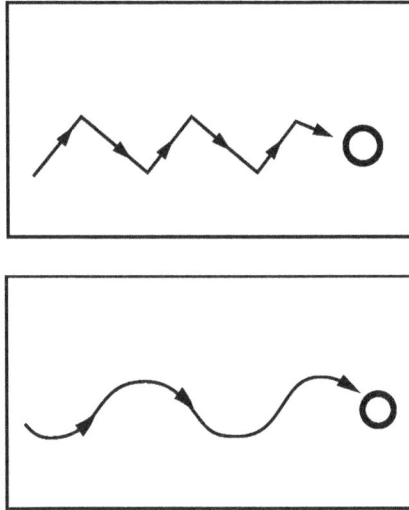

Evidence

The following extract came from the NORAD (North American Aerospace Defense Command) Command Director's logs, released on March 26, 1976 by the U.S. Freedom of Information Act. It relates to UFO activity near North Bay, Ontario, Canada, in a region of the Falconbridge Air Force Station:

"November 12, 1975 0715 23rd NORAD Region. UFO Reported from Radar site at Falconbridge Ontario, Canada (Sudbury). Reported by Mr. Julian Prince of Sudbury thru Ontario Provincial Police (also observed by 2 OPP constables ZADOW & BRETT) 2 objects seen appeared to be artificial light fading on and off with jerky motion. Broken cloud layer with no estimated base. No radar contact made and no request for fighter scramble initiated."[19]

18 Luminosity of glowing region changes near certain objects / substances.

The luminosity of the glowing effect varies over certain materials and substances. For example, the glowing effect will dim down as the UFO moves over metal surfaces, large bodies of water, and in close proximity to other UFOs.

Evidence 1

Seventy-seven-year-old Alfred Burtoo of Aldershot, England, revealed one interesting observation concerning the UFO he had seen on the night of August 11, 1983. Burtoo said:

> "…[I] saw a vivid light coming towards me from the south, which is over North Town. It wavered over the railway line and then came on again, then settled down. The vivid light went out, though I could see a light through the boughs of the trees."[20]

Evidence 2

In Chapter 4, we learned of the UFO case from France, where on the sunny afternoon of Friday 17 October 1952, hundreds of witnesses from Oloron-Sainte-Marie and nearby villages in southwest France saw numerous small UFOs and a larger, cylindrical white object moving across the sky.

This artist's impression of the UFOs in action reveals a fascinating observation: when pairs of these small UFOs approached each

19   Fawcett & Greenwood 1984, p.46.
20   Good 1997, p.88.

other and then moved away, a kind of "electric arc" was produced between the bottom end of one of the UFOs traveling higher up and the top end of the second UFO traveling below it. This strongly indicates electric charge is present on the UFOs with some kind of positive and negative charge difference over the top and bottom ends of the objects.[21]

19   UFOs can mimic behavior of electrons in a magnetic field.

UFOs can exhibit bouncing, drift and spiraling motions that are reminiscent of an electron moving in a magnetic field.

Evidence 1

On a night in July 1975, not far from the Hobart Airport in Tasmania, Australia, the driver and passengers of a car were forced to stop at around 9.30 p.m. after an unexplainable electrical malfunction and a sensation of heat was experienced in their car. After the car rolled to a halt, the witnesses suddenly saw a spherical object suspended motionless 20 metres above the road ahead of them. It was estimated to be about 5 metres in diameter and at a distance of 200 to 300 metres. The witnesses described the color as white-grey, with a metallic, iridescent appearance. Still transfixed by the sight, they continued observing the object until it took off at high speed, moving in a spiraling fashion. Understandably anxious to get to their homes, the witnesses managed to restart the car, and they immediately drove off.[22]

----

21   Brookesmith 1980–84 *The UFO Casebook*, pp.37–38.
22   Tasmanian UFO Investigation Centre, *Australian Centre of UFO Studies Bulletin* 12, p.11.

Evidence 2

On the night of July 27, 1965, at approximately 7.40 p.m., near Carnarvon, Western Australia, a Mr. Kulka and a Mrs. Lawrence saw in the sky a light changing from green to orange. The light grew bigger until it was clearly seen to be a glowing disc. Apparently, the object then hovered in the air and began rocking (or bouncing) from side-to-side and drifting slowly westward at the same time. It eventually went out of sight without making a noise.[23]

Evidence 3

The best UFO case the Grand Canary Island authorities have on record is from the evening of June 22, 1976. The UFO was witnessed by many people, and some photographs were taken of it. The highly active UFO apparently performed interesting flight behaviors, including a spiraling motion, in front of the witnesses.

One particular report from three men in a taxi cab who were at close range gave explicit

---

23   *Telegraph*, Brisbane, August 18, 1965. *Australian Flying Saucer Bureau.*

details of the object. The men reported how their radio faded out when confronted by this object while it hovered only a few metres above the ground not far from where the men were positioned. It was described as a giant "perfect sphere—as perfect as though marked out with a compass". It had a pale-greyish-blue rim and a yellow-white centre. The protective outer sphere was transparent, and the witnesses could see two humanoids wearing large black helmets and deep-red tight-fitting suits, both standing on a platform. Only their profiles could be seen, as they were facing each other at a control panel at the centre.

As soon as one of the men turned the car's headlights on the humanoids, the sphere immediately rose until the witnesses could see a transparent tube inside, which was emitting a blue gas or liquid. This gas, which swirled inside, expanded the sphere, but the aliens and instrument console remained at their normal size. The gas was finally switched off, and then the object made a high-pitched whistle and took off at high speeds, causing the spherical bubble to bend into a thin, elongated form with a halo of light surrounding it.[24]

24   Brookesmith 1980–84 *The UFO Casebook*, pp.30–32.

20   UFOs can cause metal objects to vibrate.

When UFOs approach nearby metal objects attached to the ground or another object, those metal objects are likely to vibrate.

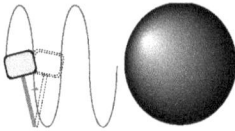

UFOs that emit some kind of invisible vibrational energy into the environment have been observed to cause nearby road signs to vibrate from side-to-side, presumably in time with the frequency of the emitted energy. Antennae attached to UFOs have been seen to vibrate in the same way.

Evidence

At 3.00 p.m. on April 14, 1957, a Mrs. Garcin and a Mrs. Rami were walking along Road D24 (not far from an intersection) in the direction of the French village of Vins, about 1 kilometre away, when all of a sudden they heard a deafening noise. They turned around, and descending before them was an object described as "like a big top" about 1.5 metres high and 1 metre in diameter. Mrs. Rami said that "on the upper part there were numerous metallic rods that looked like car antennae". Apparently, the noise they heard was a result of a road sign near the intersection vibrating furiously. The women also noticed the metallic antennae on the object strongly vibrating. No longer intrigued by its presence when the object touched the ground, the two women screamed.

Louis Boglio, who was working in a field only 300 metres away from the UFO, also heard a great noise. "I thought there had been a car crash," Louis said in his report. "I ran to the site of what I thought was an accident; and I saw a metallic craft, which made an enormous leap through the air."

The object then briefly hovered near the ground not far from where Louis Boglio was standing, made another leap in the air and stopped at a point near a second road sign—it too began to vibrate. A short while later, it took off into the distance.

Two other witnesses in the neighbouring village, La Moutonne, reported seeing the object in flight.[25]

21   UFOs can heat up nearby metal objects.

UFOs suddenly shooting across a road can cause steering wheels of nearby vehicles to heat up. Witnesses have noticed the interior of cars getting very hot when UFOs hover nearby.

Evidence

On March 13, 1980, a 31-year-old man was driving his car between Stratford-upon-Avon and Worcester at 7.50 p.m. when, without warning, an unusual object—described as cigar-shaped and glowing white with red lights at its ends—flew right across the road in front of him at high speed. The result of the object's passage was a sudden burning sensation in his hands; he realized then, after unwittingly making a few unmentionable remarks to himself, that the car's steering wheel had suddenly become very hot. His burns were severe enough to require medical treatment. The incident occurred not far from Haselor, a village in Warwickshire, England, and the object was reported to have headed in a north-easterly direction.[26]

25   Jacques & Janine Vallee 1966, pp.19–21.
26   Randles 1983, p.86. The case was investigated by Tony Green and published in *Flying Saucer Review*, Volume 26, Number 5, 1981.

22   UFOs leave behind colored glows in the air.

UFOs have been seen leaving behind aurora-like glows in the air. Also worth noting is how witnesses have reported the smell of ozone in the presence of UFOs.

Evidence 1

As British UFO investigator and researcher Jenny Randles stated in her study of UFOs:

> "Witnesses do sometimes describe the presence of a smell in association with a UFO....When a witness is asked to compare it with a known odour...it is found that two are most common. One is ozone. Another is nitrous oxide. Both are possible by-products of ionisation in our atmosphere (whose chief constituents are oxygen and nitrogen)."[27]

Evidence 2

There is an interesting case in Russia of a military pilot observing how a UFO, as it accelerated, left behind an unmistakable greenish-blue cloud. Dr. Felix Zigel, professor of cosmology at Moscow's Aviation Institute, spoke with Captain Vladimir Dubstov, the pilot who observed the UFO over the Arctic Ocean. While flying his patrol bomber on 22 October 1980, Dubstov noticed a rounded object below him. He decided to investigate.

"He told me it was truly immense," Zigel said, after having conversed with the captain. "A cone of light protruding down from it gave it an

27   Randles 1983, p.104.

eerie appearance, but it showed no sign of life. Then Dubstov's instruments went haywire, and he lost altitude. The UFO took off vertically and soared past him, leaving behind a greenish-blue cloud. Dubstov nursed his crippled jet home and reported the incident."[28]

23   UFOs can masquerade as cloud-like structures.

UFOs seem to favor foggy conditions to conceal themselves. They will also create clouds around the object.

Faint halos of light or mist surrounding the circumferential edges of UFOs have been commonly seen[29].

On closer examination of these halos, witnesses have reported a gap between the outer surface of the UFO and the ring, suggesting a reduction in air pressure on the very surface of the UFOs. Perhaps this reduction in air pressure helps to explain how UFOs achieve high speeds through the air?

Besides the usual white or grey misty, cloud-like colours, these "halo effects" can be red, though blue or green are not uncommon. In fact, such colours are remarkably similar to auroras that we see in the sky (which are known to be caused by air molecules being electrified by collisions with solar radiation, only to regain their lost electrons and in return emit the coloured lights).

Where there is a reduction in air pressure, and humidity is present in the atmosphere, these halos may sometimes expand to encompass the entire object in a glowing haze or fog-like structure. Hence, some UFOs have been observed to masquerade as clouds.

28   Blundell & Boar 1984, p.172.
29   Story 1980, p.163.

Evidence 1

For example, in the story of Moses as described in the Bible, a mysterious cloud with a bright glow in the centre was observed by a large number of people as it allegedly "guided" Moses and his people out of Egypt.

Evidence 2

The Finland UFO case discussed in Chapter 4 reveals a modern-day version of a UFO immersed in a cloud-like structure as it flew into range of two witnesses.

Evidence 3

On November 10, 1961, the Yugoslav news agency *Tanyug* officially published the following interesting report:

Mysterious Radioelectric Incident in
Croatia

Belgrade: A few days ago, the transmitter at the local radio station of the Croatian town of Vukovar was suddenly blacked out. In the studio the light went out, and then flickered on and off for thirty or forty seconds. Recording instruments showed a sharp increase in voltage. At the same moment, a strange dark-grey cloud passed over the town. According to the Belgrade newspaper 'Politika', a radio technician witnessed another extraordinary phenomenon. Several sodium lamps on a shelf completely isolated from any electric apparatus or electric cable began to glow. So far, no scientific explanation has been found that would account for the connection, if there was one, between the cloud and the phenomena observed at the radio station.[30]

24   UFOs can cause disruption to man-made electrical circuits.

UFOs have been known to cause electrical disruption to cars and aircraft. Where UFOs are seen hovering over power lines or power stations, blackouts are not uncommon.

Evidence 1

Experts Convinced ...

Dazzling UFOs Caused Mysterious Power Blackouts

Dazzling UFOs which lit up the night skies over Honduras have been linked by experts to two mystifying power blackouts which plunged the capital city into darkness.

Hundreds of eyewitnesses — including government officials — were terrified by the strange glowing objects.

And a Honduran professor openly admits he believes the power losses are due to a "controlled force" that could have come from "extraterrestrial life."

On Oct 14, 1968, a bizarre boomerang-shaped UFO was sighted swooping through the sky only a few seconds before power was knocked out for 25 minutes.

Just two weeks later, on October 27, a brilliant octopus shaped UFO appeared — and the sightings of this strange craft also coincided with a massive electricity failure of 1 hour and 10 minutes.

The Honduran government officially attributes the power losses to children's kites getting tangled in the power lines. But many people — even some in government — scoff at that theory.

"That's absurd," snorted Dr. Salvador Pardo, dean of the school of engineering at the National and Autonomous University of Honduras. "Kites could not have caused this damage.

"The power went out and then came back — without repairs, he stressed. "This to me makes it obvious that a controlled force caused this blackout. It is perfectly logical to assume that extraterrestrial life could have caused the blackouts."

Added Lieut. Alexander Her

By MEL LUNA

nandez, commandant of the Police Officers Academy at Tegucigalpa, the capital city:

"The kite theory is a joke. I personally have seen UFOs in the past here."

Prof. Jose Berrian, director of the program for technical education for the Honduran government told THE ENQUIRER that he, too, had seen a UFO — on September 16. "What appears to me is that a UFO swept down the high tension lines for whatever reasons and sucked up the power," he said.

As for his own encounter, Prof. Berrian added that while driving at about 1 a.m. September 16 he and his wife had spotted a "glowing globe" that descended rapidly and landed just off the road. "I slowed down and wanted to take a close look," he recalled. "But my wife was terrified and wouldn't let me."

Major Honduran newspapers and TV and radio stations have

been flooded with hundreds of reports of UFO sightings dating all the way back to September.

Radio America, one of the country's most popular radio stations, received as many as 200 calls the night of October 27.

"All described the object in exactly the same manner — an octopus-like object showing a

brilliant light," commented Radio America news director Rodrigo Wong.

The first blackout, on October 14, was linked to a very different UFO.

"I noticed a V-shaped or boomerang-like object hovering over the airport about a kilometre (half a mile) away at an altitude of 300 meters (1,800 feet)," remembered Rogelio Berrian, director of publicity for La Tribuna, and a brother of Prof. Berrian.

"Suddenly I saw the UFO dive over the airport at an incredible speed and the entire city went black I later saw this object soaring up in figure-eight maneuvers before it disappeared."

But even more spectacular was the incredible glowing ob-

ject with tentacles of blinding light that coincided with the power outage of two weeks later.

"It looked like an octopus with moving tentacles," remembered taxi driver Robert Aguilar. "As it swept down into the valley — boom! — all the lights went out."

Hernan Badgette, press secretary to the Honduran military junta, was standing on the terrace of the Maya hotel in down town Tegucigalpa with Associated Press correspondent Tom Fenton.

"I saw a large, bright ball of light," Badgette remembered. "From the white center of the object multicolored rays of light descended downward.

"Some were blue and red. They looked like balls of lighting. It disappeared at great speed."

SHE SAW IT: Danatila Hernandez Mojan points to where UFO hovered over substation.

BERCIAN   LOPEZ   AGUILERA

Evidence 2

Another good example supporting this common observation is the case of the glowing oblate

30   Vallee, Jacques and Janine 1966, p.220.

spheroidal object observed by a number of witnesses near Levelland, Texas, on the night of November 2-3, 1957 (as discussed in Chapter 1). The object apparently cut out, rather inconveniently, all electrical power to motor vehicles and trucks just on the outskirts of Levelland before disappearing. At least two dozen drivers observed this effect and complained to police at Levelland about their sighting.

25   UFOs can bend light beams.

Beams of light from car headlights or emanating from a UFO can bend for no apparent reason. This will only occur in the presence of UFOs at close range.

Light beams (e.g., car headlights) have been observed to bend at right angles in the presence of UFOs. Where the UFO emits a light beam, it can appear to bend at right angles or stop at a distance through the air, as if it is being bent back on itself by 180 degrees.

Evidence

A respected businessman, 38-year-old Ronald Sullivan, was driving his car along a straight stretch of road between Bendigo and Saint Arnaud in the state of Victoria, Australia, on the clear, moonlit night of April 4, 1966. In a statement he made to the police at Maryborough, near Melbourne, he said his car's headlight beams suddenly bent sharply to the right for no apparent reason. The strange effect on his lights occurred at a point on the road some 15 kilometres east of a small town called Bealiba.

Sullivan swerved his car back onto the road and narrowly avoided a potentially fatal crash. After stopping his car, he looked out at what had caused the light beams to bend.

He could see nothing that could have reflected, refracted or absorbed the light. However, in the field alongside the road, he could see a display of "gaseous lights" emitting different colours of the spectrum. Then an object rose from the ground, and as it reached 3 metres into the air, the coloured-light spectacle and object suddenly disappeared.

Mr. Sullivan returned to the scene on April 8, when he learned of a fatal accident the night before at exactly the same spot where he had his encounter with the mysterious object. The motorist and only occupant of the vehicle, Gary Turner, was killed. Police investigated the accident but could find no logical reason for the crash. The only peculiar thing police discovered near the crash scene was a 1.5 metre diameter circular depression, between 5 and 13 centimetres in depth, in the field off to the side of the road, some 20 metres in from a fence. When Sullivan learned of this fact, he went to the police to report what he had seen.

Since the death of the motorist, the object

has not returned.[31]

26 UFOs can render themselves invisible.

UFOs can render
themselves invisible in a
periodic or constant way.

As researcher Kenneth W. Behrendt has noted in a number of reports:

"There is now considerable evidence that some UFOs are able to become invisible. There are many cases in which an observed hovering UFO fades away or alternates between visibility and transparency. Very occasionally, one finds a case wherein the UFO is detected on radar, but cannot be observed visually."[32]

Evidence 1

The photograph[33] shown below is considered genuine by scientific experts. It was taken by Stephen Pratt at around 8.30 p.m. on March 28, 1966 over the skies of Conisbrough, South Yorkshire, England. Also observed by his mother, the UFO appeared as a "throbbing" orange light as it moved slowly westward lasting 10 minutes. When the negative film was developed, three black objects appeared in the sky. This is either a flight of three similarly shaped UFOs, or a single UFO rendering itself invisible in a periodic way as it moved across the sky.

---

31   *London Daily Express*, April 12, 1966; Brookesmith 1980–84, *The UFO Casebook*, p.80.
32   Evans & Spencer 1988, pp.296–297.
33   Jackson 1992, p.42.

Evidence 2

In December 1979, Wang Dingyuan and Wang Jianming were driving together in separate trucks early one morning, when suddenly they were confronted by a beam of light on the road ahead. The truck drivers stopped. From the safety of their cabins, Dingyuan and Jianming could make out two humanoids standing in the light. Then, without warning, the light, together with the humanoids, disappeared. But as soon as the drivers were about to continue on their way, the entities reappeared. Jianming, now frustrated, decided to get out of his truck and walk over to them with a crowbar in one hand. However, the entities and the light beam disappeared for good.[34]

Evidence 3

On September 30, 1954, in a remote, rural region near Nouatre, France, a team of seven construction workers—Mr. Beurrois, Mr. Lubanovic, and Messrs. Sechet, Villeneuve, Rougier and Amiraut, and their supervisor, Mr. Georges Gatay—were walking towards an excavation site when they encountered something a little out of the ordinary. All had been expecting the day to be quite uninspiring, until around 4.30 p.m. when Gatay found himself irresistibly drawn away from the group and compelled to walk in a different direction for some unknown reason. He later recalled feeling particularly drowsy at this time. Then, all of a sudden, to his and his co-workers' amazement, they saw a strange humanoid creature dressed in a grey suit standing about 30 feet away from where Gatay stood. Just behind the creature was a large, shiny, metallic saucer-shaped object with a cupola on top, hovering only 3 feet above the ground.

Feeling somewhat apprehensive of his position near the uninvited and odd-looking visitor, Gatay cautiously tried to move, but discovered that his muscles would not respond. His co-workers, standing some distance behind him, could do nothing for him, as they too were paralysed. As the team watched the spectacle helplessly, Gatay recalled what happened next:

> "[Then] suddenly, the strange man vanished, and I couldn't explain how he did, since he did not disappear from my field of vision by walking away, but

34   Spencer 1991', p.210.

vanished like an image one erases suddenly.

Then I heard a strong whistling sound which drowned the noise of our excavators; the saucer rose by successive jerks, in a vertical direction, and then it too was erased in a sort of blue haze, as if by miracle."

Following this, Gatay and the others were "released" from whatever paralysis they had been in. Gatay then ran down to the others to confirm what he saw.

All the men, being understandably of the pragmatic sort, believed the whole incident was nothing more than a secret experiment performed on them by a nation on Earth (perhaps France), because they thought it had been too real to be anything but a terrestrial event.[35]

27 UFOs exhibit unusual acceleration behavior.

UFOs often show an unusual accelerating behavior, in which they move slowly at first but quickly build up to rapid and extraordinary acceleration beyond any man-made aircraft.

There is an intriguing case currently on record in Canada where it is alleged that the accelerating motion of a UFO behaved in an unusual manner not seen in ordinary aircraft. For some details of this case, read the above abstract[36], available from several UFO websites as of January 2008.

---

35   Vallee 1988, pp.92–93.
36   An example source for this Canadian UFO case is http://www.ufobc.ca/Sightings/Saved%20V1/sights2006.htm.

28   UFOs emit a buzzing/humming sound at close range.

UFOs are generally quiet, but at close range they produce sounds reminiscent of an electrical generator.

UFOs usually produce little or no sound unless witnesses are in close proximity to them. When a sound is heard, it is usually described as a hum or buzz that reminds witnesses of the sound associated with an electrical generator. When the UFOs start up, witnesses have noticed rotating parts on the UFOs increasing in speed, the flashing of lights intensifying in brightness and increasing in frequency, and a humming or buzzing sound emanating from the UFOs.

Evidence

Alfred Burtoo described the UFO he saw land near the Basingstoke Canal, not far from the Gasworks Bridge (now called the Government Road Bridge), as "…a large object, about 40 to 45 feet [12 to 14 metres] across…", and it had intriguing occupants described as small, unusually thin humanoids "dressed in pale green overalls…and helmets of the same color with a visor [covering their facial features] that was

114

blacked out".

One feature of the UFO he noticed was the noise it made. In interviews with UFO investigator Timothy Good, Burtoo claimed that as the object was about to take off, he heard a noise that reminded him of the sound of an electrical generator. In his own words:

> "...And then I heard this whining noise, just as if an electric generator was starting up, and this thing lifted off and the bright light came on again."[37]

29 UFOs may come with antennae sticking out of the central body.

It is common for UFOs to have an antennae sticking out of the central body. For large UFOs, the antennae is usually on top. For small UFOs looking like drones (usually 1 metre in height and width), a number of antennae can stick out the bottom.

Evidence

The following two UFO photographs taken in the United States and France are considered genuine by scientific experts. They reveal what looks remarkably like an antenna attached on top of the central symmetrical body:

One of several UFO photographs taken by Mr. Paul Trent (and first witnessed by his wife) on a farm close by the Salmon River Highway about 12 kilometres southwest of McMinnville, USA, at 7:45 a.m., on May 11, 1960.

---

37   Good 1997, p.90.

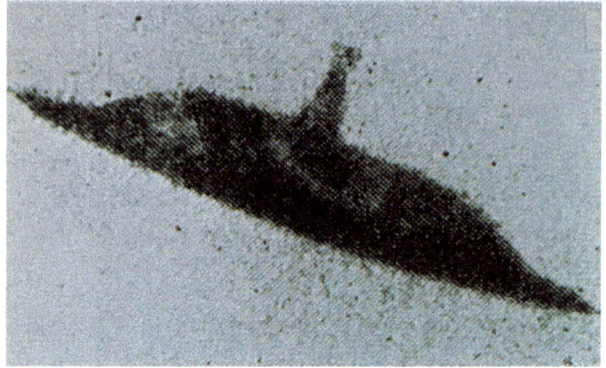

A UFO photographed by an anonymous French air marshal near Rouen, France, at 8:13 a.m., on March 5, 1957.

Computer analysis has failed to show any signs of trickery. In relation to the Trent UFO case, the Condon report stated:

> "This is one of the few UFO reports in which all factors investigated, geometric, psychological and physical appear to be consistent."

Notice the uncanny resemblance between the shapes of the two UFOs photographed at different times and places.

30 UFOs can have rotating outer sections.

It is not unusual for UFOs to come with rotating parts of UFOs, either on the upper or lower sections or as outer revolving rims.

Evidence

The classic case of Antonio Villas Boas from Brazil revealed a rotating copula on top of the UFO.

31 UFOs near metal objects may cause radiation poisoning to witnesses.

Witnesses have experienced symptoms strongly suggesting radiation poisoning in the presence of UFOs.

Witnesses of UFOs have experienced symptoms of radiation poisoning, including loss of hair soon after the UFO encounter. The radiation received appears highest when metal objects, such as cars and road signs, are close to the UFO.

The symptoms never occur at large distances from UFOs (unless humans are aggressive towards UFO occupants, see Observation 32) and when there are no metal objects near the witnesses.

Evidence 1

We have read the case in Chapter 4 of Mr. Aarno Heinonen and Mr. Esko Viljo, who observed a large metallic disc on January 7, 1970. At close range, a light beam was observed emanating from the UFO, and an entity floated down. It carried a small black box that emitted a light beam of its own in the direction of one of the witnesses. After the encounter, the witness who was hit by the light beam reported symptoms consistent with some form of radiation poisoning.

Evidence 2

A classic example is the UFO case of Betty Cash and Vickie Landrum of December 29, 1980 (see Chapter 4 for details). Briefly, the witnesses were exposed to ionizing X-ray radiation emitted by the outer stainless-steel car body as they stood outside in plain sight of the glowing, diamond-shaped UFO. As the UFO moved away, it was escorted by black U.S. military helicopters.

32 UFOs may retaliate against aggressive witnesses using light beams.

If humans make the unfortunate decision to take out a gun and shoot at a UFO occupant, the UFOs will retaliate with an intense beam of light.

UFOs and their occupants have never been known to attack humans or other lifeforms on the Earth. However, if a UFO occupant is attacked by humans, there is a high probability the UFOs will defend themselves if it cannot runaway quickly enough from a dangerous situation.

Evidence

A rather extreme case of radiation poisoning leading to death from leukemia soon after a UFO encounter can be seen in the case of 41-year-old Brazilian Inàcio de Souza. The incident occurred on August 13, 1967, and further details can be read in Chapter 4.

Artist impression of the UFO with its beam of light emerging from it. A UFO occupant was injured after being fired upon by a Brazilian man.

33   UFOs have remarkable maneuverability in flight.

UFOs are highly agile in flight. UFOs are characteristically known for their high maneuverability as if devoid of inertia or as if they possess very little mass.

Evidence

As UFO researcher Ed Zivkovic noted:

"No known aircraft is capable of the extreme manoeuvres displayed by UFOs…. No known aircraft, even to this day, can stop in mid-flight, change directions, and speed away at thousands of miles per hour or dart up and down, back and forth while changing colours."[38]

34   UFOs are built using the most lightweight and toughest materials.

When the materials from certain crashed disc cases from the late 1940s were analyzed, they were described as extremely lightweight and tough.

Part of the ability of UFOs to show great agility in the air is the high probability that they are constructed using the lightest materials and using the least amount of materials to keep the objects lightweight.

Evidence 1

More details will be revealed in Chapter 7, but in Frank Scully's *Behind the Flying Saucers*, people who have secretly studied crashed disc cases in the United States have been consistent in their claims the UFOs are made with the lightest and toughest materials.

---

38   Quote available at http://www.ezau.com/latest/articles/068.shtml.

Evidence 2

The Roswell case of early July 1947 is another interesting, high-probability UFO crash case worthy of consideration in the light of current UFO evidence. Here observations revealed a super-strong, tough and extremely lightweight newspaper-thin shape-memory alloy allegedly used as the external hull of the disc-shaped flying object. In this, the world's most detailed case of an alleged crashed disc, in which small, thin bodies were found in the Plains of San Agustin, along with additional materials dropped over a local rancher's property near Roswell, New Mexico, all the materials were described as extremely lightweight. For example, a dark-grayish metal foil was found that had the ability to return to its original shape and to withstand extraordinarily high temperatures from a blowtorch, as well as impacts from a sledgehammer, and was described as being newspaper-thin and extremely lightweight.

As for the plastic-like beams that provided structural support to the flying object and separated the outer metal foil from the internal components, these too were described as extremely lightweight. Its toughness was beyond the norm, as these plastic beams, together with some brown, insulating, parchment-like material, could not be burnt or torn by ordinary means.

The USAF has made the unlikely claim that the materials found were from a secret weather balloon made of newspaper-thin and rather flimsy aluminium foil, ordinary plastics, and balsa wood frames.

The front page of the *Roswell Daily Record*, July 8, 1947.

35 UFO occupants are described as thin and lightweight individuals.

With the exception of the head region, it is a universal fact that UFO occupants are thin creatures, short short in stature.

In keeping with the lightweight aims of the UFOs themselves, the occupants are invariably lightweight too. By staying thin and usually choosing individuals who are short, it seems to aid in the agility of the UFOs when in flight.

UFO occupants are most often reported as small humanoid beings with large, usually slanted or wrap-around eyes and large bulbous heads that converge markedly to a narrow chin. Apart from the size of the head, which suggests a large cranial capacity, UFO occupants are of extremely slender appearance, and the most common height is between 1 and 1.5 metres. Exceptions to this rule are some tall "attractive" and more human-like beings, almost always said to be 1.8 metres in height with whitish skin and blond hair, often with large blue or pink, almost Asian eyes. They will usually appear with broad faces that converge to narrow chins. Despite

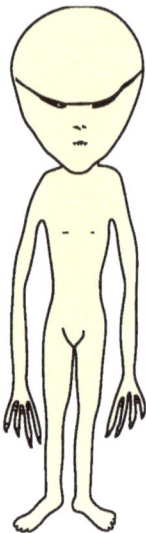

their height, all are very slender, as if to minimise mass in order to operate their UFOs with the least amount of energy when accelerating to high speeds and to withstand inertial forces during acceleration. Only a very few appear to be more exotic, such as having large ears and eyes and being extremely slender, small and highly agile, playful and curious creatures, especially when testing humans for similar characteristics.

Evidence 1

The so-called Aztec crashed disc case first mentioned in the book by Frank Scully (but likely to be an amalgamation of information from different genuine crashed disc cases, or a direct link to the Roswell case), alien bodies were allegedly found inside a recovered disc: two charred remains were found sitting on dull-bronze-colored, "bucket-like" seats and slumped over an instrument panel. Another 12 bodies were found in disarray on the floor of a chamber within the cabin. The aliens had died as a result of sudden decompression when a porthole in the disc was somehow punctured. On closer examination, the aliens were described as small (between 90 and 108 cm in height) and very thin.

Despite the smallness of their noses and mouths and a lack of ears extending beyond the sides of their disproportionately large heads, their eyes certainly made up for an otherwise empty face—they were very large and slanted. The creatures' average weight was around 18 kilograms.

Further indications on how these occupants achieved their incredibly low body-mass were revealed by the alleged autopsies performed by several unnamed scientists working for the U.S. military. Although there were some conflicting reports, witnesses claimed some could not observe a digestive tract (though patches were noticed on the skin, possibly to transfer nutrients directly to the body). There was also a lack of agreement on the existence of reproductive organs. The only thing they could agree on was the appearance of the blood, which was colourless and smelled similar to ozone.

Evidence 2

More likely to be referring to the Roswell UFO crash case, a former member of the U.S. Department of Defense Research and Development Board, the late Dr. Robert I. Sarbacher, claimed to have been "invited to participate in several discussions associated with the reported [crashed disc] recoveries" in the United States in the 1950s before revealing some details to Dr. Wilbert B. Smith. In a letter to author and UFO investigator Mr. William S. Steinman, dated November 29, 1983, Sarbacher confided that:

> "Although I could not personally attend the meetings, about the one thing I remember at this time was that certain materials reported to have come from the flying saucer crashes were extremely light and very tough. I am sure our laboratories analysed them very thoroughly...people operating these machines were also of very light weight, sufficient to withstand the tremendous deceleration and acceleration associated

with their machinery....I still do not know why the high order of classification has been given and why the denial of the existence of these devices."[39]

36   UFO occupants wear skin-tight and smooth metallic suits.

A majority of UFO occupants have worn silver or grey, metallic one-piece suits, usually with no zippers, buttons or pockets and covering almost all parts of the body.

There are some rare examples where it is hard to be sure if anything metallic is woven into some clothing worn by UFO occupants, but on the whole, the majority of occupants will wear metallic one-piece suits having no signs of sharp edges or points formed by buttons, zippers or pockets.

Evidence 1

Examples of occupants wearing metallic suits are mentioned in various UFO cases in this chapter, such as the occupants observed by French lavender farmer Maurice Masse.

Evidence 2

The Antonio Villas Boas UFO abduction case is another classic example of UFO occupants wearing metallic suits. The witness provided considerable details of the clothing worn by his captors. This is also a rare case in which one of the crew members, notably a female, had chosen to wear no clothing for the purposes of acquiring certain biological samples from the

39   Good 1988, pp.525–526.

male witness.

37 UFO occupants are universally humanoid in shape.

There have been no UFO reports emerge of exotic creatures coming out of a UFO that can be described clearly as a non-humanoid shape. Practically every genuine UFO report that mentions occupants have a humanoid shape to their bodies.

UFO occupants are described as universally humanoid. In other words, the occupants always have two limbs for manipulating controls inside the UFOs, using instruments carried around to collect samples, or aiming a pen-like device to immobilize human witnesses coming into close proximity, while the other two limbs are naturally for mobility (i.e., walking and running). While there may be superficial differences, UFO occupants have eyes, ears, mouths and noses in the same numbers and positions as we on our heads. And nearly all have heads that are larger in size compared to their bodies than ours, suggesting a large cranial capacity.

Why UFO occupants appear almost always in human form is not precisely known. Perhaps, as astronomer Dr. Harlow Shapley (1885–1972) commented:

> "A mixture of pure chemical elements will always under the same physical conditions produce the same result, whether it be an odour, an explosion, or a colour. Perhaps we should expect that a mixture of starshine, water, carbon, nitrogen and other atoms, when physical conditions are fairly similar, will everywhere produce animals that are much alike in structure and operation and plants that have certain standard behaviour, notwithstanding great morphological differences. If we should

visit a planet essentially identical with ours in mass, temperature, age and structure, we would probably not find the biology queer beyond comprehension....Therefore we surmise that the biology on Planet X and Planets Y, Z and so forth might have much in common with the living forms on Planet Earth just because the carbon compounds will have it so, and because the same chemistry and the same natural laws prevail throughout the universe we explore."[40]

Dr. Paul Davies, professor of mathematical physics at the University of Adelaide, Australia, supports this belief. Despite the natural tendency for the stupendous genetic diversity of life, Davies says:

"I believe that life is channeled by the [same] forces of nature and evolution, so it would not surprise me if life forms somewhere or other in the universe were very similar to ourselves."[41]

While most UFO occupants tend to walk on two legs, there are occasions when the smaller variety of occupants are seen floating in the air, but only in the vicinity of the UFO and sometimes while immersed in a beam of light emitted by the UFO. Alternatively, a special belt with a flashing light may be worn for the purposes of helping with this weightless movement through the air, but only in close range of the UFO.

40   Holmes 1966, pp.78–79.
41   Hough & Randles 1991, p.123.

São Francisco de Sales, Brazil
16 October 1957
Witness: Antonio Villas Boas

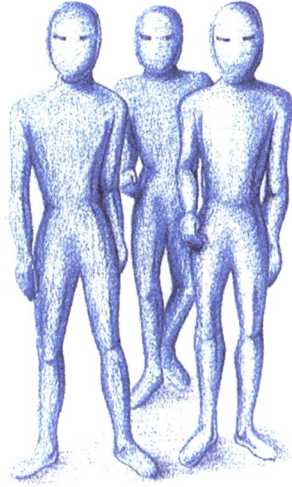

Monte Maiz, Argentina
12 October 1963
Witness: E. Douglas

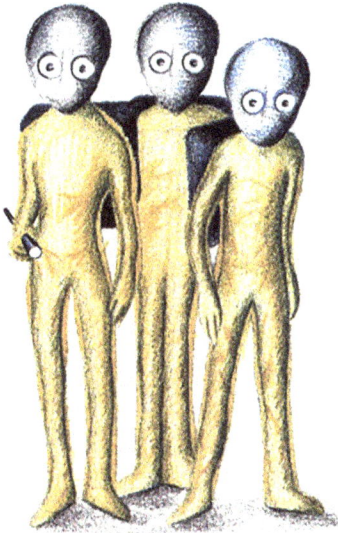

Sao Paulo, Brazil
23 July 1947
Witness: Jose Higgins

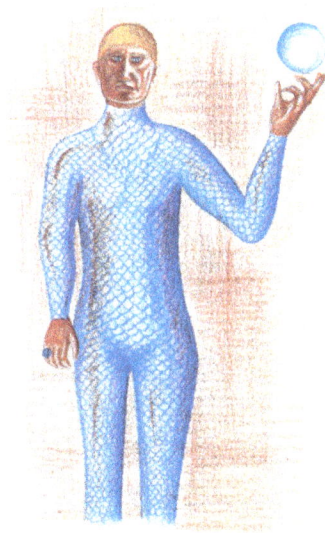

Cordoba, Argentina
13 June 1968
Witness: Maria Eladia

Valensole, France
1 July 1965
Witness: Maurice Masse

Vaienciennes, France
10 September 1964
Witness: Marius Dewilde

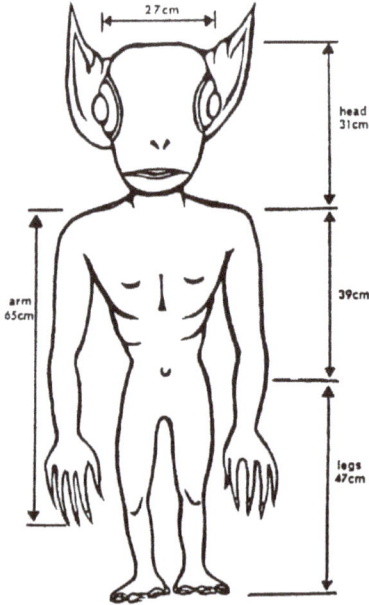

27cm

head
31cm

arm
65cm

39cm

legs
47cm

Kelly-Hopkinsville, Kentucky
21 August 1965
Witness: Five adults and seven children

On a farm near Rio Piedras Puerto Rico
3 March 1980
Witness: Vivian and Jose Rodriguez

CHAPTER 6

Are We Dealing with Ball Lightning?

If you put together inexplicable atmospheric phenomena, maybe of an electrical nature, with human psychology and the desire to see something, that could explain a lot of these UFO sightings.[1]

—Dr. Stephen Hughes, Australian astrophysicist

After reviewing all the common UFO observations in Chapter 5, it now looks like UFOs are indeed an electromagnetic flying object of some sort. The question is, what kind of flying object could we be dealing with here?

In the natural world we live in, we do know of the one and only flying object capable of displaying electromagnetic observations. This is the familiar, but rare ball lightning. Of course, this naturally raises the question of whether we can be sure that UFOs are not this type of natural phenomena.

---

1   Amos 2010.

Due to the rarity of ball lightning, photographs of this type of phenomenon are hard to come by. The best humans can do is draw illustrations, such as this one published in "The Aerial World", by Dr. G. Hartwig, London, 1886.

To determine how feasible this is, it is important to look at common observations of ball lightning. By doing so, we can make reasonable comparisons between observations of ball lightning and those of UFOs, and so answer the all-important question of whether or not the descriptions of UFOs do show a sufficient resemblance to ball lightning.

Below is a table summarising the observations[2] for this phenomenon as reported by witnesses.

---

2   Information on ball lightning observations can be found in Cooray & Cooray 2008, p.101, and http://wiki.wunderground.com/index.php/Educational_-_Ball_lightning.

Common Observations of Ball Lightning

Appearance

Shape	Generally spherical in shape, although occasionally a teardrop or oval shape has been reported.
Size	Usually grapefruit size but anywhere from 10 to 40 cm in diameter
colour	The most common colours are red, orange, yellow and white. Green has been reported for some ball lightning but is considered rare.
Brightness	Not exceptionally bright but can be seen during the daytime and usually maintains its brightness and size until the very last moment when brightness suddenly increases before it disappears explosively. Or the brightness may simply fade gently until the ball lightning is gone.
Sound	Usually quiet except for in about 10 percent of cases, in which a hissing noise has been heard. When the ball lightning disappears explosively, a much louder noise like a bang is not uncommon.
Odour	If a smell is detected, it is usually sharp and acrid, resembling ozone, burning sulphur or nitric oxide.
Attraction	Ball lightning usually heads for, or when attached follows along, metallic objects such as wire fences, power lines and telephone wires, or bodies of water.
Rotation	Many reports tend to describe ball lightning as either a solid core structure surrounded by a translucent envelope, or a structure that emits spark-like phenomena. In either case, the structure may be seen as rotating or spinning.

Occurrence

Where	Ball lightning can take place anywhere in the world.
When	Any time when thunderstorm activity takes place, usually simultaneously with a cloud-to-ground lightning discharge, or during other stressful natural conditions such as earthquakes or tornadoes. However, ball lightning can appear during good weather with absolutely no connection to a thunderstorm or lightning.
Height above ground	While ball lightning can appear out of a cloud and fall to the ground, it is not unusual for ball lightning to hover high in the air. In a few instances, ball lightning may occur within a few meters from the ground and

	sometimes without a lightning discharge to initiate it.
Duration	The lifespan of ball lightning is often less than 10 seconds, but in a small number of reports ball lightning has lasted for just over a minute.
What happens when it disappears?	Usually fades away over time or rapidly disappears in an explosive manner.

Motion

Direction	Usually moves horizontally or vertically in a downward direction (rarely does it travel upward). Will also follow along metallic objects such as fences. Occasionally, some ball lightning will stop motionless in the air. On other occasions, a zigzag motion may also be seen.
Speed	When moving horizontally, the speed slows down dramatically to a walking speed of about 1 to 2 meters per second. When moving vertically in a downward direction, the speed can be very fast. Sometimes there is no speed at all, as ball lightning can occasionally stop mid-flight.
Speed and direction affected by wind?	Usually not, as ball lightning has been reported to move against the wind.

Other

How far back has ball lightning been seen in history?	Ball lightning has been observed since antiquity.
Can ball lightning be reproduced in laboratories?	No.
Can we rely on photographs of ball lightning?	Few photographs exist of ball lightning and most, if not all, are questionable. Most details about the properties of ball lightning have to be extracted from eyewitness records.

What Have We Learned?

Already from these brief tables summarising what we know of ball lightning, we see some interesting similarities between UFOs and ball lightning. For instance, the spherical shape is not entirely an uncommon observation for both UFOs and ball lightning, and both types of phenomena will glow. Furthermore, the colour of the glow, the presence of some kind of rotating structure, a smell reminiscent of ozone, how both UFOs and ball lightning can be seen at any time during the day or night and need not necessarily be associated with a thunderstorm activity, and have been observed all the way back since antiquity, does give the impression that ball lightning could explain a number of UFO cases.

Philip Klass gave support for the ball lightning theory. According to *The UFO Encyclopaedia* by Margaret Sachs:

"Writer PHILIP KLASS is the leading proponent of the theory that plasmas [or ball lightning] account for a large number of UFOs. He believes that an observer, seeing a cloud of illuminated particles moving through the air, might assume it to be some sort of flying machine. Conditioned to accept certain structural features as integral parts of aerial craft, the mind might distinguish dark areas as part of the structure, and lighter areas as windows. Many properties ascribed to UFOs are also attributable to plasmas. Not only are they good RADAR targets, but they can actually produce a stronger echo than a solid metal object. Low energy plasmas can sometimes be detected by radar without being seen visually. Sudden or gradual disappearance of plasma-UFOs occurs when their energy is dissipated. Plasmas can change size and shape, and divide and merge just as UFOs are reported to do. Abrupt stops and sharp directional changes are easily accomplished by an essentially weightless and electrically propelled plasma. The reddish-orange and bluish-white colours frequently reported by UFO witnesses are the characteristic colours of ionised nitrogen and oxygen, the principal constituents of air. Ultraviolet radiation generated by plasmas could cause sunburn and bloodshot eyes, PHYSIOLOGICAL EFFECTS often associated with UFOs. PHYSICAL EFFECTS characteristic of UFOs, such as radio and television interference, could be attributable to the agitated electrified particles which can serve both as generators of radio-frequency energy and as a screen to block other signals. An automobile battery might be short-circuited if the terminals were to be surrounded by plasma or if the plasma's proximity to the distributor were to change the shape of the spark sufficiently to cause engine failure."[3]

This is the best man-made attempt at creating a ball lightning type of effect under controlled laboratory conditions. Here, scientists at the Max Planck Institute discharged a high voltage capacitor in a body of water, leading to a glowing plasma effect to rise above the water's surface.

However, when we closely examine UFO cases, as in Chapter 5, we discover the following discrepancies:

- Daytime disks show a metallic surface.
- The size of the UFOs at close range when on the ground often exceeds three meters in diameter.

---

3   Sachs 1980, pp.252–253.

- UFOs at close range have been observed with portholes, doors, and interesting movable and fixed protrusions around the circular bodies.
- UFOs at close range are likely to reveal occupants.

When faced with these observations, it makes ball lightning a rather difficult phenomena to rely on as a suitable explanation for all UFO cases.

Even the short duration of ball lightning seems to contradict a number of UFO cases in which the sightings have lasted far longer than a couple of minutes. Take, for instance, the Oloron-Sainte-Marie UFO case in France. Witnessed by hundreds of people, multiple UFOs were observed for nearly two hours.

As Margaret Sachs writes in *The UFO Encyclopaedia*:

"Opponents of the plasma theory claim that it is an invalid explanation for the majority of UFO sightings, since plasmas normally exist for only a few seconds near high-tension lines during severe thunderstorms."[4]

Something does not add up when we compare ball lightning to genuine UFO reports, especially at close range. Of particular concern are the observations of portholes, doors and the appearance of UFO occupants—not features normally associated with ball lightning unless one wanted to add magic mushrooms from the French countryside to this mix (as mentioned in Chapter 4). However, given the numbers of genuine UFO cases worldwide from people who have acted perfectly normal and calmly reported their sightings to the authorities in a highly rational way, the UFO reports are clearly telling us that there are no hallucinatory drugs to consider here. Add to this from Chapters 4 and 5 those people who have suffered radiation poisoning in the presence of UFOs and it is clear that there must be another physical object in the environment capable of creating these common and interesting electromagnetic observations.

But again, we must not assume that UFOs are of alien origin. We still have to consider how likely it is that UFOs could be a new type of man-made electromagnetic technology. Before we can ever look more closely at the possibility of a new "alien" technology and any associated

4   Sachs 1980, p.253.

new concepts to advance science, it is important to know whether humans have built one of these flying objects. Because, if there is evidence we can and have done so, it could go a long way towards explaining the nature of these UFOs. Therefore, let us see how well the evidence for a new man-made electromagnetic technology stacks up in this analysis.

CHAPTER 7

Are We Dealing with a Secret Man-Made EM Experiment?

The phenomena reported is something real and not visionary or fictitious.[1]

—General Nathan D. Twining, Chairman of the
Joint Chiefs of Staff (1957–1960)

S THERE are strong indications that UFOs are some type of electromagnetic (EM) technology, it is reasonable to consider whether UFOs are man-made. If this is true, then we have to take note of the fact that this technology isn't one readily accessible to the public. This suggests that whoever has developed this technology seems rather shy or secretive. Yet, at the same time, we must acknowledge that there are UFO cases in which whoever is flying them

---

1   USAF memorandum dated 23 September 1947 written by General Nathan D. Twining of the Air Material Command at Wright-Patterson AFB.

is keen to give us clues about how it works. How odd? It is almost as though someone wants to give us the puzzle to figure out for ourselves. However, if the makers of this technology want to keep it a secret, it does not make sense to show us some aspect of the technology. If UFOs are meant to be man-made, there is a tussle going on between those who want to maintain the secrecy and those who want to tell the public about it.

But how likely is it that we are dealing with a man-made flying object? Is there an EM technology for flying machines available to us that could make this a reality?

To figure out the probability of UFOs being man-made, it is important to give a brief history of UFOs. Not just any history, but one paying particular attention to EM studies carried out—in secret or otherwise—by the authorities and anyone else, for that matter, in relation to projects involving the development of new flying objects. In this way, it may be possible to learn who might be involved in such a secret project and how long this EM technology may have been available to us.

So, without further ado, here is the history of UFOs as we know it today.

A Brief History of UFOs

For as long as humans can remember, we have witnessed the appearance and disappearance of countless UFOs in the sky. For those with a propensity to communicate what they saw to others, there have been various attempts to record UFOs as fervent memories of these special events.

UFOs in Prehistoric Times

In prehistoric times, humans drew numerous pictures on rocks and cave walls not only of animals and themselves (such as the outline of hands, and simplified representations of humans chasing or throwing spears at animals) but also of strange objects they saw in the sky, including pictures of flying circles, discs, and the occasional individuals dressed in strange clothing.

UFOs in Biblical Times

The Old Testament of the Bible isn't exactly devoid of possible UFO accounts.

In the Book of Exodus, we learn of an white-bearded old man named Moses and his band of fellow Hebrew men and women making their famous escape on foot from the clutches of a ruthless and power-driven pharaoh and almost certain death at the hands of the relentless Egyptian army chasing them down. After many generations of passing the story down through word-of-mouth, the written scriptures recorded a strange-looking cloud, in an otherwise clear sky, hovering during the day and glowing at night while seemingly guiding Moses and his people to safety across the Sea of Reeds. The cloud was described as able to move of its own accord and successfully kept the Egyptian army at bay by retracing its flight path and approaching the ground to frighten the soldiers. This UFO appeared within a day or two of a massive volcano blowing apart the Greek island of Santorini. The eruption led to all sorts of prophesies coming true, made by the old man based on what God had told him prior to the event (perhaps another UFO encounter?), such as the waters turning red, an infestation of frogs and insects moving through Egypt, and darkness descending with intense wind and electrical activity. The ensuing floods that wiped out the Egyptian army were also believed to be part of the volcanic event, where the waters of the Sea of Reeds receded temporarily only to return with a vengeance as the African continental plate suddenly slipped back into position following the massive volcanic eruption.

In the Book of Ezekiel, another old man claimed a similar observation of a mysterious cloud-like structure with a glowing or bright light around it. Of course, on this occasion, nothing would be complete without the appearance of four living entities just to get the old man's heart racing a bit and keep him wondering what he saw. As mentioned in Ezekiel 1:4-7:

> "As I looked, behold, a stormy wind came out of the north, and a great cloud, with brightness round about it, and fire flashing forth continually, and in the midst of the fire, as it were gleaming bronze. And from the midst of it came the likeness of four living creatures. And this was their appearance: they had the form of men, but each had four faces, and each

of them had four wings. Their legs were straight, and the soles of their feet were like the sole of a calf's foot, and they sparkled like burnished bronze."

Interesting to see the "fire flashing" observation. As we know, one of the most common observations of UFOs (see Chapter 5) is how the glowing surface is able to brighten and dim in a regular and periodic manner. As we shall see in the next chapter, there is a technological explanation for this specific observation. Until then, we have to ask ourselves, Does this mean the old man really did see a genuine UFO in biblical times? As good scientists, it is important to always keep an open-mind.

We also find another strange UFO in 2 Kings 2:9-12:

"...Eli'jah said to Eli'sha, 'Ask what I shall do for you, before I am taken from you.' And Eli'sha said, 'I pray you, let me inherit a double share of your spirit.'...And as they still went on and talked, behold a chariot of fire and horses of fire separated the two of them. And Eli'jah went up by a whirlwind into heaven. And Eli'sha saw it and he cried, 'My father, my father!'...And he saw him no more."

In such biblical accounts, the people who witnessed the events felt inclined to describe the UFOs as God, specifically "the Father", or sometimes as a "sign of God" such as one of the angels or messengers of God.

To this day, scientists are not entirely sure what to make of these UFOs. Could these be examples of overactive minds willing to hoax UFO events for the sake of a good yarn? Probably not. The Moses story is particularly more troublesome because there were too many witnesses. If it had been a fake, someone would have piped up and said so after a while.

For now it is worthy to keep at the back of one's mind these examples, especially those that mention flashing lights and cloud formation surrounding a hidden glowing object in the sky, when deciding later how likely it is that an EM flying technology was developed by humans.

UFOs Observed by the Romans

Between 218 B.C. and 43 B.C., Roman civilians and military soldiers reported seeing UFOs. Most sightings tended to get ignored, but a number of sightings were spectacular enough and seen by enough people to warrant the Roman authorities to record them for posterity.

The space agency NASA has looked at the more authentic UFO accounts in ancient times, of which those Roman UFO sightings written into history are among the ones the agency considers to be worthy of study.

Here is a summary[2] (compiled by Richard Stothers) of the main UFOs observed in Italy at around this time, with possible natural explanations included[3]:

- At Rome in the winter of 218 B.C. "a spectacle of ships (navium) gleamed in the sky" (Liv. 21.62.4). Franklin Krauss, for lack of an alternative explanation, speculated that the "ships" were probably clouds or mirages, although suggestive cloud formations had been long-understood and considered familiar features.

- In 217 B.C. "at Arpi round shields (parmas) were seen in the sky" (Liv. 22.1.9; Orosius 4.15). A *parma* was a small round shield made partly or wholly of iron, bronze or another metal; we do not know whether the lustre of these devices (and not just their shape) was intended to be an element of the description. Mock suns are an unlikely explanation, since in the Roman prodigy lists these were routinely described as "double suns" or "triple suns" (i.e., two mock suns on either side of the real one).

- In 212 B.C. "at Reate a huge stone (*saxum*) was seen flying about" (Liv. 25.7.8). The implication would seem to be that the object in question was a stony grey colour; it is said to have moved irregularly (*volitare*), leaving open the possibility that the object Livy describes was a bird or some kind of airborne debris.

---

2   https://pubs.giss.nasa.gov/docs/2007/2007_Stothers_st02710y.pdf
3   Stothers 2007, pp.82-84.

Sporadic reports of similar objects continue to appear after this in the Roman prodigy lists. The immediate sources are again Livy and his extractors Pliny, Plutarch, Obsequens and Orosius:

- In 173 B.C. "at Lanuvium a spectacle of a great fleet was said to have been seen in the sky" (Liv. 42.2.4).
- In 154 B.C. "at Compsa weapons (*arma*) appeared flying in the sky" (Obsequens 17). The term *arma* refers to defensive weapons, especially shields.
- In 104 B.C. "the people of Ameria and Tuder observed weapons in the sky rushing together from east and west, those from the west being routed." Thus Pliny (Nat. 2.148) who uses the term *arma* for UFOs is essentially the same as Obsequens' (43) version. Plutarch (Mar. 17.4) also gave similar support by calling the weapons "flaming spears and oblong shields", but may be merely glossing and expanding; since he noted the time as night, the phenomenon in question might be the streamers of an aurora borealis.
- In 100 BC, probably at Rome, "a round shield (clipeus), burning and emitting sparks, ran across the sky from west to east, at sunset." Thus Pliny (Nat. 2.100), although Obsequens (45) called the phenomenon "a circular object, like a round shield." The *clipeus* was a round shield similar to the *parma*, but bigger. Seneca (Nat. 1.1.15; 7.20.2), quoting Posidonius (1st century B.C.), referred to a class of *clipei flagrantes*, saying that they persisted longer than shooting stars. Nothing in the ancient reports forbids that these were spectacular bolides (meteoric fireballs), which move across the sky more slowly than ordinary shooting stars, but enormously faster than genuine comets, which are seen for days or weeks.
- In 43 B.C. at Rome "a spectacle of defensive and offensive weapons (*armorum telorumque* species) was seen to rise from the earth to the sky with a clashing noise". It might be possible to visualise in this report a bolide exploding while rising above the horizon.

UFOs in Early Greek Times

The early Greeks reported seeing UFOs. The Greek biographer and historian Plutarch (46–120 A.D.) recorded a historical testimony of a UFO observed by military witnesses minutes before the start of a battle during the third Mithridatic war (75–63 B.C.). The intriguing story was transferred by word-of-mouth until Plutarch made the decision to write it down.

At the time of the sighting, Lucullus and his Roman army were preparing for battle to fight against Mithridates, the king of Pontus, and his ally Marcus Marius who was dispatched by Sertorius. The sighting was enough for the armies to withdraw from the battle as it was seen as a sign of the gods' displeasure at the event that was about to unfold on the ground.

Here is the best English translation and interpretation of the extract concerning the UFO incident:

> "With these words, he [Lucullus] led his army against Mithridates, having thirty thousand foot soldiers, and twenty-five hundred horsemen. But when he had come within sight of his enemies, he was amazed by the number of people. He desired to avoid and delay the battle. However, Marius, who was dispatched from Iberia [Spain] by Sertorius to provide an army to join forces with Mithridates, got set for battle. As the armies were on the verge of commencing the fight, without any observable change in the weather, the air opened and appeared a rapidly descending and large flame-like object, which appeared like a wine jar or vase in shape and like a glowing annealed metal in colour. Both armies, astonished and frightened by the sighting, decided to withdraw. They said that this happened in Phrygia, at a place called Otryae."[4]

UFOs in Europe

As human language reached a level of sophistication in describing things, some people began to report UFOs as "coloured globes", "fiery lights", "flying shields" and "chariots of fire" to name a few. Where a

---

4   http://penelope.uchicago.edu/Thayer/E/Roman/Texts/Plutarch/Lives/Lucullus*.html and http://www.historydisclosure.com/plutarch-wrote-about-ancient-ufo-sighting/. This account of the UFO sighting compiled by SUNRISE.

mysterious column of light appeared below these UFOs or from strange clouds moving around in the sky, it may have been described as a "pillar of fire". In other observations, flying crosses and cylinders were not unheard of.

If an unusual aerial event was seen as significant, humans were likely to draw pictures of what they saw to make a permanent record of it. For example, a woodcut made in celebration of a momentous hour-long event that occurred in the sunny sky over Nuremberg, Germany, on the morning of April 14, 1561, suggested that a number of flying spheres, hollow cylinders with spheres inside, other cylinders without the spheres, as well as flying crosses made their presence known to observers on the ground. Not unheard of observations for UFOs in more modern times.

The spheres made their triumphant return at sunrise and sunset on two days in July and again at sunrise on August 7, 1566, in Switzerland, where the following picture was produced.

UFOs over Basel, Switzerland, in 1566.

A student from the Swiss town of Basel reported the sighting in the following way:

> "...[M]any large, black globes were seen in the air, moving before the sun at great speed and turning against each other as if fighting. Some of them became red and fiery and afterwards faded and went out."

UFOs over Nuremberg, Germany, in 1561. Hand-coloured woodprint by Hans Glaser

In both sightings, it would appear the UFOs were engaging in a type of aerial battle as some of the objects collided and faded off as they fell from the sky. Could this be evidence of the early introduction of fireworks into Europe from the Arab world, where such technology had been known since the 12th century? Or were these UFOs something more mysterious?

It is true that many UFOs seen during these times tend to have a natural explanation. A good example of this is the bright light that suddenly appeared in the sky over China on July 22, 1054. Today, astronomers can confidently explain that this was due to a massive supernova explosion, the remnants of which can be seen in photographs taken of the Crab Nebula. However, not all UFOs have an obvious explanation. For instance, there are strange flying objects depicted on the canvases of several medieval paintings in monasteries from Yugoslavia and other parts of Europe. They suggest that either the artists were trying to render an image of God as a human being flying a strange object, or that this was just a simple way of representing ordinary comets flying above the skies of Earth.

Electricity is Discovered

Of course, if UFOs are meant to be some kind of man-made electromagnetic flying object, it stands to reason that at some point in our history, someone must have discovered the power of electricity. This is the means by which electric charges are separated by an electrical generator and brought back together to perform work. In the case of UFOs, the work must involve moving the object and making it fly.

The earliest known device that could have been designed as a battery for storing electric charges appeared in the first century A.D. In 1936, archaeologists working for the Iraqi Antiquities Authority unearthed a jar roughly 14 cm tall. On removing the asphalt seal, they found a tube of copper surrounding an iron rod. The two metals did not make contact. They were held in place by more asphalt. Wilhelm Koenig, an Austrian who served as director of Baghdad Museum at the time, looked at the strange jar. Apart from the fact that it was constructed by someone living in the Parthian empire, the Parthians were not noted for any special engineering or technical prowess. Yet, somehow, this jar excavated from a site of Khujut Rabou near Baghdad looked remarkably like a battery.

Experiments to replicate the design and add an acid such as salt water, copper sulphate, citrus juice or vinegar, can produce a measurable voltage. Salt water was the least useful as it quickly corroded the iron, causing the voltage to drop after 60 seconds. Copper sulphate was better, allowing a voltage of 0.45 volts to be maintained for several hours until the copper plated the iron rod, which in effect stopped any further production of electricity, assuming this was the purpose. However, a mixture of acetic acid (or vinegar) and grapefruit juice produced 0.49 volts for several days.[5]

After uncovering more "batteries" from the Sassanid Period, archaeologists remain baffled by what use the Parthians had for these little jars. Definitely no motors were found near these jars, and certainly no thin filament metals to heat up at night to become electric lights (a common feature of glowing UFOs). The only clue one can fathom is that a number of the jars were stuffed with papyrus, and may have come with magical amulets.[6]

The most likely explanation is that these jars were either a means of preserving papyrus scrolls and anything else the Parthians thought would stand the test of time once the jars were sealed, or it is possible the Parthians may have discovered electricity and used it to treat certain medical conditions. Certainly the Egyptians were known to have used (and still do to this day) the electric catfish from the River Nile to treat headaches and nerve pain. The voltages need not have been high to treat pain. Modern medical literature claims anywhere between 0.8 and 1.4 volts and 0.2 to 1.0 milliamps is sufficient. This is roughly in the range of how much electricity the Baghdad jars could technically produce if an acid was added to them.

Perhaps the Parthians had accidentally found another way to create electricity for this very purpose?

Moving forward quite a few hundred years, we learn the true and deliberate effort to separate electric charges was successfully achieved by humans thanks to the invention of the electrostatic generator. The earliest type was friction-based using a sulphur globe that was rubbed by hand—first discovered in 1663 by Otto von Guericke. The design was improved, with Sir Isaac Newton recommending a glass globe,

---

5   Keyser, "The Purpose of the Parthian Galvanic Cells," pp.88-92.
6   Keyser, "The Purpose of the Parthian Galvanic Cells," p.82; James and Thorpe, *Ancient Inventions*, pp.148-149.

while his assistant Francis Hauksbee (1660–1713), whom Newton later appointed curator, instrument maker and experimentalist of the Royal Society, used an electrical machine, enabling the rapid rotation of the glass globe against a woollen cloth.

As for realising that negative charges flow toward the positive end forming an electric current, this was first observed in 1781 when Luigi Galvani, professor of anatomy at Bologna University, Italy, saw the effect of electricity on the nerves and muscles inside the legs of a dead frog. It was an accidental discovery. Galvani's assistant just so happened to be playing with an electrostatic machine in the same room as he was dissecting a frog with his steel scalpel. A spark was generated by the machine, which managed to complete a circuit, whereupon Galvani observed the muscles in the frog's legs twitch significantly. As Galvani said:

> "…[S]uddenly all the muscles of the frog's limbs were seen to be violently contracted just as though they had been seized with a violent cramp."

In 1800, Alessandro Volta became credited with inventing the battery. It may not have been the world's first battery in the strictest definition, if the Baghdad jars are anything to go by, but it would be the moment when people actually recognised the usefulness of batteries as a means of storing electrical energy.

Electricity would be exploited further following the development of magnets and coils for generating power. The prolific inventor Thomas Edison made the most of this power by redesigning the light bulb in 1879 to ensure a reliable source of light was achieved using lower electric currents thanks to an improved vacuum established inside the globe.

Despite these interesting advancements in electromagnetism and the development of the humble light bulb, there are no indications that humans had the means of developing an electromagnetic UFO by the late 19th century. If we did, people would have been reading by the light of these glowing UFOs as well as traveling the globe in whatever special way these objects could fly. Instead, at the time, the electric light bulb was the best-known method for illuminating dark places, and certainly did not generate movement and flight.

UFOs in the 19th Century

During the 19th century, reports of mysterious UFOs proliferated in the scientific journals and publications of several Western countries.

John Staveley of Hatton Garden, London, wrote one such account where he assumed he saw a swarm of meteors "dancing" in and out of clouds on August 10, 1809:

> "I saw many meteors moving around the edge of a black cloud from which lightning flashed. They were like dazzling specks of lights, dancing and traipsing thro' the clouds. One increased in size till it became of the brilliancy and magnitude of Venus, on a clear evening; but I could see no body in the light. It moved with great rapidity, and coasted the edge of the cloud. Then it became stationary, dimmed its splendour, and vanished. I saw these strange lights for minutes, not seconds. For at least an hour, these lights, so strange, and in innumerable points, played in and out of this black cloud. No lightning came from the clouds where these lights played. As the meteors increased in size, they seemed to descend."[7]

The Era of the Airship

More detailed descriptions of UFOs would emerge nearly at the turn of the 19th century when humans developed the first machine-powered objects to fly in the air.

It all began on November 21, 1773, when Joseph Michel (1740–1810) and Jacques Etienne (1745–1799) Montgolfier became the first humans to lift their feet off the ground in a high-quality wallpaper-and-linen hot-air balloon[8]. Although the entire 12-kilometre flight across the French countryside took 26 minutes, it gave humans a brand-new perspective on the surrounding land below.

As early as 1852, Henri Giffard designed and flew a balloon-like device called an airship over Paris; it was made of rigid steel frames and driven by a steam-powered engine attached to a propeller[9]. A similar machine was devised by the American inventor Solomon Andrews, who flew it near New York in 1865. And by 1869, an English expatriate by

---

7   Reader's Digest *Mysteries of the Unexplained* 1982, p.211.
8   Graham 1993, p.21.
9   Hough & Randles 1995, p.15.

the name of Fred Marriot flew a steam-powered, cigar-shaped balloon prototype with wings on either side of it in San Francisco.

Following the audacious development of balloon-like, cigar-shaped flying devices, an intense wave of aerial sightings soon appeared over the United States in the period between November 1896 and May 1897.

According to *The UFO Encyclopaedia* by Margaret Sachs:

> "The sighting which sparked the mystery occurred on November 22, 1896, when a group of streetcar passengers in Oakland, California, saw a winged, cigar-shaped UFO emitting a stream of brilliant light. As the phenomenon began to spread across the country, the diversity of reports revealed that more than one airship was involved. Descriptions varied considerably from an object eighteen inches in diameter and twelve to thirty feet long, to a seventy-foot long structure with wings and propellers. Sometimes hissing or humming sounds accompanied the craft [i.e., UFO] but generally no sounds were heard. coloured or bright white lights plus red or white searchlights were common features. The objects sailed through the skies, often against the wind, at speeds estimated to range between five and two hundred miles per hour."[10]

Similar phantom airships were also reported by citizens of Britain, New Zealand, South Africa, Japan, Australia and across much of Europe during 1909 and 1912–13.[11]

The Tunguska Explosion

On June 30, 1908, an object was seen hurtling from space at approximately 7:15 a.m. The object entered the Earth's atmosphere and turned into a bluish-white ball of fire as it raced across the summer sky, leaving behind it a trail of multi-colored smoke[12]. Soon afterwards at 7:17 a.m. (Siberian time), it exploded at an altitude of 16 kilometers above the ground with a blinding flash of light and released the energy of a thousand atomic bombs in a desolate region near the Podkamennaya Tunguska River in remote central Siberia.

---

10   Sachs 1980, p.8.
11   *Mysteries of the Unknown: The UFO Phenomenon* (Time-life Books) 1987, p.25 and Spencer 1991 *The UFO Encyclopaedia*, p.10.
12   This was first observed by travellers in the Gobi Desert according to Sachs 1980, p.326.

As a result of the massive explosion, over 3000 square kilometres of pine forests ignited and continued burning for days, while ferocious winds rattled doors and windows of people's homes up to 600 kilometres from the epicenter, and tremors were recorded at Irkutsk's seismographic centre nearly 900 kilometres to the south. The explosion was of such tremendous proportions that almost all trees within a radius of 64 kilometres around the blast site (approximately 80 million trees) were flattened outwards like match sticks.

The accompanying thunderclaps could be heard up to a distance of 80 kilometres. Dirt and burnt debris was sucked up and thrown 20 kilometres above the Tunguska region to fall as a shower of "black rain" within 24 hours. Massive glowing "silvery clouds" suddenly appeared over northern Europe and Siberia, which became so bright during the next few nights that in some places it was possible to read a book at midnight without the aid of artificial lights[13].

When Russian scientists finally investigated this remote and mostly uninhabited area—notably Leonard Kulik, a mineralogist—nearly 20 years later, no crater or meteor rock was found.

Further discussions with witnesses near the scene of the explosion reported a blinding flash, expanding shockwaves, black rain of debris, and an apparently mushroom-shaped cloud[14] formed immediately after the blast. This made some scientists think that perhaps it was a nuclear explosion of some sort. However, measurements of radioactivity in the Tunguska soil using sensitive equipment were performed some 50 years after the incident. Scientists found little sign of radiation, but radioactive caesium-137 was traced in much higher quantities than normal in the inner rings of living trees, which would coincide with the 1908 explosion. Consequently, some people have proposed that a nuclear-powered spaceship went out of control and crashed in this sparsely populated area of swamps and forests[15].

Today scientists have opted for a natural explanation: that an icy comet fragment weighing between 100,000 to more than a million tons and measuring up to 70 meters across (or possibly a small asteroid covered with ice), previously lost in the glare of the sun, collided with

---

13   Sachs 1980, p.326.
14   Sachs 1980, p.326.
15   *Mysteries of the Unknown: The UFO Phenomenon* (Time-Life Books) 1987, p.24; Sachs 1980, pp.326–327.

Earth at a speed of perhaps 100,000 kilometres per hour, leaving no trace of itself after impact.[16]

The World's First Man-Made Airplane

American inventors Wilbur (1867–1912) and Orville (1871–1948) Wright designed, constructed, and flew the first successful motor-driven "heavier-than-air" piloted airplane. The historic flight was made on 17 December 1903, at Kitty Hawk, North Carolina, USA.

With Orville piloting the biplane, the brothers covered a distance of 36.5 meters in 12 seconds under the machine's own power. Later, with Wilbur at the controls, they made the longest and last flight on this aircraft, which lasted 59 seconds of flying time and covered 260 meters of ground.

The Wright brothers' success encouraged others to devise innovative new aircraft designs of greater power, with materials chosen to provide extra strength and increased lightweight properties, and by 1910 many had been built and test-flown.

Of these aircraft designs, all incorporated a battery and some electrical circuits to help perform basic functions, such as turning on lights in the cockpit and in the cargo bay areas, as well as solenoids to activate mechanical devices or to measure and display data on a cluster gauge for the pilots to see. Eventually, radio devices would be added to the pilot's instrument panel as a form of communication to the outside world. However, electricity was not used as the fundamental energy source for propelling the aircraft; rather, burning non-renewable fossil fuel for an engine to rotate propellers—and later as thrust for jet engines—was preferred. The presence of electricity was merely for a side benefit. Furthermore, the central bodies of all aircraft were non-symmetrical and utilised non-movable protrusions called wings for lift, using the flow of air moving over them. This fundamental design of man-made airplanes has remained unchanged to this day.

The industrial manufacturing of this type of aircraft intensified just prior to and during the First and Second World Wars, where they were put to destructive use.

---

16   This latest information about the Tunguska event is available from the University of Bologna's Department of Physics website at http://www-th.bo.infn.it/tunguska.

The Mystery Ghost Rockets

Before the commencement of the Second World War in 1939, strange flying objects haunted the skies of Sweden, Norway and Finland during the winter months of 1933–4 and 1936–7.[17] Approximately 15 (10 visual sightings and 5 sounds[18]) of the 111 sightings reported in Sweden alone after the 1933–4 period remain inexplicable in terms of known natural or human-made phenomena. Those that were explainable tended to be sightings of lawful civilian or military aircraft[19].

High-ranking military officer Major-General Reutersۚärd of Sweden gave the following press statement on April 30, 1934 regarding the 1933–34 UFO wave of sightings:

> "Comparison of these reports shows that there can be no doubt about illegal air traffic over our secret military areas. There are many reports from reliable people which describe close observations of the enigmatic fliers. And in every case the same remark has been noted: no insignias or identifying marks were visible on the machine....It is impossible to explain away the whole thing as imagination. The question is: Who or whom are they, and why have they been invading our air territory?"[20]

Foo-Fighters

First appearing over the Pacific (e.g., the Sea of Japan by Japanese pilots) in 1943 and later re-appeared in greater numbers toward the end of World War II (between November 1944 to February 1945) over Belgium, France and Germany by pilots of Allied forces, a number of mysterious glowing metallic discs or luminous balls were observed usually ascending towards their planes from the ground, chasing them for several kilometers and flying about in incredible maneuvers. Foo-fighters, as airmen have called them, have been observed on a number of occasions to change color, usually from red to orange to white and

---

17   Evans & Spencer 1988, pp.53–55; Brookesmith 1984 *The Age of the UFO*, p.109; Good 1997, pp.xviii–xx.
18   Evans & Spencer 1988, p.54.
19   Evans & Spencer 1988, p.55.
20   Good 1996, p.xviii; Brookesmith 1984 *The Age of the UFO*, p.109..

back to orange again, and many have been known to cause the aircraft's ignition system to malfunction.

A grey foo-fighter chases Japanese planes over the Sea of Japan between Japan and Korea in 1943.

As *The New York Times* reported on January 2, 1945:

"BALLS OF FIRE STALK U.S. FIGHTERS IN NIGHT ASSAULTS OVER GERMANY

The Germans have thrown something new into the night skies over Germany – the weird, mysterious foo-fighter, balls of fire that race alongside the wings of American Beaufighters flying intruder missions over the Reich."[21]

World War II fighter pilot Lieutenant Donald Meiers of Chicago recalled seeing these strange lights at the time. He said:

"There are three kinds of these lights we call foo-fighters. One is red balls of fire which appear off our wing tips, and [another] fly in front of us, and the third is a group of about fifteen lights which appear off in the distance—like a Christmas tree in the air—and flicker on and off."[22]

In a *Daily Operations Report* dated 17 February 1945 for the 416th Night Fighter Squadron (NFS) stationed in Pisa, Italy, the following

---

21   Flammonde 1976, p.134.
22   Flammonde 1976, p.134.

excerpt acknowledged the existence of these enigmatic UFOs near La Spezia as observed by American pilots George Shultz and Frankie Robinson:

"At 21:30 saw reddish white light going off and on in spurts about 6 or 8 miles away, near La Spezia at 10,000 ft. going NE., chased it at 280 MPH for 1 1/2 minutes. It took erratic course and faded out. At 21:40 saw some type of light 10 miles South of La Spezia and it went North and turned East of La Spezia at 9000'. Faded near La Spezia. Pilot came within 5 miles of La Spezia, suspected Ack Ack trap. At 21:55, 10 miles south of La Spezia chased another and it went across La Spezia and pilot followed. Faded 10 or 15 miles North of La Spezia. Our aircraft at 300 MPH couldn't catch it. No ack ack at La Spezia. At 22:50, 5 miles south of Pisa, saw same light from distance of 10 miles. Chased it for 2 or 2 1/2 minutes. It took north course, disappeared over mt. this light 10,000'. Light described as glow that alternates between weak and bright. No contacts on AI. Apparently no jamming."[23]

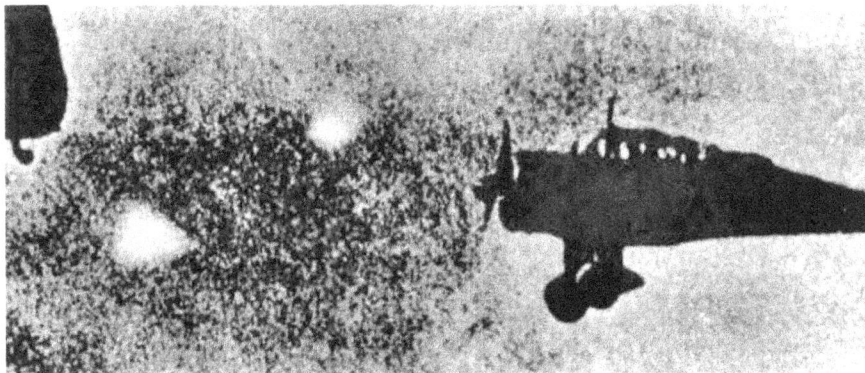

Two luminous foo-fighters chasing Allied aircraft in Europe.

These uninvited visitors eventually disappeared from the skies soon after Allied forces captured a key region in Germany at the closing stages of the war in Europe.

As Lieutenant Colonel Jo Chamberlin, an air crew member of the 415th squadron based in France who observed one of these mysterious

---

23   *Daily Operations Report*, 416th NFS, 12th AF-SCU-01, 17 February 1945.

foo-fighters, confirmed in a December 1945 article he wrote for *The American Legion Magazine*:

> "The foo-fighters simply disappeared when Allied ground forces captured the area East of the Rhine. This was known to be the location of many German experimental stations."[24]

The foo-fighters returned over Japan before eventually disappearing at the time of the first atomic bomb tests in the United States. As Chamberlin said:

> "The lights, or balls of fire, appeared and disappeared on the other side of the world, over Japan."[25]

Besides the possibility of "intelligently controlled alien spacecraft", more earthly explanations for foo-fighters, such as ball lightning or perhaps reflections from ice crystals forming on the cockpit windows and even mass hallucinations, have been suggested. As Curtis Peebles, author of *Watch the Skies! A Chronicle of the Flying Saucer Myth*, wrote on page 2:

> "...the Foo-Fighters were judged to be electrical or optical phenomena or mass hallucination."[26]

Interestingly, not all pilots were fully convinced by these natural explanations.

The Return of the Ghost Rockets

Only six months after the United States dropped two atomic bombs in Japan to officially end World War II, those mysterious aerial objects reappeared over the Baltic Sea and Scandinavia as well as Western Europe between 1946 and 1948, and this time people began to observe strange rocket-like shapes and bright, meteor-like objects careening overhead.[27]

---

24   Chamberlin 1945, p.47.
25   Chamberlin 1945, p.47.
26   Peebles 1994, p.2.
27   Brookesmith 1984 *The Age of the UFO*, p.111.

Subsequent investigations by the Swedish government into 997 of their own "ghost rocket" sightings between May and December of 1946 managed to dismiss about 80 percent of them as conventional aircraft, meteors, stars, planets, or clouds[28]. However, once again they were forced to admit that at least 200 reported sightings had defied explanation, including some described as metallic disc-shaped objects and others as "...mysterious spool-shaped objects with fiery tails..."[29].

Following a press release by the Swedish government in October 1946, *The New York Times* stated on October 11:

> "Swedish military authorities said today that they had been unable to discover after four months of investigations the origin or nature of the ghost rockets that have been flying over Sweden since May.
>
> A special communiquè declared that 80 percent of 1,000 reports on the rockets could be attributed to 'celestial phenomena' but that radar had detected some [200] objects 'which cannot be the phenomena of nature or products of the imagination, nor can be referred to as Swedish airplanes'.
>
> The report added, however, that the objects were not the V-type bombs used by the Germans in the closing days of the war."[30]

UFOs before June 1947

Prior to June 1947, a military pilot, a meteorologist, and a field engineer observed three UFOs in separate incidents and mentioned the sightings to the authorities.

As recorded in the Project Blue Book files, the first sighting occurred on January 16, 1947 in the North Sea, 50 miles from Holland. An RAF pilot flying a Mosquito aircraft tried to pursue the UFO after radar detected it taking an efficient and controlled evasive action. It headed off toward Norfolk at a speed that was considered either the same as or exceeding the pilot's own aircraft, making it impossible to keep up with it.

In the second UFO sighting, a meteorologist noticed a silvery disc through a theodolite while tracking a weather balloon. The disc was

---

28   Peebles 1994, p.3.
29   Flammonde 1976, p.139.
30   Good 1997, p.xxxiii.

traveling east to west at an estimated height of 15,000 feet and looked larger in size than the weather balloon. This event allegedly occurred near Richmond, Virginia, in April 1947.

In the third UFO sighting, not far from Oklahoma City, a field engineer claimed he saw a "frosty white" disc (apparently flatter in height than it was in diameter by a ratio of 10 to 1, based on the witness estimate). It was the size of a B-29 bomber traveling north at a height between 10,000 and 18,000 feet and moving at least 3 times the speed of a jet with a slight swishing sound as it passed overhead. The witness reported the sighting to authorities on the day it happened, which was May 17, 1947.

The Kenneth Arnold UFO Case[31]

Then the civilian interest in the United States over UFOs and number of sightings dramatically increased after June 1947 when reporters talked to a highly rational and respected man with many hours of flying experience who claimed to have seen UFOs over the Cascade Mountains.

It began on the warm, sunny afternoon of Tuesday June 24, 1947. Thirty-two-year-old American airman Kenneth Arnold (1915–1984), a businessman and private pilot with more than 4,000 hours of flying experience, was preparing to travel home by flying over the Cascade Mountains of Washington state. Before he left, he remembered hearing reports of a plane that had gone missing somewhere in these mountains and decided he would spend about an hour searching for it. Sadly, he did not find the plane, but he did spot something else much more unusual and intriguing.

A few minutes before 3:00 p.m., as Arnold was flying toward Mount Rainier at an altitude of 9,200 feet[32], a blue-white flash seemed to light up his cockpit. He looked around frantically to see where it was coming from.

"The only actual plane I saw," he later recalled, "was a DC-4 far to my left and rear, apparently on its San Francisco-Seattle run."

---

31   Full details of this case available from Evans & Spencer 1988, pp.26–31; Sachs 1980, pp.207–208; Flammonde 1976, pp.28 & 162–166; *Mysteries of the Unknown: The UFO Phenomenon* (Time-Life Books) 1987, pp.36–37; Peebles 1994, pp.8–9.
32   Peebles 1994, p.8.

Relief turned to surprise once again when another brilliant flash struck the surface of his plane. This time the flash emanated from the north, and when he turned his head, he noticed, in his own words:

"…a chain of nine peculiar-looking aircraft flying from north to south at approximately 9,500 feet elevation and going, seemingly, in a definite direction of about 170 degrees."[33]

The flash of light was apparently due to two or three discs dipping and tilting slightly to correct their flight course and causing sunlight to be reflected off their highly polished metallic surfaces. At first it was difficult to make out how far away they were until they passed in front of snow-covered Mount Rainier.

"They were flat like a pie-can, and so shiny that they reflected the sun's rays like a mirror," said Arnold. He added afterwards in a radio interview that "I was baffled by the fact that they did not have any tails…I judged their wing-span to be at least 100 feet [30 meters] across. The sighting did not particularly disturb me at the time, except that I have never seen planes of that type [before]."

Kenneth Arnold estimated their distance from him to be about 32 to 40 kilometres, and he tried to time their flight as it passed the peaks of Mount Rainier and Mount Adams as reference points. According to his measurements, the objects took about one minute and 42 seconds to pass both peaks. At 75 kilometres apart, that meant they were moving at 2665 kilometres per hour. Allowing for error, he subtracted some 800 kilometres per hour off this numerical figure. In spite of this, it was still too fast for an ordinary jet aircraft to accomplish at that time; only a rocket could have achieved such a speed.

At about 4:00 p.m., Arnold arrived at Yakima, Washington, where he hurried to tell his story privately to a friend named Al Baxter, manager of Central Aircraft Company, who in turn called in another pilot for his opinion. The pilot listened and suggested the objects were probably guided missiles from a military base at Moses Lake. "I felt satisfied that that's probably what they were," Arnold said. "However, I had never heard of a missile base at Moses Lake, Washington."[34]

Still a little puzzled by the whole affair, Arnold continued his flight plan. He flew to Pendleton, Oregon, where upon landing his plane on a

33   Peebles 1994, p.8.
34   Flammonde 1976, p.28.

small airfield, he learned that his amazing story had arrived ahead of him, as a number of journalists started to gather around his plane, waiting to hear more. He was surprised by the commotion created by his story, saying:

> "I never could understand why the world got so upset about nine discs, as these things didn't seem to be a menace. I believed that they had something to do with our Army and Air Force."

Gradually, as reporters asked a barrage of questions, many expressing doubt, they soon learned of Arnold's sincerity and his flying experience, and they realised his standing as a reputable citizen—even the most skeptical were impressed.

The official U.S. Air Force (USAF) report on this incident, when it was finally released, mentioned nothing about any aircraft or missiles aloft near the Cascade Mountains. Instead, USAF authorities concluded that Kenneth Arnold saw an optical illusion generated by a mirage due to a temperature inversion[35]. Arnold disagreed on the basis of his experience, saying:

> "I observed these objects not only through the glass of my airplane but turned my airplane sideways where I could open my window and observe them with a completely unobstructed view."

Other investigators argued that Arnold's estimate of both the object's size and distance was questionable. For instance, in order for him to make out any details at a distance of 32 kilometres, the objects must have been either considerably larger, which would make them more of a UFO since anything larger than the largest man-made airship had never been seen to travel that fast, or more likely, they were much closer, perhaps only 10 kilometres away, in which case ordinary aircraft could have been responsible. However, a complete check of all pilot records shows that no other aircraft was around on that clear and sunny Tuesday afternoon.

After much heated debate and controversy over Arnold's sighting, it is evident that this case still stands to be reckoned with.

---

35   Sachs 1980, p.207.

The Modern UFO Era Officially Begins

Following Arnold's highly publicized sighting of UFOs over the Cascade Mountains, the incident triggered an avalanche of other unusual sightings throughout the USA and the rest of the world, which still persists to this day, in what is considered the beginning of the modern UFO era. So popular was the topic of UFOs to the general public immediately after Arnold's famous UFO sighting that one brave soul, apparently an anonymous headline writer in the United States, coined the now famous term "flying saucer" in an attempt to describe the shape of the more unusual UFOs being reported locally, nationally and around the world.

Some of the UFOs reported at this time were obviously hoaxes by people wanting to get their names in the newspapers. But a growing number of reports were providing unusually detailed information never seen before by the scientific community. Among the observations getting mentioned in newspapers were the common "flying saucers skipping on water" due to their apparent high manoeuvrability, rounded symmetrical shape, terrific speed and bizarre flight behaviour. In other reports, glowing spheres, dark cylinders and flying crosses started to make a comeback (apparently not unique to the 15th century) and this time the number of developed nations to report such sightings had increased.

UFO researchers today would describe the period from 1947 through to the mid-1960s as the "Golden Age of UFOs" because of the richness and numbers of UFOs reported. Of course, from a totally skeptical point-of-view, it could also be described as the "Ass Age of UFOs" because people were making asses out of themselves for misrepresenting common natural and man-made phenomena as UFOs. Take your pick.

Whatever your perspective on the UFO phenomenon, it is clear that something in the air spooked the American and worldwide public, and something had to be done. As Arnold noted:

> "…if I was to go by the number of reports that came in of other sightings, of which I kept a close track, I thought it wouldn't be long before there would be one of those things in every garage."[36]

---

36   Evans & Spencer 1988, p.31.

Did the U.S. Military Capture a UFO?

Suddenly, one of these UFOs did arrive in the "garage" of the United States Army and Air Force, within a month of Arnold making his famous UFO sighting. Whatever it was the military had found, if the captured UFO was indeed nothing out of the ordinary, the remains of the object would probably be in an American history museum by now; however, this is not the case. What it did do was change the attitude among top military brass and certain personnel involved in the recovery operation as soon as the object was discovered and they understood what it was. After that important day of discovery, the attitude of the U.S. military has been one of continuous total secrecy behind the scenes while presenting a persona of public denial about the existence of UFOs. Not even the slightest possibility that some UFOs might advance our scientific knowledge or support a potential alien presence in the reports.

Because of how dramatic the change in attitude towards UFOs was at this time by U.S. authorities, it is necessary to discuss what had allegedly occurred.

The Crashed UFO near Roswell, New Mexico

If we can rely on the testimonies of civilian and military witnesses gathered by various U.S. investigators since 1978 while the witnesses were alive and from family and friends, then it would appear that on the night of July 2–3, 1947, something crashed in the desert not far from Roswell, New Mexico. The earliest reports indicate that a glowing disc-shaped object allegedly flew over the city of Roswell as it headed in the direction of a large thunderstorm that was sweeping over New Mexico. Later that night, a rancher named William "Mac" Brazel and his family suddenly heard multiple lightning strikes over one area of his property. The streaks of lightning appear to be focused on one particular spot in the desert. Initially, Brazel thought underground mineral deposits were attracting the lightning, but then, in the midst of the lightning strikes was an odd explosion, unlike an ordinary thunderclap. As soon as the explosion was heard, the lightning quickly died down.

The next morning, the rancher and the son of a neighbour went on horseback to check on his sheep in the general vicinity of the odd explosion. What he found later that morning, in a shallow area below an escarpment, was something that would leave an indelible mark on the rancher's memory due to the way the remarkably lightweight materials revealed unheard of physical properties not seen in any known man-made material at the time, or even to this day. Scattered over a wide area with most of the materials concentrated in one spot and fanning out to a great distance were large amounts of super-tough and extremely lightweight materials, consisting mostly of a metal foil, together with plastic insulating sheets and balsa-wood-like plastic I-beams (some with hieroglyphic-like writing on them). Clearly, something had flown in the air and exploded as it tried to speed away from the area for whatever reason. Was this the cause of the odd explosion heard during the night?

Among the materials found was a newspaper-thin yet extremely tough, dark-greyish foil, definitely a metal of some sort, but with the uncanny ability to return to its original shape.

The rancher was also intrigued by an insulating, thin, plastic-like sheet, which could not be torn by hand or burned with a cigarette lighter. Neither he nor the young lad sitting on his horse could explain the materials they found, let alone the type of object that flew over the property.

Major Jesse A. Marcel

Excited by the find, the rancher rushed to speak to his local neighbours. They could not explain what he found. After talking to friends in the town of Corona, it was recommended that he go to the authorities in the city of Roswell to report his discovery. Eventually he did, carrying with him two boxes containing the mysterious debris. Incredibly, neither the sheriff nor a USAF expert who visited the office at the sheriff's request named Major Jesse A. Marcel (1907–1986)—an experienced airman who knew every known man-made aerial device and aircraft in existence at the time—could explain what the materials were or the type of object that flew over the rancher's property. In fact, Marcel was so astonished and excited by the find that he had to take one box to the Roswell Army Air Field where he was stationed to show his commanding officer, Colonel William Hugh Blanchard (1916-1966). Again even his superior was stumped as to exactly what these materials were.

A decision was made. Not knowing the nature of the object and its importance, Blanchard ordered Marcel to go with the rancher to the debris site to conduct a further inspection of the materials and to gather it all up and return to the base. Marcel was permitted to get help with this task. A colleague, CIC Agent Sheridan Cavitt, agreed to assist. Both men and the rancher rode in their own vehicles, making the long journey to the debris field, staying overnight at a small hut just a few kilometres from the debris. The next morning, the rancher left the two men to inspect and collect the materials. It was assumed the two men would be enough to pick up the remaining debris, but as it turned out, it wasn't.

It was nearly the end of the day when Marcel said he would head back to Roswell and leave Cavitt to do some more collecting.

Before reaching the base, Marcel took a turn off to his home in Roswell. He was eager to show his wife and son the materials he had gathered. Actually, no one could miss the excitement in Marcel's voice and his animated appearance as he tried to piece together the foil pieces to get an idea of the size and type of object it was. He even showed some of the physical properties of the materials. Marcel's son could not forget that moment and later recalled his experiences that night to U.S. investigators.

Despite several hours effort, Marcel was unable to get an estimate on the size or type of object it was. With time running out, and too excited by the find to get some sleep, Marcel decided to put the materials back in his vehicle. It was getting close to sunrise and he had to be back before anyone else arrived.

As soon as Marcel and Cavitt returned to base with as much of the materials as they could carry in two vehicles (the two men took a moment to have coffee together and chat about the previous day's activities and materials they found), a meeting was held. Various military men started to arrive and inspect the materials. Further inspection took place during the meeting. One person had the bright idea of using a sledgehammer to try to put a permanent dent in the newspaper-thin metal foil. He wasn't able to succeed. The foil simply thumbed its proverbial nose at the man by returning to its original shape without a scratch. Another man spoke of having used a blow torch to see what would happen. Remarkably, the foil would not melt, and within seconds became cool to the touch. As the meeting progressed, talk of a new type of flying object soon started to enter the minds of everyone who was there.

Could this be evidence of one of those mysterious flying discs witnessed and regularly reported in the media? The men started talking. Even Blanchard could not rule out this possibility. So he ordered his men to take several trucks and make the long distance to the debris field to collect all the materials as thoroughly as possible as it appeared the materials were an important find and he didn't want one piece of the debris to be missed.

Colonel William H. Blanchard

As it was unlikely that a major military operation of this kind, with numerous military trucks going through the town heading north, would not be noticed by the public and lead reporters to ask questions, Blanchard ordered First Lieutenant Walter Haut, the public information officer at the base, to distribute to the media, in an attempt to keep curious reporters away from the recovery operation, a news release claiming that a flying disc had been recovered —an interesting choice of words to describe the object and perhaps not the most ideal for some military officials higher up the chain of command. However, no one in the top brass knew what was happening. Everything remained quiet. Blanchard was merely acting on considered and reasoned advice from his men, as well as those with more knowledge and experience that a new flying object not made by anyone in the world might have been found and could represent one of those mysterious flying saucers getting mentioned regularly in newspaper articles and news broadcasts of various radio stations around the country. It had to be a new flying object not manufactured by any organisation on the planet given the nature of the materials. This was a new find. He thought he had a significant find based on what he saw and the advice he received. Despite making what he thought was a reasonable decision, Blanchard would be later chastised for this after failing to seek advice higher up the chain of command (he would later be replaced by another commanding officer even though the USAF claimed, on the record, there was nothing out-of-the-ordinary in making this decision). Unfortunately, far from quelling the curiosity of the press, it had the opposite effect. The news release had merely heightened the interest among reporters in the military recovery operation and wanted to learn more.

One eager reporter, who had heard about the military commotion, attempted the long drive to the site to see what it was the military were picking up. Unfortunately, he was prevented from doing so. Almost immediately he was taken to the Roswell AAF to be held against his will while sitting in an office until the recovery operation was complete. However, it was not without overhearing from some military officers at the base of so-called "bodies" having been recovered some distance from the main wreckage site. The journalist was so excited about the information that he attempted to make a secret call to his radio station to state that "little green men" had been found. Almost immediately, the line was tapped and military personnel at the base were informed about what was happening. Since then, the journalist kept quiet and told his colleague at the radio station not to say anything. It would appear that he was threatened with his own life, as did the rancher who was eventually brought in for questioning and held against his will for several days. All forms of interrogation and bad treatment of the rancher were handed out in the hope that he would reveal everything he knew. Eventually, he had to be released; however, before doing so, the rancher was ordered to go to a local radio station in Roswell, accompanied by two Air Force men, to state it was a weather balloon, and was later allowed to visit another radio station on his own in the belief that he was complying with the military directive. However, the rancher changed his mind. He stated that what he saw was not a weather balloon and that if he had his time again, he wouldn't tell the authorities, even if it was an atomic bomb. He was left distrusting of the military ever since and only spoke to his son years later about what he found.

So what happen to the main body of the object comprising the lost pieces on the ground? Where did it end up landing?

The only clue we have of the possible whereabouts of the stricken object came from a civil engineer who was driving in the Socorro region in the "summer of 1947". He would not give a specific date to his closest friends and employer thinking that he might get into trouble with the authorities. At the same time, he was an old man with not many years left in him. He made the decision to speak frankly and honestly about his experience and cautioned his listeners not to mention it to others, at least not while he was alive.

If we can rely on this witness' testimony, then it would appear that early one morning (presumably on 3 July 1947) as the sun was just starting to rise above the horizon, a civil engineer named Grady Landon "Barney" Barnett noticed to his right as he was driving along a highway heading to the city of Socorro some light reflecting off an object sitting in an area he described as the Plains of San Agustin. This was approximately 200 kilometres away in a west to north-westerly direction from the initial debris found on the rancher's property, except Barnett was not aware of the other debris site. All he could see was a metallic object in the desert some distance away. Since he could not see what it was, he decided to take a closer look.

When he arrived at the scene, Barnett and a small group of people, who arrived a few minutes later and appeared to be part of an archaeological team from an unnamed university, started to look closely at the object and bodies. What the people were witnessing wasn't anything they had expected to see. Not the average looking man-made aircraft and pilots, but rather something more unusual. Barnett claims they saw unusual-looking bodies and a badly damaged and symmetrical object ripped open by an explosion or impact (yet there were no materials spread out from the object). The object was described as a dark greyish or "dirty stainless steel"[37], disc-shaped metallic object approximately 9 meters in diameter.

The bodies were described as small with large heads, and each was wearing a grey metallic one-piece suit without zippers, buttons, or belts of any kind. Barnett didn't see any gender differences. He assumed the bodies were all males. A total of six bodies were found scattered inside and outside the object.

---

37   Good 1997, p.474.

Not long after this, the U.S. Army arrived in several trucks and a jeep after allegedly having observed the object in flight on several radars in the area and used the method of triangulation to estimate the landing site. Upon arriving at the scene, the witnesses were told in very strict terms not to talk about what they had seen—that it was their patriotic duty not to do so. As Barnett recalled:

> "...We were told to leave the area and not to talk to anyone about what we had seen...that it was our patriotic duty to remain silent..."[38]

Unbeknownst to Barnett was another location for the debris that could have been part of the object he was looking at. The question is, are they linked? The only things linking to two debris locations are that the materials at both locations were found in the summer of 1947 in New Mexico, that there is U.S. military involvement, and that the metal skin of the disk, with its dark-greyish colour, is highly reminiscent of the dark-gray metal foil found on the rancher's property. Unfortunately Barnett and the others did not check the metal to see if a shape-memory response was present.

It would be several days before the USAF finally cleared the first wreckage site from the local rancher's property (as if the military had no knowledge of what occurred), but not without more civilian and military witnesses noticing the wreckage up close. At any rate, all of the materials on the rancher's property (and presumably the disk in the Socorro region) were picked up and sent in large wooden crates to Wright-Patterson AFB to be scientifically analysed (as confirmed by media outlets at the time, and military personnel who claimed to have been involved in the recovery operation on the rancher's property) before eventually being sent to various secret locations around the country for further analysis. From the way the materials were being treated, they were considered sufficiently unusual to require extra special attention to understand them as if the USAF had not manufactured this object or been part of what could be a secret military experiment. Despite the effort, scientists at Wright-Patterson laboratories were perplexed by at least one of the materials: the dark-greyish foil. As if the military had no idea what it was or who created the foil, the people at the base needed outside help to access the world's

---

38   Berlitz & Moore 1988, pp.61.

first vacuum furnace developed by the Battelle Memorial Institute. The aim was to create and test its own small samples based on the composition details uncovered from the Roswell foil using the equipment first installed at the New York University in 1947. Later, the military needed more help from Battelle to find ways to attain a higher level of purity for titanium and its alloys and study the properties of selected and notable titanium-based shape-memory alloys, including one distinctly dark-greyish and most powerful shape-memory alloy known to science at the time: NiTi.

A declassified USAF/Battelle "Second Progress" report in 1949 and classified "First Progress" reports commencing in March 1948 together with a few metallurgical articles published in scientific journals confirming a handful of scientists have requested a viewing of the second progress report for their own scientific work have revealed an interest in titanium-based shape-memory alloys by the USAF, including NiTi.[39] In the scientific literature, according to a study by SUNRISE on titanium-based shape-memory alloys, of all the shape-memory alloys the USAF looked at (e.g., ZrTi, NiTiCr, NiTiCo, NiTi, etc.), only NiTi (or it could be NiTiX, where X is another element added in small quantities of roughly less than three percent to enhance its engineering properties) could match the observations from the witnesses, even right down to its colour.

In fact, NiTi (or NiTiX) is the only distinctly dark-greyish and most powerful titanium-based shape-memory alloy known to science at the time, and even by today's standards.

The choice of using titanium to make a shape-memory alloy is no coincidence. Brigadier General Arthur E. Exon, who was a Lieutenant-Colonel stationed at Wright-Patterson AFB in 1947 and was aware of the recovery of the Roswell materials, learned that the Roswell shape-memory foil probably contained titanium to help explain its toughness[40]. However, for something to be dark-greyish in colour and have the most pronounced shape-memory effect known to science, while also supporting the witnesses' claims of the same properties in the Roswell foil, only NiTi fits the bill, based on the alloys studied by the USAF so soon after the Roswell crash.

---

39   Craighead, Fawn & Eastwood. Battelle Memorial Institute Second Progress Report on Contract
      AF 33 (038)-3736 to Wright-Patterson Air Force Base, 1949. On closer examination of the
      section mentioning NiTi, it stated the work into this alloy appeared in the First Progress Report
      (with all work commencing in March 1948), which have yet to be declassified by the USAF.
40   *Roswell UFO Crash Update*; Kevin Randle, 1995; transcript of interview, 18 June 1990.

The alloy of interest here is known as NiTi. It is the most powerful shape-memory alloy at an approximately 50:50 ratio, and at the time the USAF first studied it, it was the toughest shape-memory alloy in the world. More amazingly, the USAF was the first in the world to properly look at a very pure sample (as necessary to reveal the shape-memory property) of NiTi in the latter half of 1947 by a young man named Dr John Nielson. An intriguing chap in his own right, Nielsen suddenly emerged from the USAF at Phillips Laboratories to become a full professor in metallurgy at NYU in "the summer of 1947" (apparently uncontested) in order for him to have unfettered use of the world's first and only vacuum furnace to make small and highly pure samples of NiTi for his study. Nielsen's move to NYU suggests that he wanted to find out if purity was an important and critical first step to understanding the shape-memory property of the alloy.

This research into NiTi would not stop. It continued into 1948 at the request of certain individuals at Wright-Patterson AFB, but this time, Battelle was asked to make NiTi samples at the titanium-rich end of the NiTi composition (where the least information was known due to the impurities of man-made titanium), and most likely without revealing its shape-memory property to the scientists. All this work was necessary because the USAF did not understand the shape-memory property and how it worked, nor did they have the technology to make highly pure NiTi in 1947 to reveal this property, or even in 1948, without outside help, and with no means to mass-produce the alloy into sheet form to cover the skin of a large flying object. All the USAF had at the time was the help of the only scientific institution in the world to have developed the vacuum furnace capable of producing small samples of highly pure titanium-based alloys.

Hence, this begs the question as to how did the USAF could have manufacture a tough, NiTi-like, newspaper-thin foil in the quantities found near Roswell by July 1947? No technology existed anywhere in the world to manufacture sufficiently pure NiTi or any titanium-based alloy, and certainly not in the required quantities. And no other shape-memory alloy existed with the required colour and toughness characteristics as NiTi, and with such a significant shape-memory effect.

Indeed, today, we understand this extreme shape-memory property to be part of NiTi's super-elastic room temperature phase, but only if the purity is there to reveal it. Otherwise, no one in the world could have known NiTi or any titanium-based alloy of the right combination of elements would show a shape-memory response. You need the extreme purity to have a hope of seeing this property in action.

Furthermore, to resist scratching and permanent bends in NiTi requires considerable preparation and ageing to cold work the alloy through regular bending to help enhance its hardness over time. However, no one knew at the time that this was important. In fact, the USAF were actually in a quandary for a long time as to exactly how the Roswell foil attained its incredible hardness and yet remarkable flexibility.

And now we know the Roswell foil has to be a metal and that it probably contained a significant amount of titanium to explain the toughness properties.

Due to the continuing secrecy surrounding this incident, efforts were made by former New Mexico congressman Steven Schiff to get to the bottom of the mystery. Without realising the importance of NiTi and other shape-memory alloys to the USAF, the USAF was effectively able to get away with its own explanation as to why the incident remains secret.

At first, the USAF was in damage control when the initial news release from Lieutenant Walter Haut arrived at several radio stations and newspapers in the city of Roswell. There was a need to quickly convince people that there was nothing unusual about the materials found. Without mentioning bodies to the media (as the rancher and the news release mentioned nothing about this), and at the request of Major General Clemence McMullen, vice-commander of the Strategic Air Command, Brigadier Maxwell Roger Ramey, commanding officer of the Eighth AAF at Fort Worth AAF, Texas, was ordered to concoct the weather balloon explanation to explain the whole situation. The torn up remains of a weather balloon was brought into Ramey's office. Just prior to this, Marcel arrived with some of the original Roswell materials to set up in Ramey's office and thought these would be shown to reporters. However, Ramey asked Marcel not to take part in the media discussions. While Marcel was out of the picture, someone else substituted the materials with a standard and familiar metallic weather balloon that had been torn up to make it look like it had crashed to the ground. James Bond Johnson, the only media representative from the *Fort-Worth Star-Telegram* (and with military credentials) was permitted to enter Ramey's office to ask questions, and take several photos of the weather balloon.

In order to make the story stick, Ramey ordered Warrant Officer Irving Newton, a weather officer on duty at the time, to attend the meeting. With no first-hand experience of the original wreckage and with the principal military witness (Marcel) and others involved in the viewing of the original wreckage and recovery process taken out of the office, Newton looked at the tangle of aluminium foil and balsa wood and confirmed to Mr Johnson that it was a weather balloon. Newton was permitted to leave the room.

After Mr Johnson left Ramey's office convinced that the weather balloon was probably the cause for all the commotion in media circles following the USAF's unfortunate and premature news release, Marcel was ordered to enter the room, sit on the chair next to the weather balloon, and have his picture taken by USAF photographer Major Charles A. Cashon. It was at this point that Marcel realised what had happened. A cover-up had begun.

After the photos were taken, Marcel left the office. However, Ramey thought it was important to have one more photograph taken, this time of Newton squatting next to the weather balloon. Newton was ordered back into the room and posed with the balloon for Major Cashon, Later this photograph and those of Marcel were quietly slipped in with the rest of Mr Johnson's pictures. Incredibly, until his death in 2006, Mr Johnson was never aware of the possibility that additional photographs were taken and included in his collection. He always assumed he was the only photographer in Ramey's office. He could also not recall how many photographs he took in the office on that day.[41]

Despite Mr Johnson not raising the question of bodies (and hence one less headache for the USAF to deal with), an analysis of one photograph taken by Mr Johnson showed that Ramey had inadvertently held part of the contents of a secret memo he was carrying in his hand to the camera as if, for some reason, he knew the historical importance of the event and wanted himself to be in the pictures for posterity. Today, researchers have identified the word "victims" in the memo and the suggestion that if the photographer was to ask about bodies, Ramey would quickly mention dummies as having been used by the military. Fortunately for Ramey, no such question was forthcoming.

---

41   Email communications with Mr Johnson in January 2006 by SUNRISE just prior to his death.

However, despite successfully quelling the interest among media representatives in the crashed object (as the journalists had nothing more to go on and the witnesses were suddenly going quiet on the matter), the continuing secrecy over the incident and what was found to this day even when the USAF has been asked about the incident in more recent times had caught the eye of Mr Schiff in the early 1990s. He wanted to find out more. So, in turn, he asked the General Accounting Office (GAO) to audit the Department of Defense of all documents relating to the incident. Realising the situation was heating up in an unfavourable way, the USAF had to find an explanation for the continuing secrecy.

In 1994, the USAF found something that could be used to explain what happened. Taking on the view that close enough is good enough to explain the event, it is alleged the USAF was conducting a secret military experiment known as Project Mogul using weather balloons to send instruments high up into the atmosphere in an attempt to detect nuclear explosions in Russia. Although the timing and location for the closest downed weather balloon is well outside the early July period and nowhere near the original debris location, it was assumed that the rancher never knew about this debris for a long time, and weather conditions may have assisted in blowing some of the debris onto the rancher's property only for him to discover the materials by accident during his round to check on his sheep.

If this is true, then why use so much of an expensive and impossible to build alloy in 1947 to create the weather balloons? Why not use aluminium foil or a simple metallic-looking plastic? Surely it would be cheaper and easier to make.

Well, it appears the USAF has taken the latter position for the materials in its final report on the matter published in 1994. In other words, no exotic materials were created for this experiment. Therefore, the Roswell foil must be aluminium—end of story.

However, there is a big problem for the USAF in choosing this official position on the materials. Scientists know aluminium foil does not return to its original shape. It stays crinkly or crumpled when a deforming force is applied, just as we see in Mr Johnson's photographs of the man-made weather balloon in Ramey's office. And if that is not enough, aluminium foil is a bright silvery metal, not dark-grey as in the original Roswell foil. It is as if the USAF has somehow conveniently forgotten all the work on shape-memory alloys done in 1947 (and, therefore, the application of those alloys to build whatever flying object had crashed in the New Mexico desert must still remain a secret), which is hard to imagine given the significance of the find and the amount of work that took place at Wright-Patterson AFB to understand the Roswell materials. Surely, this cannot be true. The witnesses are certain they had observed a distinct dark-grey metal foil, as thin as a sheet of newspaper, that could not be burned or cut, was extremely lightweight, and showed a distinct ability to return to its original shape. Now, at last, the scientific literature not only highlights the connection to the USAF's research at Wright Patterson AFB so soon after the incident, but it also shows support for the witnesses' claims with a direct link through a special class of alloys; namely, the world's most powerful shape-memory alloy of dark-grey appearance known as NiTi.

Furthermore, the general who worked at Wright-Patterson AFB in 1947 stated the Roswell shape-memory foil contained titanium. Titanium? How interesting. We know NiTi contains titanium, and its shape-memory property is evident in the titanium composition range of 49 to 51 per cent. Such a choice of element to make the foil is significant.

To add to this interesting discovery, the purity of titanium is crucial to revealing the shape-memory property of NiTi and other titanium-based shape-memory alloys. We are talking about extreme purity approaching 99.995 or higher.

It is no wonder the USAF were interested in getting help to improve the quality of titanium as needed to study specific titanium-based alloys, of which NiTi was among the list.

All this is unlikely to be a coincidence.

And can we be sure the USAF was looking at the shape-memory effect at around the time of the Roswell crash? Actually, the military had no choice but to consider it once the original Roswell foil arrived at Wright field. Firstly, the witnesses who saw and touched the foil were certain it was a metal of some sort, not a plastic, and yet it exhibited a plastic-like shape-memory effect. Those military personnel involved in the recovery operation also noted the shape-memory effect. This is further reinforced by the efforts of some military personnel at the Roswell AAF to subject the foil to a sledgehammer and saw it return to its original shape, in addition to the fact that a blow torch was also used to demonstrate the foil would not melt. Hence, it is suffice to say that we are dealing with an alloy of some sort. No known plastic material made in 1947 and even to this day can resist the high temperatures of a blow torch as far as the scientific literature can reveal.

However, a titanium-based shape-memory alloy can withstand the high temperatures of a blow torch.

Not only that, but we should also mention the meteoric rise in the interest in titanium after 1947. As the 1969 edition of *Encyclopaedia Britannica* states:

> "TITANIUM: a metallic element, changed after 1947 from a rare metal to an important structural metal. Because of its lightweight and high strength, particularly in alloy form, it is in demand for use in structural parts in high-speed airplanes....No other structural metal has been studied so extensively nor has advanced in technical stature so rapidly."

More astonishingly, the ones who instigated this interest in titanium turned out to be the USAF. No other military establishment in the world was considering this metal due to the high costs to manufacture it. As the 1987 edition of the *Encyclopaedia Americana* states:

> "The high cost of titanium metal often limits its use to military purposes. Because of its lightness and strength, titanium is used as a structural material in high-speed aircraft, rockets, guided missiles, and recoil mechanisms for artillery."

A titanium-based shape-memory alloy must be connected to this Roswell object and forms part of the hull or exterior skin of the object. It is a newspaper-thin sheet covering a very large area, much larger than the USAF could do after 1947, and virtually impossible to make in 1947.

Yet no one in the scientific community knew of the existence of a titanium-based shape-memory alloy. The only time scientists would realise this fact came in 1958 when U.S. Navy scientists discovered the shape-memory effect of NiTi from a very pure sample of the alloy, and from there did the world of shape-memory alloy research actually begin.

Further details of what has been uncovered can be found in the SUNRISE book, *Roswell Revealed: The New Scientific Breakthrough into the Controversial UFO Crash of 1947.*

As for the bodies allegedly found in July 1947, the USAF took on the assumption that dummies were used in some secret high-altitude weather balloons, even though the official use of dummies by the military did not begin until the 1950s. In which case, the actual experimental test of a new symmetrical flying object carrying a significant amount of a shape-memory alloy on its surface, and the use of at least half a dozen diminutive and thin dummies in an object (it is more likely to be as many as nine individuals if the discovery of bodies on the rancher's property and military personnel testimony of the recovery operation are anything to go by) that successfully flew in the air and travelled at high speeds to escape a thunderstorm has never been revealed to this day. Either there are no records available to support the existence of this object (which is unlikely, considering the amount of work done on shape-memory alloys), or the matter still remains secret to this day. In other words, the USAF's official conclusion on the Roswell case, written into a report in 1994 to explain the continuing secrecy of the Roswell object, is not the final word.

Indeed, what exactly is the USAF hiding from the public after all this time? Why can't the military come clean on exactly what it found that required so much of a shape-memory titanium-based alloy to form the outer skin of the large object? And why the lingering rumours of aliens allegedly having crashed to earth in this part of the world? Were there indeed bodies recovered from the wreckage? If so, who are these "people" that flew the stricken flying object? Why do we not know the names of these people who died in the crash? And what is so sensitive about these people that the world cannot be told who they are?

Even if we are to assume the bodies were merely dummies, the subsequent interest in the shape-memory Roswell foil and method of production strongly indicates that this object that crashed to earth was not manufactured by the USAF or anyone else in the world for that matter. Knowing how long it took for the USAF to realise what had happened before cleaning up the initial wreckage site, it looks very much like the USAF was not prepared for this event, nor did it have prior knowledge that such an event would take place or how to build the highly-advanced materials. Whoever manufactured this object had advanced material engineering knowledge and production technology that wasn't part of the USAF's area of expertise.

General Nathan F. Twining (1897–1982)

Thus, two prominent questions remain unanswered: who made the object, and how did it end up in the state of New Mexico if it was not part of a USAF military experiment?

After Roswell

However, without this intimate knowledge of shape-memory alloys and the chain of events that unfolded in this field as instigated by the USAF after July 1947, no one in the public or media knew what was going on. So, luckily for the USAF, the only thing it could do while people remained ignorant was to apply a silencing tactic to prevent the local rancher and the sheriff at Roswell, as well as anyone else who saw the wreckage up close, from speaking out about the incident and the shape-memory effect of the dark-greyish foil.

This silencing tactic was reasonably successful; the rancher and sheriff only spoke to their sons briefly about the discovery many years later, whereas Marcel's son would recall seeing the original materials when they were brought into the house by his excited father. The rancher's son, Bill Brazel had opportunities to revisit the debris field and after some rains to pick up some more of the original materials. These included a piece of a plastic optic wire-like material that he couldn't break or burn, some of the insulating sheets, and several pieces of the shape-memory foil. Unfortunately, he spoke too much at a bar about his materials, and was later visited at his home by two USAF personnel who demanded that he turn over the materials to the military. Bill complied. The story would have ended there, but in the late 1970s when U.S. investigators spoke with a number of remaining first-hand witnesses—including the original USAF witness, Major Marcel—and their friends and families. This was the time when interesting and unusually detailed information was revealed about what had happened and about the materials that were found.

Despite the event being kept under wraps by the U.S. Military to this day, more information would later be revealed about the Roswell object. And this time, the electromagnetic nature of the object would begin to emerge.

More Details of an Electromagnetic Technology Emerge from the Roswell Case

Leaving aside certain electromagnetic side-effects being observed of UFOs by witnesses, the strongest indication that UFOs could represent a new electromagnetic technology came from a high-ranking military official, General Nathan Farragut Twining (1897-1982), in charge of the Air Force Matèriel Command based at Wright–Patterson AAF in Dayton, Ohio. It began when Twining was stirred into action by the event near Roswell.

On July 7, 1947, Twining suddenly changed his travel plans to attend a matter of great importance in New Mexico. The official letter[42] mentioning the sudden change in plans was written to Mr. Julius Earl Schaefer (1893-1974) of the Wichita Division of Boeing Airplane Company. It showed that Twining had canceled his scheduled meeting for July 10, 1947 with the Boeing company because, in his words, of "a very important and sudden matter that developed here".

The U.S. Air Force later officially explained the visit by Twining as a "routine inspection".

An urgent "routine inspection" of a weather balloon? Must have been a very special weather balloon to get Twining off his feet so fast to visit New Mexico and with such urgency. Or, as the USAF suggested in their 1994 *Roswell Report*, it was to attend a Bomb Commanders' Course on July 8, 1947:

> "An example of activity sometimes cited by pro-UFO writers to illustrate the point that something unusual was going on was the travel of Lt. General Nathan Twining, Commander of the Air Matèriel Command, to New Mexico in July, 1947. Actually, records were located indicating that Twining went to the Bomb Commanders' Course on July 8, along with a number of other general officers, and requested orders to do so a month before, on June 5, 1947."[43]

This must have probably been one of the most important Bomb Commanders' Courses ever attended by the General, and would appear to have been organised at the very last minute for the General to

---

42   Hesemann & Mantle 1997, p.58. A copy is available in Appendix H.
43   *USAF Executive Summary: Report of Air Force Research Regarding the 'Roswell Incident'*, September 8, 1994, p.14. *The Roswell Report* 1995, p.21.

suddenly change his travel plans. Either that or Twining must love attending these types of courses.

However, a secret memo[44] dated July 15, 1947 and written by General Twining, released to the public under Freedom of Information (FoI), seems to reveal the true reason for the sudden urgent and rather important visit and possibly the type of object that was recovered. On page 2, it states:

1. As ordered by Presidential Directive, dated July 9, 1947, a preliminary investigation of a recovered "Flying Disc" and remains of a possible second disc, was conducted by the senior staff of this command [Air Material Command at Wright Patterson AFB]....

2. It is the collective view of this investigative body, that the aircraft recovered by the Army and Air Force units near Victorio Peak and Socorro, New Mexico, are not of U.S. manufacture for the following reasons:

   (a) The circular, disc-shaped 'planform' design does not resemble any design currently under development by this command nor of any Navy project.

   (b) The lack of any external propulsion system, power plant, intake, exhaust either for propeller or jet propulsion, warrants this view.

   (c) The inability of the German scientists from Fort Bliss and White Sands Proving Ground to make a positive identification of a secret German V weapon out of these discs. Though the possibility that the Russians have managed to develop such a craft, remains. The lack of any markings, ID numbers or instructions in Cyrillic, has placed serious doubt in the minds of many, that the objects recovered are not of Russian manufacture either.

The document is highly revealing in the sense that it discusses the initial examination of the wingless disc-shaped craft. It reveals quite remarkable and interesting insights into the construction, and likely

---

44   For original document, see appendix I.

propulsion system, stating "...the craft itself comprises the propulsion system", "...the reactor [or engine] to function as a heat exchanger and permitting the storage of energy into a substance for later use", "... storage battery", "...no moving parts discernible within the power room", "...this activation of a electrical potential is believed to be the primary power to the reactor [engine]", "...air outside the craft would be ionised", "...[the] crew compartments were hermetically sealed via a solidification process", "...[the] craft components appear to be molded and pressed into a perfect fit", "...no weld marks, rivets or soldered joints".

Other interesting snippets of information include a doughnut-shaped tube for circulating a clear fluid:

"A doughnut shaped tube approximately thirty-five feet in diameter, made of what appears to be a plastic material, surrounding a central core....This tube was translucent....This tube appeared to be filled with a clear substance, possibly a heavy water. A large rod centred inside the tube was wrapped in a coil of what appears to be of copper material, run through the circumference of the tube. This may be the reactor control mechanism or a storage battery."

There was also a metal ball turret and four circular cavities sitting below the tube and virtually at the base of the disc:

"Underneath the power plant, was discovered a ball-turret, approximately ten feet in diameter. This turret was encompassed by a series of gears that has an unusual ratio not known by any of our engineers. On the underside of the turret were four circular cavities, coated with some smooth material not identified. These cavities are symmetrical but seem to be movable."

All these quotes are indicating something electromagnetic in the technology. So, far from carrying a nuclear technology, the disc could quite easily have incorporated a purely electromagnetic technology designed to generate a high electric potential (another term for voltage or charge), and the symmetrical "circular-cavities" evenly positioned around the base could be the means by which one could concentrate and distribute the charge to one end of the disc to help achieve some yet unspecified purpose. The doughnut-shaped tube filled with a movable "energy storage" fluid and a central rod and coil presumably intended to charge and move the fluid continuously could be a method of generating a magnetic field designed to deflect ionised materials outside the disc for protection, as well as storing the energy at a cool enough temperature.

Who knows?

One thing is certain. We should be careful not to rely on some nuclear-like descriptions (e.g., "reactor") in the memo when explaining how this disc might work.

Assuming this astonishing document is genuine, it suggests that by July 15, 1947 a second crash site suddenly came to the attention of the USAF (probably through the Army) where a disc was found, and that this disc had unusual construction techniques and an unusual propulsion system not recognised by any military or scientific establishment at the time. The information had presumably been passed on to General Twining after he made an unplanned visit to Roswell AAF in New Mexico.

Also, within the document the "remains of a possible second disc" are mentioned. Is this a mistake? Or did the military discover a second disc?

It is feasible to suggest that the first crash location on the local rancher's property was presumed to be the second disc at this early stage of the military's investigation given the amount of debris found scattered over a large area, despite not revealing any clear evidence of its shape. The fact that the document mentions nothing about the internal structure of the second disc to give a sense of comparison through discussion of similarities and/or differences suggests that the military had assumed a "two discs" theory (as they were probably not told by the rancher about the "odd explosion" during the lightning episode), with the second disc being sufficiently disintegrated over the rancher's property to the point where it wasn't worth discussing its internal structure, according to General Twining. Certainly, if we look more closely at the document, we do see the words "...remains of a possible second disc".

In addition, to stay in the air for 200 kilometres (for a *single* badly damaged disc) before crashing would be unheard of for normal military or civilian aircraft. It would seem natural to assume the possibility of two discs.

Disregarding for the moment the number of discs, it is beginning to look like the U.S. Army and Air Force found at least one disc. We are led to believe it was a symmetrical, wingless "aircraft....not of U.S. manufacture", requiring a high electrical potential or voltage to move the object, ionisation of the air to perhaps reduce air pressure near the surface, with no sharp metal points, and no identifying marks. This is one very special type of weather balloon designed to move and protect itself using obscure electromagnetic principles. Or maybe it wasn't meant to be a weather balloon? The fact that the document describes the disc as a "craft" suggests it could carry instruments and possibly people inside.

Does this mean the military have found a new type of high-speed electromagnetic aircraft as well as the remains of pilots, if any, during their recovery?

A Change in Attitude Toward the UFO Phenomenon by the U.S. Government

Despite the USAF claiming an ordinary weather balloon (which would imply the use of materials familiar to everyone in the 1947 period, or else we have a new type of electromagnetic aircraft built like a weather balloon for extreme lightweight but using much tougher materials) as the cause of the crash near Roswell, one fact cannot be denied. After the Roswell incident, the U.S. military and government suddenly changed their attitudes toward UFOs in a way never seen before or since.

At first, like the Swedish authorities, the U.S. government and military appeared to be bewildered by the number of UFO reports coming in on a daily basis prior to 1947. Natural explanations would be given for certain unusual sightings. At other times, the U.S. authorities had no idea what was going on. If anything, the USAF just wanted the whole situation to disappear, thinking that perhaps UFOs were natural phenomena.

Then, immediately after 1947, the new official attitude toward UFOs from the U.S. authorities suddenly changed to one of tremendous significance and treated UFOs as something that had to be kept secret at all costs, but at the same time, they were extremely keen on gathering

UFO information worldwide with the help of the CIA on a covert basis.

One can see the change in behaviour from newspaper reports on UFOs and in observing U.S. military personnel and certain government officials prior to and immediately following the Roswell incident in 1947.

Even more telling is how the U.S. military would later denounce to the public any possibility of an alien explanation for UFOs, as if it did know what UFOs were but didn't want to be open-minded and conduct an impartial investigation. Certainly it would not be scientific to say that there were no aliens visiting the Earth because there is no way the military would know for sure, unless it did find the evidence and wanted to keep it secret. At the same time the change in attitude and decision to keep quiet on the subject would not be indicative of a natural phenomena. How could a meteorite or something similar warrant such secrecy, especially as long as it has been going on for, which is to this day. It simply does not make sense. It suggests something else more artificial was found. Perhaps a secret military experiment? But even if it could be man-made, the USAF seemed totally unaware of what it was that was discovered as if the Americans had not been the ones to build and test it. The only problem is, we still do not know who created this object. And if we look at the technology of titanium-based shape-memory alloys to make one of these UFOs, there was no one in the world who had this technology. Only the USAF started looking at these alloys first after July 1947. But as we have said, the USAF acted as if it didn't own the object that crashed in the New Mexico desert. The fact that it couldn't manufacture pure enough titanium in 1947 to make the world's most powerful shape-memory alloy as the witnesses reported in the Roswell foil tell us the object could not be made by the USAF. Therefore, if we cross the USAF off the list and so too the rest of the world as well, we are inevitably forced to consider the alien explanation.

Now that would be a disturbing thought.

Conspiracy theory or not, this change in attitude would also be proven with the release of thousands of pages of U.S. government documents under the Freedom of Information (FoI) Act in 1976. While the release of such documents was supposed to show that the U.S. government had nothing to hide about UFOs, it is clear there had been a change in behaviour. And in case the Roswell event was nothing out of the ordinary, there is one document that somehow got through the censors which claimed a disc was recovered in New Mexico. In addition to this, it has been proven in the law courts that tens of thousands of documents about UFOs still remain hidden from the public to this day. The U.S. government claims on the one hand that it has no interest in UFOs but argues on the other how the remaining documents pose a risk to national security in the United States if they were to be released.

Not so for the Australian and UK governments, which have seen no need to maintain secrecy about UFOs. Thinking all UFOs must be natural or man-made IFOs, and with no crashed disc of their own to warrant a need to follow in the footsteps of the U.S. in maintaining secrecy, being open about releasing all their UFO government documents for public scrutiny was seen as the best way to quell public interest in UFOs at the time.

But for the U.S. authorities, this is not the case. Something has spooked the Americans, and they are not willing to tell the world what it is they found.

Well, how long can the U.S. government maintain this position? If there are no UFOs as the U.S. military would like the public to believe, then why all the secrecy? Something doesn't make sense—unless something significant and unexpected did occur near Roswell, New Mexico, in early July 1947.

U.S. President Harry S. Truman Allegedly Briefed on the Crashed UFO Situation

Another person who changed his attitude toward UFOs rather quickly, at least on a secret level, was former U.S. President Harry S. Truman (1884–1972).

A high-ranking military officer named Brigadier General Arthur E. Exon who worked at Wright-Patterson AFB at the time, where the original Roswell materials ended up for analysis, not only suggested the Roswell metal foil probably contained titanium, but he heard from the people involved in top-level meetings with Mr. Truman that the object had not been manufactured by the USAF or anyone else in the world. Kind of makes sense given what we know of the titanium-based shape-memory alloy NiTi. Such an alloy made to such high purity to reveal the shape-memory effect and in such quantities to cover the hull of a large flying object would have been impossible to achieve for anyone in the world in 1947. Sounds like Exon was telling the truth.

During an interview held in 1990, Exon said top military and intelligence officers debated with Truman over what to do with the Roswell materials. Perhaps they were discussing in detail whether to release the evidence? At any rate, it would appear a decision was made not to tell the world.

This would seem to make sense. It was possible by this time for preliminary analysis of the crashed disc to have revealed to Truman a lack of weapons on board, suggesting that UFOs were probably no threat to national security. But to be absolutely sure, and so avoid panic among his citizens if the military admitted it could not defend the nation against these potentially marauding UFOs, Truman decided to reorganise all the intelligence organisations in various departments within his country. The role of Director of Central Intelligence was created after September 1947 and given powers by the president to centralise all intelligence information. Part of this intelligence-gathering involved the collecting of UFO reports in his country and elsewhere on a covert basis.

Since then, Truman would officially claim he had no interest in flying saucers. But as General Robert B. Landry, the USAF Aide to the President, said:

> "I was directed to report quarterly to the President after consulting with Central Intelligence people, as to whether or not any UFO incidents received by them could be considered as having any strategic threatening implications at all. The report was to be made orally by me unless it was considered by intelligence to be so serious or alarming as to warrant a more detailed report in writing. During the four and one-half years in office there, all reports were made orally. Nothing of substance considered credible or threatening to the country was ever received from intelligence."[45]

In 1949, Truman commissioned a report into the mysterious "Foo-Fighters" that had menaced military aircraft on both sides during World War II. General Jimmy Doolittle, who headed the official study, concluded that both German and Allied forces had seen them; all the

---

45   http://www.presidentialufo.com/harrys.htm.

witnesses claimed the objects were not secret German or Allied weapons, and the objects were not caused by natural phenomena. When Doolittle told Truman of his results and suggested the objects were "most likely of extraterrestrial origin", it is claimed that he received a message from Truman on April 4, 1950, saying:

"I can assure you that flying saucers, given that they exist, are not constructed by any power on earth."[46]

The Idea that UFOs are Electromagnetic Flying Machines Reaches the Public

Despite the decision by one unusually well-informed U.S. president on the UFO situation and others to keep things under wraps from the public, information surfaced soon after the Roswell event from at least one anonymous source claiming that UFOs were real and that up to three had been allegedly recovered by U.S. military officials. The results of such potentially explosive information—if it could be substantiated with physical evidence, such as a piece of the alien technology, or an admission from the government on the reality of the UFO situation— first appeared in an article by Hollywood gossip columnist Frank Scully (1892–1964) published in October 1949 by the entertainment trade paper *Variety*. Scully obtained this information from an anonymous "scientist".

Readers found the story sufficiently intriguing. So much so that Scully had to quickly finalise his second and more widely-read article in November 1949. In it, he stated that the UFOs allegedly travelled along "magnetic lines of force" and that the scientists involved in the study of downed UFOs expressed outrage over a military decision to have the discs "...dismantled by the Air Force over the protests of magnetic research scientists" and sent to various secret government research centres around the country.

Assuming Scully's claims were correct, this would suggest that the U.S. government was covering up on an important find and knew the truth about UFOs. To the FBI tasked by the U.S. government via the Office of the Attorney General in the State of California to learn more about Scully's mysterious source, it was viewed differently in an FBI file dated 28 October 1951 as:

---

46   http://www.presidentialufo.com/harrys.htm.

Harry S. Truman (1884-1972)

"...the so-called 'flying saucers' are really manifestations of secret and sinister scientific research by the government."[47]

Due to the popularity of the two articles, Scully wrote another follow-up piece on 11 January 1950 explaining the results of sending "twenty questions for the Air Force". Among the questions he asked included:

Q5. Weren't all the saucers found on the western hemisphere magnetic rather than jet jobs?

Q6. Wasn't the small one, which was 36 feet in diameter, equipped with landing gear which had steel-looking balls instead of wheels and which when moving could not be tipped over by ten men but when not moving could be tilted by one man?

Q10. Did you ever find the secret of how these flying saucers were hermetically sealed so as to show no outside crack when the door was closed?

Not surprisingly, the Air Force avoided answering these questions so early in its investigation of the electromagnetic flying object retrieved from Roswell.

On 8 March 1950, the anonymous "scientist" held an unannounced lecture at 12:30 pm in front of 90 first year science students at the University of Denver in Colorado to reveal more about his controversial claims. As news of the lecture reached other students, the audience quickly swelled to 350. In the audience were Professors of astronomy and engineering.

After the lecture, Scully claimed two U.S. intelligence agents attempted to identify the lecturer by talking to George Koehler (who also confirmed what happened), the man who escorted the lecturer, without success. However, the concerns expressed by the agents was made more apparent when they asked Koehler to hand over tape recordings of the discussions with the agents that he was secretly recording.

---

47   https://vault.fbi.gov/silas-newton/silas-newton-part-02-of-06/view

As for the mystery lecturer, those students who attended the presentation eventually determined who he was by 17 March. His name turned out to be Silas Mason Newton.

After the lecture and with the U.S. government now fully aware of the man's identity, Professor Francis F. Broman quickly tried to downplay the interest shown by the students and academic staff by saying the lecturer was brought in as part of a test to see how well the students could evaluate the veracity of the lecturer's claims, and at the end of the lecture the students opinion was that the lecturer's information was "of little value"[48].

If the students had thought so little of the lecturer, why the interest by students to find out who the man was? And if it was of little value, why the visit by the intelligence agents? Doesn't the government know the information was meant to be bogus and harmless to the students, and part of a supposed "test"?

Several newspapers reported what had happened at the university. Far from seeing the lecturer as a crackpot not worthy of a mention, within a week, the public got wind of the mystery man and his story in several published articles.

Scully could not help notice those articles too. He quickly read them with interest.

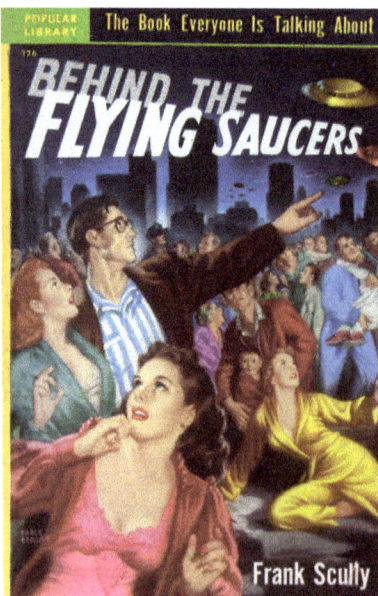

POPULAR LIBRARY
The Book Everyone Is Talking About
BEHIND THE FLYING SAUCERS
Frank Scully

Intrigued to learn whether this was his mysterious source, Scully tracked down the man. Newton confirmed that he had passed information to Scully that led to his ground-breaking and highly popular column pieces in late 1949. As Newton had already mentioned a mysterious magnetic engineer in his lecture as the source of his claims, an opportunity arose for Scully to meet up with this "expert". Newton invited Scully to visit Mojave to check on an oil exploratory operation he was involved with and said the engineer would be there. Scully agreed.

48   https://www.saturdaynightuforia.com/html/articles/articlehtml/anatomyofahoax-part4.html

After discussions with the engineer, Scully decided he would compile all the claims into a book, *Behind the Flying Saucers*, published in September 1950. At the request of the two men, names were kept out of the book. To keep the two men separate, Scully used "Dr Gee" for the magnetic engineer, and "Scientist X" and "Dr X" for Newton. In newspapers that covered the story of Newton's lecture, he was initially called "Mr X".

Following the book's release, and unbeknownst to everyone connected to the book, the FBI secretly developed a profile on Newton and his connections with anyone mentioned in the book. Included in the profile were criminal records that could be used against the man. We see the first FBI report written on the man on 6 August 1951, with the earliest request for the bureau to investigate Newton's activities made on 1 May 1951.[49]

Looking at the FBI files, we see Newton was issued with a warrant for his arrest in 1931 on conspiracy charges, and another charge in 1934 of "false stock statements". He was eventually tracked down and "arrested by the New York Police Department in September 25, 1935, for grand larceny". Afterwards, things went quiet for the man until perhaps a tip-off was made by someone together with a letter written by the office of the Attorney General in the State of California dated 14 June 1951 requesting that the FBI take a look into Newton's business affairs and connections to people he had met, especially those mentioned in Scully's book, as well as a complete look at his criminal record.

Scully was oblivious to the intelligence gathering work by the FBI. He was more interested in gathering further information about the UFOs and the pilots driving these machines from Newton's mysterious source.

While Scully painted a glowing picture of Newton and the magnetic engineer as great scientists performing an important national duty of informing the public and other scientists about UFOs, other researchers have uncovered a little more information about the two men.

In the case of Newton, researchers learned (as did the FBI) that he was the president of Newton Oil Company and a graduate of Baylor University and Yale University. This makes the claim of Newton being a scientist correct. However, his only claim to fame, if one may call it

---

49   https://vault.fbi.gov/silas-newton/silas-newton-part-01-of-06/view.

that, was in drilling "three small [oil] producing wells" in Kansas in 1941. According to an FBI file maintained on the man after 1950, this apparently earned him $200 per month, but not enough to pay the bills. The picture painted by the FBI was of a man who tried immensely to give the impression of wealth by wearing an expensive business suit, and living in a hotel of some reasonable repute among socialites, but in truth he was not doing all that well. In fact, his business was largely unsuccessful. Eventually, Newton was broke after spending too much money drilling numerous dry holes in search of new oil reserves in Arizona and Colorado.

It was about this time that Newton met up with a man involved in the oil exploration business. He had something to tell, but was afraid to give precise details. Given the incredible nature of the claims being made by the man, a decision was made to release the information to the public while retaining anonymity for all concerned. If it were a complete hoax, selling the story to the highest bidder would have been the order of the day. However, strangely, Newton and his mysterious colleague only wanted to tell the story to Frank Scully in late 1949. Later, Newton went on his own to present his university lecture without asking for a fee. In fact, the only one to make money from the situation was Scully, who took the opportunity of publishing his book. The men knew what Scully did and there was no complaint about it. It seemed the primary aim of the men was to get the information out to the public.

By 1952, the financial plight of Newton became more dire, and his mysterious informant was not earning enough from his daytime job to make ends meet. Then the two men did something that would forever damage their reputation and the credibility of their claims. It was likely the men noticed Scully's financial success from his book. The men probably thought they should do the same in some capacity. So an unfortunate decision was made by the men to work together on a scheme to cash in on the 'alien technology' theme. The idea was to sell electronic devices (later found to be relics of old wartime devices) to unsuspecting investors allegedly designed to detect oil or gold in the ground.

This was the men's downfall. However, it need not necessarily mean that what was said by the men is one hundred per cent untrue. It just makes their claims less credible unless someone else can come along to highlight something important in what was said that no one else could have known.

Leaving aside the men's financial predicament and folly, Scully tried all he could to get the mysterious Dr Gee to tell him where he obtained his information. He would not say. The most Scully could do was write his book while respecting the wishes of the men to remain anonymous (even despite Newton having been identified by a number of students) as the way to avoid getting into trouble with U.S. government and military officials.

With no indications in the early stages of the men trying to make money from the story, it looks fairly clear that the primary aim for Newton and his informant was to let the public know of something important and to inspire science students to study interplanetary travel. Seems like a reasonably noble cause. A less obvious secondary aim concerns where the students should focus their studies in science— electromagnetism. Given the way the crashed disc stories were presented, it seems as if the intent of whoever gave this information to the informant was to weave together key information about the way these discs were constructed and some electromagnetic concepts. The presence of odd-looking occupants was just an interesting side-note to make the whole thing sound a little more "alien", but it is interesting to see how lightweight and small the aliens were described. Not entirely an unheard of observation from other UFO reports (see Chapter 5). By creating the stories and adding crucial pieces of information, the identity or identities of those allegedly involved in the analysis of at least one crashed disk (possibly more) could be avoided and so protect those working for the U.S. government. Otherwise we must assume the men had somehow made up the stories with remarkable realism, right down to presenting these key observations.

Or did a group of scientific whistleblowers create the stories for someone else to leak to the public? If Dr Gee was indeed a group of scientists who saw something and had trouble understanding how the electromagnetic technology worked and wanted to tell the world about it, it is probably not so far fetched for Scully to describe Dr Gee as having:

> "...more degrees than a thermometer and had received them from such diverse institutions as Armour Institute, Creighton University, and the University of Berlin."

The exact number of scientists appears to be in the order of eight. As Scully said:

"...Dr Gee and seven of his group of magnetic scientists were called in to examine this strange ship."

Multiply the degrees for each of the scientists and it is probably not too outrageous for Scully to state what he did.

If this is true, it would suggest that one scientist was designated as the group's spokesperson, who in turn leaked some of the results of a preliminary examination of crashed discs in story form to a person on the outside. It seems the aim of the leak was to get other scientists to recognise the electromagnetic nature of UFOs and hopefully pursue it until a solution is found. Then maintaining secrecy on UFOs would become ineffective. That could have been the intent of releasing this information.

Whatever the truth, due to the sensational claims, Scully's book proved to be a bestseller when it hit the stands mainly because no one else knew if the claims were true or what would be possible. And indeed, for a while, no one could figure out whether the information was true.

Eventually, a San Francisco journalist John Phillip Cahn was commissioned by the editors of *True* magazine to look more closely into Scully's claims and the people responsible for supplying the information. After some hard work, Cahn managed to track down the mysterious magnetic engineer, or so he thought. Prior to Cahn getting involved, the FBI already knew the name of the man—Leo Arnold Julius GeBauer. Now that Cahn was aware of this too, he confronted GeBauer and asked if he had been involved in passing on information to Scully and Newton about the alleged "crashed discs". The Texas oilman suddenly denied all involvement in the matter. When asked whether he was Dr Gee, he denied it. In fact, he told Cahn he was not a scientist.

GeBauer has signed a letter written by him to Cahn confirming he was not Dr Gee, but gave no indications where he got the source of his information. He also denied meeting Scully in Mohave and had no knowledge of Scully's book or any connections with the author, but he did state that he was acquainted with Newton. The sudden decision to deny where he obtained his information suggests that he has been frightened off from saying anything more now that things were getting out-of-hand.

Today, researchers believe the informant that gave GeBauer his unusually knowledgeable insights into UFOs was probably from a geologist who had done some work for GeBauer, and who had previously worked for the U.S. military where the information was obtained. But to protect the original source (probably one or a number of scientists) involved in the study of at least one recovered UFO (was this from Roswell?), up to three stories were created of supposedly additional crashed discs with the aim of presenting key observations obtained from a preliminary examination of the original recovered UFO.

Or were there genuinely additional crashed discs retrieved by the U.S. military?

However, the more skeptical researchers and journalists (of which Cahn was one of them) were quick to discount GeBauer's claims when further investigations uncovered a newspaper article titled "'Saucer Scientist' in $50,000 Fraud" in the *Denver Post* dated 14 October 1952. It stated that Newton and GeBauer had been involved in a scheme to defraud an industrialist named Herman A. Flader of $50,000 by selling what Newton described to investors as a supposedly alien technology consisting of magnetic devices called "doodlebugs", allegedly designed to detect underground oil or gold deposits. Authorities at the time discovered the machines were nothing more than old electronic war relics.

With the confidence trickster status now firmly established for GeBauer and Newton, it was enough for most researchers to label their information given to Scully as unreliable and not worth looking further into it.

David Thomas summarised the investigative work into these two men in the following way:

"Turns out that these two gentlemen were oil swindlers. They were con men. GeBauer worked as a tech at an auto parts supply place. He was not a top-secret government scientist. Newton, also a con artist, would tell investors of the recent crash of an alien ship, the analysis by top-secret government labs, and the discovery of magnetic devices that could find oil and gold. Then came the pitch. Newton told the investors that he could get them access to this incredible 'alien' technology—for a price."[50]

Moving on from those difficult financial times for the two men, GeBauer became the chief manager of AIResearch Company's laboratories in Phoenix, Arizona, and Los Angeles, California.. With no evidence of any further wrong-doing and holding down a stable and well-paying job until he retired, GeBauer had every opportunity to confess on whether the three "crashed discs" stories were a hoax. When he died, there was no admission of guilt on this matter. The only thing he was willing to admit was not being the mysterious "Dr Gee" talked about in Scully's book or the magnetic engineer in Newton's lecture.

THE VOICE OF THE ROCKY MOUNTAIN EMPIRE

DENVER POST HOME EDITION

DENVER, COLO.—Climate Capital of the World—TUESDAY, OCT. 14, 1952   52 PAGES

'Saucer Scientist' n $50,000 Fraud

Rocky Mountain Rodeo Ticket Offices Listed

Tickets for all performances of the Rocky Mountain Empire rodeo may be obtained at the following places: The coliseum, from 10 a. m. to 5 p. m. Sunday, and 9 a. m. to 8 p. m. on weekdays; The Denver Post and the downtown J. C. Penney company store; Engelwood Men's store, 3163 South Broadway; Reinert's Clothing stores in Boulder and the Cheyenne Travel Service in Cheyenne.

Military personnel can get tickets at special services offices at Lowry air force base and Fitzsimons Army hospital.

Telephone reservations may be made by calling AComa 4784. Tickets still may be ordered by mail through use of the coupon printed on page 2.

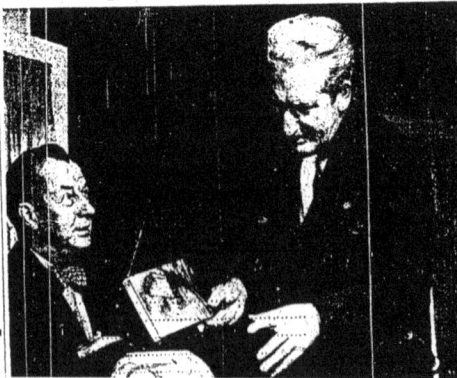

Ending War

Swindle Alleged In Oil Tests

By CHARLES ROOS.
Denver Post Staff Writer.

Silas M. Newton, the "Mr. X" lecturer of flying saucer fame, and a Phoenix, Ariz., radio parts merchant were charged Tuesday by District Attorney Bert M. Keating with operating a $50,000 confidence game swindle.

Keating accused Newton, an oil promoter, and Leo A. GeBauer of Phoenix of defrauding Herman A. Flader, Denver industrialist, out of $50,000 in a swindle involving oil well exploration tests with electronic "doodlebugs," one of them represented as costing $800,000.

Two similar machines have been examined and declared to be war surplus items worth about $3.50, the district attorney said.

BOTH MEN SOUGHT.

Newton died on 15 December 1972. Again we see no indications of a confession, verbally or otherwise, to friends or family.

50   https://alibi.com/news/30634/Skeptic-on-Skeptic.html

Scully was offered $25,000 to admit the story was a hoax by Cahn. The author refused the money and steadfastly stuck by his belief in the story as it was relayed to him by Newton and his mysterious "Dr Gee". No written confession could be found when Scully died on 23 June 1964.

Despite this unfortunate blemish on the reputation of the two principal men, a closer examination of the information provided by GeBauer from his unknown source reveals remarkable and rather consistent scientific information in the manner in which he described the construction and materials used to build the discs, especially in relation to the laws of electromagnetism and Einstein's special theory of relativity. The specifics involving purely magnetic propulsion may be out-of-whack, but this might reflect the early stages of the analysis into a potentially genuine crashed disc (or were there more than one?) and a lack of understanding of how UFOs worked. To give as an example of how little was known of UFOs and extraterrestrial life in general, some scientists in the 1950s thought it was possible for alien life to exist on Venus or Mars, and if so, it would not have required a massive amount of energy for aliens to reach the Earth. Interplanetary travel is more feasible than interstellar travel. Not even with the help of nuclear power can the distances between the stars be covered quickly enough to permit UFO occupants to participate in the flight. Moreover, the recovered UFO(s) looked too small and lacked a fuel tank or recognisable combustion engine and yet numerous occupants were found inside, all apparently healthy enough until time of death. It seemed the most logical thing to do for scientists involved in the examination of the disc(s) was to consider life on other planets as the likely location for these diminutive beings. Furthermore, the lack of a familiar engine and fuel system with only a few moving parts and gears underneath the floor suggested that the energy must have been derived from the surrounding environment. As the UFO clearly operated on some unknown electromagnetic principle, it stands to reason for the hapless scientists to consider any electromagnetic energy in space as the source of fuel. However, the only substantial electromagnetic energy available in space had to be the magnetic fields generated by planets, but the radiation in space seemed too little to be of much use. It made more sense in the 1950s to consider interplanetary distances for the UfO technology and to somehow imagine the technology as being able to concentrate the magnetic energy and use it to achieve some kind of motion quick enough for the occupants not to starve to death during the travel. The only problem for the scientists was, how could this be achieved? Well, that is why the technology was called "alien". There was something the scientists had not understood and were not familiar with. The scientists knew the crashed disc(s) worked on some unknown electromagnetic principle. But to figure it out, someone had to start somewhere. Must as well consider the possibility of magnetic propulsion, even if it seems highly unlikely by today's standards and available technology.

What matters here is the bigger picture being presented by the men, which is:

(a) An electromagnetic technology apparently exists in UFOs.
(b) Something was found in the southwest of the country.
(c) There is a very good chance someone with knowledge of the UFO situation was trying to find a way to leak information to the public to show how important electromagnetism was to the understanding of UFOs, but exactly how it worked remained a work in progress by 1950.

For example, we learn from Scully's book how the discs were constructed in a very smooth way (presumably to prevent electrical sparks emanating from sharp points), employed a symmetrical Faraday cage design for the disc itself because of its perfect circular shape and for the cabin internally, used tough and extremely lightweight metals and plastics that could withstand high temperatures and impacts (highly reminiscent of the materials used in the Roswell object), and that the amount of materials used to construct the discs was kept to an absolute minimum (of which the Roswell materials were extremely lightweight), with signs of a newspaper-thin outer skin (just like the Roswell foil), a couple of bucket seats, a control panel and certain rotating mechanisms beneath the floor.

Interesting to hear the tough and lightweight nature of the materials used to construct the discs. It looks like Scully's book could be describing the Roswell case. Does this mean the stories of three crashed discs in Scully's book are meant to be a throwback to this original Roswell case? Or have there been other crashed discs subsequent to the Roswell event?

As an additional scientifically reasonable observation to support the smooth outer appearance of the disks, Scully asked the Air Force based on what he heard from Dr Gee:

"Did you ever find the secret of how these flying saucers were hermetically sealed so as to show no outside crack when the door was closed?"

This makes reasonable sense if the aim is to remove any sharp edges on the outside. One should expect an alien civilisation to understand the importance of engineering precision to make the parts fit snuggly and perfectly, especially in applying an electromagnetic technology involving the use of electric charge and generation of electromagnetic fields. Otherwise, any charge build up could result in electrical sparks that could affect the operation of the discs and cause problems for the occupants.

Apart from the disc, the occupants were also described in some detail. Here, the most notable features of the occupants were their small and very thin bodies with large heads. Unless these individuals were skinny dwarf humans suffering from a disease that made their heads larger than usual (e.g., macroencephalopathy), it would suggest both the occupants and disc did not originate on Earth. About the only other thing that could be agreed upon by those participating in the study was the considerable efforts by the occupants to reduce their own mass (with the obvious exception of the head region) and the mass of the disc.

Indeed, such an attempt to achieve extreme lightweight measures in both the occupants and the discs is not just highly reminiscent of the Roswell case, but it is scientifically consistent with trying to achieve high speeds approaching that of light. Einstein's special theory of relativity states that no object can reach or exceed the speed of light because mass increases with speed and because of how mass reaches an infinite value at the speed of light. However, nothing in the laws of physics tells us we cannot approach the speed of light as closely as we would like. It is all a question of reducing the mass so that the object and occupants can be accelerated to near the speed of light using the least amount of energy and as quickly as possible. Given the low-mass nature of the occupants, one can fathom how they could minimise inertial forces and withstand a much higher level of acceleration.

Furthermore, it is scientifically feasible for the occupants to have participated in such great journeys, because Einstein's theory states the distance along the direction of motion is perceived as being dramatically reduced, and so it allows the occupants extremely short journey times.

However, all these extreme measures in mass reduction are suggesting the distances travelled by the occupants could be much greater than originally thought by the scientists in the 1950s. It could pave the way for the possibility that we might be dealing with interstellar travellers, rather than planetary travellers. In which case, there has to be another electromagnetic concept to allow for more rapid acceleration to make this possible (as planetary magnetic fields can only extend so far into space and another energy source has to be used), and a means by which occupants can withstand the inertial forces. In Chapter 8, we will learn whether there is a way we can achieve this using current scientific knowledge.

In summary, everything that has been said by the mysterious source is not inconsistent with scientific concepts of electromagnetism and the special theory of relativity. The question is: why would electromagnetism be so important, especially since no solid fuel or exhaust system or obvious engine to process the fuel and achieve propulsion could be found, as if electromagnetism somehow plays a central role? Surely, the mysterious source could have spoken about nuclear power, as this was of great interest to scientists and the U.S. military after 1945. Even the scientists themselves thought nuclear power to be the most advanced knowledge available to humans at the time. This view hadn't changed right up to 1973, when scientists of the British Interplanetary Society relied solely on nuclear power[51] as the only effective solution to interstellar travel, and that was for an unmanned flight. Interesting to hear this in 1973. In 1949, more people had been left in awe by the power of nuclear energy through the atomic bomb. Yet, remarkably, GeBauer's mysterious source—whoever he or they were—did not fall into this trap. Electromagnetism was seen by the source as somehow fundamental to understanding flying saucer technology. Why?

Given what we have seen so far, we should be careful not to ignore this information. It is possible that what was said by the unnamed source could provide important information for helping scientists to

---

51   Sternbach 1983, p.52.

understand the modus operandi of genuine UFOs and potentially advance science.

The Independent UFO Investigation by Major Donald Edward Keyhoe

Whatever it was about the UFO phenomenon that stirred the top echelons of the U.S. government and military (understandable if it indeed involved the recovery of at least one crashed UFO—we presume at Roswell?), the seriousness and super-secret nature of the attitude of top U.S. military officials towards the UFO situation became evident after an investigation was conducted by freelance journalist and retired U.S. Marine Corps pilot, Major Donald Edward Keyhoe (1897–1988).

Major Donald E. Keyhoe

The story began when Keyhoe received a telegram from Mr. Ken W. Purdy, the editor of *True Magazine*[52]:

---

52   Peebles 1994, p.37.

HAVE BEEN INVESTIGATING FLYING SAUCER MYSTERY. FIRST TIP HINTED GIGANTIC HOAX TO COVER UP OFFICIAL SECRET. BELIEVE IT MAY HAVE BEEN PLANTED TO HIDE REAL ANSWER. LOOKS LIKE TERRIFIC STORY. CAN YOU TAKE OVER WASHINGTON END?

KEN W. PURDY, EDITOR, TRUE MAGAZINE

Upon receiving the telegram, Keyhoe agreed to investigate the story. The result of Keyhoe's effort over a six-month period was a sensational and widely read article, titled *Flying Saucers Are Real*[53], published in January 1950[54]. He attacked the USAF and other government agencies for hiding UFO evidence; he thought they strongly feared nationwide panic if people knew the truth about flying saucers.

Vice Admiral Roscoe H. Hillenkoetter

During his research (which he later expanded and published in a book of the same title), Keyhoe's military background gave him access to high-level military personnel, and he interviewed several military witnesses. Included among his contacts was the first director of the Central Intelligence Agency (CIA), Vice Admiral Roscoe H. Hillenkoetter (1897-1982), who spoke frankly about the UFO issue not long before his reign as CIA director suddenly ended in 1950 (soon after Keyhoe published his article). Director from 1947 to 1950, Hillenkoetter was reported in *The New York Sunday Times* on February 28, 1960 as saying:

"It is time for the truth to be brought out in open congressional hearings. Behind the scenes, high-ranking Air Force officers are soberly concerned about the UFOs. But

---

53   Keyhoe's 1950 book was a best seller and is now a classic in ufology. This book is an accurate look at the facts and how Keyhoe investigated the UFO business as a freelance journalist.

54   *Mysteries of the Unknown: The UFO Phenomenon* (Time-Life Books) 1987, p.46.

through official secrecy and ridicule, many citizens are led to believe the unknown flying objects are nonsense."[55]

Keyhoe discovered from his investigation that none of his high-level sources except Hillenkoetter wanted to talk about UFOs, while others did confirm to him the existence of so-called flying saucers, which led Keyhoe to deduce that perhaps there was substance to these stories after all.

New Public Opinion on UFOs and the Founding of NICAP

Soon after the famous article was published, there was a change in public perception about UFOs, as the U.S. Gallup poll results showed in May 1950. Now, people favored the "secret military experiments" explanation for UFOs. We can see something similar from the Gallup poll of August 1947 except it was specifically described as a "secret weapons experiment". However, by 1950, it seemed nearly a third of the American population were generally following the USAF in its explanations for various UFOs observed in their country (i.e., 32 percent for mirage, optical illusions, weather balloons, etc.) and around 16 percent for U.S. or Russian secret weapon experiments, although a large proportion (33 percent) couldn't answer or simply didn't know.

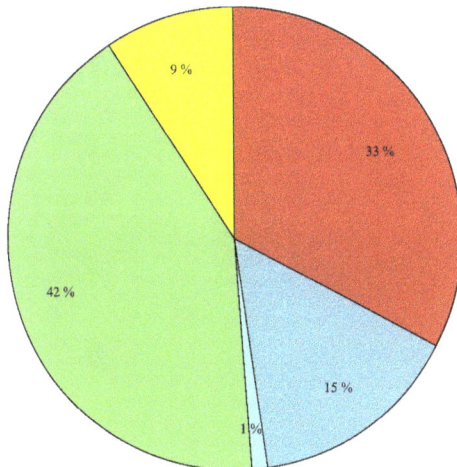

Gallup poll, August 1947 (simplified)

Original	%

55   Good 1997 pp.360–361.

No answer, don't know	33
Mirage, optical illusions, etc.	29
Hoax	10
U.S. secret weapon experiments (part of atomic bomb, etc.)	15
Russian secret weapon experiments	1
Weather forecasting devices	3
Other explanations	9

Simplified	%	
No answer, don't know	33	
Mirage, optical illusions, weather balloons, hoax, etc.	42	
U.S. secret weapon experiments	15	
Russian secret weapon experiments	`1	
Miscellaneous (Other explanations, etc.)	9	

Then, by May 1950, the public was favouring the secret military experiment or some kind of new airplane as a possible explanation for UFOs.

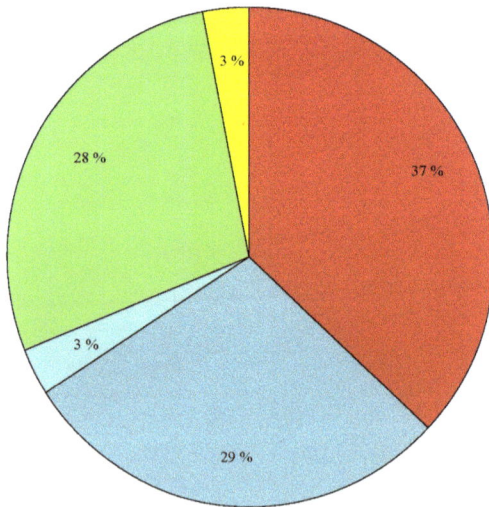

Gallup poll, May 1950 (simplified)

Original	%
No answer, don't know	32
Secret military experiments	23

Optical illusions, hoax, pipedream, etc.	16
Some kind of new airplane	6
Comets, shooting stars, something from another planet, etc.	5
Weather devices	1
Russian secret weapon experiments	3
Miscellaneous	3
No such thing	6
Haven't heard of them	6

Simplified	%	
No answer, don't know, or haven't heard of them	37	
Mirage, optical illusions, weather balloons, hoax, pipedream, no such thing, etc.	28	
Secret military experiments, or some kind of new airplane	29	
Russian secret weapon experiments	3	
Miscellaneous (Other explanations, etc.)	3	

Not long after his revealing investigation into UFOs in the military and U.S. government, Keyhoe was instrumental in the founding of the National Investigations Committee on Aerial Phenomena (NICAP), the world's largest civilian UFO organisation. Due to Keyhoe's strong connections with the military and willingness to get to the truth, some UFO researchers have claimed that the CIA later infiltrated into the organisation to hinder quality UFO research work[56] in key areas necessary to solving the UFO mystery, as this excerpt from Margaret Sachs' 1980 book, *The UFO Encyclopaedia*, indicates:

NATIONAL INVESTIGATIONS COMMITTEE ON AERIAL PHENOMENA (NICAP), 5012 Del Ray Avenue, Washington, D.C. 20014; telephone number: 301-654-8091.
The stated goals of this investigative organisation are

---

56   SUNRISE has noticed a number of former or current USAF and Battelle individuals participating in MUFON in the United States either as so-called UFO/Roswell researchers or directors of these organisations. However, when the information contributed by these people to the public is inspected more closely, or when SUNRISE unexpectedly receives information from other indirect means such as MUFON newsletters that just so happen to coincide with current SUNRISE research, it shows examples of UFO cases that are clearly hoaxes, Roswell witnesses' statements that are different from the original testimony, and/or simply not relevant to UFO research. And the UFO cases one may receive show a poverty of observations relating to electromagnetic effects.

scientific investigation and research of reported unidentified flying objects, and encouragement of full reporting to the public by responsible authorities of all
information which the government has accumulated. The
group adheres to no specific theories regarding the
identity of UFOs. Although beset with financial difficulties throughout its existence, NICAP has maintained a
conservative and dignified reputation in the field,
partially because of the credentials of the members of the
board of governors. Over the years, these have included
high-ranking military officers, former officials of the
CENTRAL INTELLIGENCE AGENCY (CIA), politicians, businessmen, college professors and clergy-men.
In recent years, some UFOLOGISTS have claimed that
NICAP has been infiltrated by the CIA as part of a
conspiracy to hinder UFO research and COVER UP
UFO information.

Dr. Wilbert Brockhouse Smith and Project Magnet[57]

The possibility that UFOs were electromagnetic flying objects started
to ramp up thanks to the work of one man. Across the border, a senior
Canadian radio engineer from the Department of Transport in Ottawa,
Ontario, Dr. Wilbert Brockhouse Smith (1910–1962), was authorized by
Commander C. P. Edwards, then Deputy Minister of Transport for Air
Services, to investigate UFOs using the department's laboratory and
field facilities from December 1950 under the auspicious code name of
Project Magnet. As the name suggests, the Canadian government would
officially describe the project as a study of "those phenomena resulting
from unusual boundary conditions in the basic electromagnetic
theory"[58].

The project went ahead on the basis that Smith knew a thing or two
about electromagnetic theory and he was expressing confidence in a
link between UFOs and electromagnetism, based on certain
information he received.

Smith, who held a master's degree in electrical engineering from the
University of British Columbia, Vancouver, was studying radio wave
propagation at the time. This eventually led him into the areas of

57   Good 1997, pp.158 and 184; Sachs 1980; Spencer 1991 *The UFO Encyclopaedia*, p.252.
58   Good 1997, p.184; Spencer 1991 *The UFO Encyclopaedia*, p.252.

cosmic radiation, auroras, atmospheric radioactivity and geomagnetism. It was in the field of geomagnetism that Smith became increasingly more interested in the possible development of power sources utilising the energy in Earth's natural magnetic field.

By chance, he then discovered Frank Scully's book. It was here, to his amazement, that he learned of the possibility of UFOs actually having crashed on Earth and later being picked up by the U.S. military, and that Scully's book described the potential use of magnetic principles as the motive force behind these downed disc-shaped flying objects.

Later, Smith understood a number of genuine UFO reports were indicating the presence of a strong magnetic field in association with the objects in flight (e.g., compasses and watches worn by witnesses were magnetically affected at the time of observing UFOs). Did UFOs derive their energy for propulsion from geomagnetic fields or some other source?

With help from Lieutenant Colonel Bremner, Smith interviewed a scientist named Dr. Robert Sarbacher, a consultant to the Research and Development Board of the Washington Institute of Technology, with regard to UFOs and the U.S. government's involvement with them. After the interview, he made handwritten notes of what he had learned (revealed to Arthur Bray by Smith's widow):

September 15 - 1950
Notes on interview through Lt/C Bremner with Dr. Robert Sarbacher.

WILBERT B.SMITH(WBS): I am doing some work on the collapse of the earth's magnetic field as a source of energy, and I think our work may have a bearing on the flying saucers.

ROBERT SARBACHER(R/S): What do you want to know?

WBS: I have read Scully's book on the saucers and would like to know how much of it is true.

R/S: The facts reported in the book are substantially correct.

WBS: Then the saucers do exist?

R/S: Yes, they exist.

WBS: Do they operate as Scully suggests on magnetic principles?

R/S: We have not been able to duplicate their performance.

WBS: Do they come from some other planet?

R/S: All we know is, we didn't make them, and it's pretty certain they didn't originate on the Earth.

WBS: I understand the whole subject of saucers is classified.

R/S: Yes, it is classified two points higher even than the H-bomb. In fact it is the most highly classified subject in the U.S. government at the present time.

WBS: May I ask the reason for the classification.

R/S: You may ask, but I can't tell you.

WBS: Is there any way in which I can get more information, particularly as it might fit in with our work.

R/S: I suppose you could be cleared through your own Defense Department and I am pretty sure arrangements could be made to exchange information. If you have anything to contribute, we would be glad to talk it over, but I can't give you any more at the present time.

Note: The above is written out from memory following the interview. I have tried to keep it as nearly verbatim as possible.

On November 21, 1950, Smith sent a formerly top-secret memorandum to the controller of telecommunications in Ottawa, recommending that action be taken to initiate a special research group to study the magnetic and gravitational properties of UFOs.

Furthermore, according to the memorandum that instigated Project Magnet, he stated, "…the matter is the most highly classified subject in the United States government"[59]. No doubt this made him wonder why the U.S. government had an intense interest in UFOs in direct contravention to its public persona of stating that UFOs were nothing out-of-the-ordinary.

In a speech he gave to the Vancouver Area UFO Club on March 14, 1961, Smith confirmed his interest in UFOs when he said:

"Being of a scientific background myself, my own interest was directed largely towards the scientific end. I wanted to know how these craft were built, what their motive power was, where they got their energy and how come they were able to do such interesting things that our craft were unable to do."[60]

In terms of the electromagnetic anomalies alluded to by his department, Smith confirmed his interest in this field when he said:

"…studies of Maxwell's equations did, however, raise certain questions regarding the nature of the various fields involved and the manner in which such fields come into being. The generally accepted hypotheses were scrutinised carefully and a number of inconsistencies were detected. These inconsistencies and their experimental investigations form the subject matter of this report."[61]

Thus, by gathering more information about UFOs, applying his knowledge of electromagnetism, and performing experiments, he hoped an analysis of these inconsistencies and coming up with improvements in current knowledge would help to better understand how these enigmatic objects worked or at least to determine the area of physics needing closer attention.

There were two parts to the study. The first involved the collecting of information from a wide range of sources, analysing the information, and forming appropriate conclusions where possible. The second consisted of a careful look at the basic scientific concepts to see

59   Good 1997, pp.181–183.
60   http://www.presidentialufo.com/march_14,_1961.htm.
61   http://www.presidentialufo.com/march_14,_1961.htm.

if there was an anomaly or area needing closer attention by the scientists.

On June 25, 1952, Smith provided a provisional report on the current situation with Project Magnet. In it, he said:

"If, as appears evident, the Flying Saucers are emissaries from some other civilisation, and actually do operate on magnetic principles, we have before us the fact that we have missed something in magnetic theory but have a good indication of the direction in which to look for the missing quantities. It is therefore strongly recommended that the work of Project Magnet be continued and expanded to include experts in each of the various fields involved in these studies."[62]

It is clear Smith was finding something interesting about UFOs through the laws of electromagnetism. But he couldn't quite put his finger on what it was precisely. He knew further study was required.

Smith filed another report on August 10, 1953, again emphasising the importance of the study and the kind of interesting results he was uncovering.

In December 1953, Smith's study group (consisting of two other engineers and two technicians working part-time) accrued a small laboratory inside a wooden building equipped with measuring devices including a gamma ray counter, a magnetometer, a radio receiver and a recording gravimeter, loaned to them by the Canadian Defense Research Board. The laboratory, located near Shirleys Bay outside Ottawa, became the world's first scientific UFO sighting station.

Then, on August 8, 1954 at 3:01 p.m., a clear and distinct magnetic disturbance—among other activity on all the measuring devices—was recorded. Unfortunately, owing to the heavy fog on that day, none of the scientists could observe anything overhead. Smith revealed his latest findings in a secret government meeting, stating that whatever had caused the magnetic disturbance at Shirleys Bay, scientists should expect some kind of magnetic or radio noise disturbance in association with UFOs. Furthermore, the gravimeter suggested that gravity waves might also exist. However, further research was required. As Smith said:

---

62   Good 1997, p.186 .

"Whether the phenomenae [*sic*] be due to natural magnetic causes, or alien vehicles, there would probably be associated with a sighting some magnetic or radio noise disturbance. Also, there is a possibility of gamma radiation being associated with such phenomena. It has been suggested by some mathematicians that gravity waves may exist in reality....While we know practically nothing of such waves in nature, nevertheless, if the possibility exists, flying saucer phenomenae, being largely an unknown field, might be a good place to look for such waves."[63]

Just as things were starting to go well for Smith, on August 10, 1954, just two days after detecting the magnetic and gravitational anomaly at Shirleys Bay, the Canadian government publicly announced the official termination of Project Magnet. Although the reason for making such a move so soon after the anomaly was not given, Smith believed it was probably on the grounds that his study was receiving adverse publicity as more and more members of the general public became aware of his activities (which in turn may have reached the ears of the U.S. military across the border).

This might explain why Keyhoe, who lived in the United States, was reported as having approached Smith to discuss his UFO investigation activities[64].

The final conclusions of Project Magnet can be best summed up in the following quote:

"The conclusions reached by Project Magnet...were based on a rigid statistical analysis of sighting reports and were as follows: There is a ninety-one percent probability that at least some of the sightings were of real objects of unknown origin. There is about a sixty percent probability that these objects were alien vehicles."[65]

After that, Smith continued his work in his own time without further government support until his death on 27 December 1962.

Before he died, Smith felt sure there was something new to be discovered in UFO reports and argued that further scientific study may

---

63   Good 1997, p.189.
64   Evans & Stacey 1997, p.92.
65   Sachs 1980, p.48.

reveal (1) the existence of extraterrestrial life and/or (2) the mystery of gravity and its relationship with the laws of electromagnetism.

Since the termination of Project Magnet, the Canadian government has continually tried to play down the importance of the team's research work whenever inquiries are made about it by the general public, presumably to avoid embarrassment for getting involved in an official UFO study, thinking there is nothing to learn about UFOs despite the link between electromagnetism and UFOs.

Or did U.S. authorities influence the Canadian government to drop the study?

Lieutenant General Nathan F. Twining and the Official Beginning of U.S. UFO Investigations

Just prior to the publication of Keyhoe's famous article, UFOs continued to be reported by reputable civilians and military personnel, including descriptions of so-called flying discs with light beams. As a result, the Chief of Staff of the U.S. Army and commander of the Air Matèrial Command at Wright-Patterson AFB, Lieutenant General Nathan F. Twining, stated, in a historic memorandum originally classified secret:

> "The phenomenon is something real and not visionary or fictitious....
> ...The reported operating characteristics, such as extreme rates of climb, manoeuvrability and evasive action when sighted or contacted by friendly aircraft and radar lend belief to the possibility that some of the objects are controlled either manually, automatically, or remotely."[66]

His memorandum led to the beginning of a string of official U.S. government UFO investigations—the largest and most expensive study ever conducted by any nation on Earth—the first of which began on 30 December 1947.

---

66   The original 3-page *Air Matèriel Command Report*, September 1947 is reproduced in Good 1997, pp.313–315. Quote also appears in Blundell & Boar 1984, p.164.

Project Sign

On 30 December 1947, the USAF initiated the nation's first official study into UFOs, called Project Sign (also unofficially called Project Saucer), with the approval of Major General L.C. Craigie[67], Director of Research and Development. Based at the Technical Intelligence Department (TID) of the Air Materiel Command at Wright Field (later Wright-Patterson AFB), Ohio, Project Sign began operations the following year on January 22, 1948.

With the lowest security classification given for the project—Restricted—its purpose was to "collect, collate, evaluate and distribute to interested government agencies and contractors all information concerning sightings and phenomena in the atmosphere which can be construed to be of concern to the national security".[68]

Astronomer Dr. Allen J. Hynek from Ohio State University was contracted by the USAF to study UFOs on behalf of Project Sign[69]. At the time, Hynek was genuinely skeptical about UFOs and appeared to be the ideal scientific expert to investigate the cases.

By July of that year, a black, legal-sized cover with the words "Top Secret *Estimate of the Situation* (EOTS)" stamped across it containing the official Project Sign report on the current UFO situation was delivered to higher military officials. In it, a scientific overview on UFOs written by chief scientific adviser to the U.S. president, Professor George E. Valley of the Massachusetts Institute of Technology, stated that any observations regarding "rays of light" or "light beams" as a form of flying saucer propulsion would be impractical as it would require more power than the world had currently available at the time[70]. Magnetic propulsion systems were also considered impractical for the same reason. At any rate, the staff of Project Sign stated that four-fifths of reports analysed were explainable as misinterpretations of clouds, weather balloons, aircraft and other natural phenomena, but nearly one-fifth of the reports remained unsolved, including observations of UFOs with light beams, and called for a more detailed study. The intelligence staff of Project Sign totally ruled out the likelihood that

67   Peebles 1994, p.16.
68   Sachs 1980, p.258; Story 1980, p.277; Peebles 1994, p.16.
69   Further details on Project Sign can be found in Sachs 1980, p.258; Story 1980, p.277. The original Project Sign report is known as *USAF Project Sign Technical Report Number F-TR-2274-1A: Unidentified Aerial Objects*, February 1949. Original Project Sign report was written by Wright-Patterson AFB personnel. In fact, many important UFO studies would emanate from Wright Patterson AFB following the official arrival of the mysterious Roswell wreckage of 1947.
70   Peebles 1994, p.31.

these unexplained objects originated from any country on Earth, and most felt inclined to believe that these "machines" were probably extraterrestrial in origin.

In February 1949, a slightly more watered-down version of the previous Project Sign report appeared. In it, one can observe a rather clear and strong divide in the views held by members of Project Sign, including those who expressed a view that they did not know what UFOs were:

> "The possibility that some of the incidents may represent technical developments far in advance of knowledge available to engineers and scientists of this country has been considered. No facts are available to personnel at this Command that will permit an objective assessment of this possibility. All information so far presented on the possible existence of space ships from another planet or of aircraft propelled by an advanced type of atomic power plant have been largely conjecture....
>
> Future activity on this project should be carried on at the minimum level necessary to record, summarise, and evaluate the data received on future reports and to complete the specialised investigations now in progress. When and if a sufficient number of incidents are solved to indicate that these sightings do not represent a threat to the security of the nation, the assignment of special project status to the activity could be terminated. Future investigations of reports would then be handled on a routine basis like any other intelligence work."[71]

Then there were other members who accepted the possibility of an extraterrestrial connection, as indicated by the following statement:

> "If there is an extraterrestrial civilisation which can make objects as are reported, then it is most probable that its development is far in advance of ours. This argument can be supported on probability arguments alone without recourse to astronomical hypotheses.

---

71   Peebles 1994, p.30 & 31.

Such a civilisation might observe that on Earth we now have atomic bombs and are fast developing rockets. In view of the past history of mankind, they would be alarmed. We should, therefore, expect at this time above all to behold such visitations."[72]

Acknowledging the possibility of something alien in the UFOs stirred up too much controversy among top USAF officials. Soon the top brass faced a dilemma as to whether it was time to prepare the American public and the world for a significant announcement on the situation.

Not so among certain individuals at the Pentagon, who were either not ready, knew the situation with UFOs very well, or were too pragmatic in considering new possibilities. Whatever the case, someone was unhappy about something in the report, and the conclusion had to change.

On April 27, 1949, a condensed, 22-page, single-spaced Project Sign report was released to the press by the USAF[73]. But, as Keyhoe discovered, after seeing the public reaction to the press release, a Pentagon group quickly countered this with a report claiming UFOs were either bunk or some form of foreign technology. It looked like someone higher up was trying to diminish the importance of the extraterrestrial hypothesis at all costs, while admitting absolutely no knowledge of any secret electromagnetic experiment being carried out by the USAF.

Well, if they were not extraterrestrial in nature and not man-made objects, what about studying UFOs to see if they could advance science and reveal new insights? Apparently people in the Pentagon were not interested. They simply wanted the whole problem to disappear as soon as possible.

The differing views coming from two official reports were enough to get Keyhoe to investigate the UFO phenomenon more closely, as he stated in his book *The Flying Saucers Are Real*:

"...the Air Force was puzzled, and badly worried, when the discs first were sighted in 1947....Project Saucer was set up to

---

72   The original Project Sign document is known as *USAF Project Sign Technical Report Number F-TR-2274-1A: Unidentified Aerial Objects*, February 1949. The quote shown here is found in Good 1997, p.328; Flammonde 1976, p.390; and Peebles 1994, p.32.

73   Swords 1997, pp.89–90.

investigate and at the same time conceal from the public the truth about the saucers. During the spring of 1949 this policy, which had been strictly maintained by [Former United States Secretary of the Navy James Vincent] Forrestal, underwent an abrupt change. On top-level orders, it was decided to let the facts gradually leak out, in order to prepare the American people. While I was preparing the article for the January 1950 issue of <u>True</u>, it had been considered in line with the general education program. But the unexpected public reaction was mistaken by the Air Force for hysteria, resulting in their hasty denial that the saucers existed…"[74]

General Hoyt S. Vandenberg

The Project Sign report continued to provoke considerable interest for some time afterward until it eventually reached the eyes of Air Force Chief of Staff, General Hoyt S. Vandenburg, who rejected its findings—especially from those Project Sign members who supported the extraterrestrial hypothesis—because it lacked any proof whatsoever. The argument by Vandenburg was used as the principal means of preventing further investigation of the extraterrestrial hypothesis in case proof could be found. Hence the report was later declassified and destroyed, and most members of the now-terminated Project Sign were transferred elsewhere for more pragmatic military duties.

In the meantime, the public was told that reports of flying saucers were misinterpretations of various conventional objects, hoaxes and other natural explanations.

---

74   Swords 1997, p.91.

Yet UFOs were not going to go away any time soon simply because a few old military men had a differing or preferred view of the UFO situation.

As UFOs continued to be reported, Project Grudge had to take the place of Project Sign on February 11, 1949, and although UFOs were now considered non-hostile, its new aims were to deny all UFO sightings specifically as nothing more than natural atmospheric phenomena or hallucinations and to set up a public relations campaign to educate and convince people that UFOs were nothing extraordinary or unusual—and certainly not alien spaceships.

It wasn't long before the Air Force terminated Project Grudge on 24 December 1949 in an attempt to counter the damaging claims from Keyhoe's independent UFO investigation[75] published in *True Magazine*. A few days later, staff members involved in Project Grudge wanted to convince the public there was nothing unusual to be found in UFOs by issuing a final report stating that 188 sightings could be explained. Despite the effort, the Air Force had to admit UFOs were still creeping into their statistics. The unknowns represented 23 percent of the 244 sightings accumulated. To get around the thorny problem, staff members decided it was best to explain these cases as being psychologically motivated on the part of the witnesses. As the Grudge Report stated:

"1. Evaluation of reports of unidentified flying objects constitute no direct threat to the United States.
2. Reports of unidentified flying objects are the result of:
    a. A mild form of mass hysteria or "war nerves".
    b. Individuals who fabricate such reports to perpetrate a hoax or seek publicity.
    c. Psychopathological persons.
    d. Misidentification of various conventional objects."[76]

The report concluded by saying that any future investigations into UFOs should be carried out with a low profile (probably without public

---

75   Mr. Curtis Peeble, the author of the 1994 book, *Watch the Skies! A Chronicle of the Flying Saucer Myth*, said on page 41 how the Air Force made the decision to terminate Project Grudge in an attempt to counter Major Donald E. Keyhoe's top-selling article in *True Magazine*.
76   Peebles 1994, p.43.

awareness) collecting only those exceptional cases "clearly indicating realistic technical applications"[77].

Project Grudge operated covertly after 27 December 1949 for the next couple of years.

UFOs Will Not Go Away

However, much to the dismay of the USAF and those in the know within the Pentagon, the conclusion of Project Grudge was not going to stop the public from reporting UFOs.

By 1952, amidst a storm of continued UFO controversy among the general public and media, and growing numbers of unexplained UFO sightings among military personnel, the poorly funded, understaffed and rather disorganised Project Grudge was unable to handle the growing weight of unexplained sightings.

Also, General Walter Bedell Smith, Director of Central Intelligence (1950-53), said in a memorandum to the National Security Council:

> "It is my view that this [UFO] situation has possible implications for our national security which transcend the interests of a single service."

It is clear there were some people in the military who did not know the full situation regarding UFOs. As a result, some felt compelled to get to the bottom of the mystery.

Captain Edward J. Ruppelt and Project Blue Book

In an attempt to get to the bottom of the mystery, USAF Intelligence Officer Captain Edward J. Ruppelt was put in charge of Project Grudge; he, in turn, brought sanity to the staff of Project Grudge by upgrading it to a separate organization called Project Blue Book in March 1952 that was based at Wright-Patterson AFB[78]. The project's aim was to uncover the true nature of UFOs: were they extraterrestrial or terrestrial in origin?

---

77   Story 1980, p.278.
78   Sagan 1996, p.79.

During his reign at Project Blue Book, Ruppelt did notice an unusual interest taken by his nation's president, Mr. Truman, in the subject following another wave of UFO sightings that had hit the country in the summer of 1952.

On July 19 and 26, 1952, UFOs were observed on the ground and on radar in Washington, D.C., flying in a provocative manner directly over the White House. The sightings made it to the front page in newspapers all over the country. As Gerald K. Haines stated in an article written for the CIA:

Captain Edward J. Ruppelt

> "A massive build-up of sightings over the United States in 1952, especially in July, alarmed the Truman administration."[79]

At first, Truman gave the order for the USAF to shoot down flying saucers on the morning of July 26, 1952. However, several prominent scientists urged the White House against such action. Truman eventually withdrew the order at 5:00 p.m. that afternoon.

The head of Project Blue Book, Captain Edward J. Ruppelt, was called into the White House by Brigadier General Landry on July 28, 1952. He was asked to explain what he thought happened in Washington when the UFOs passed overhead. As Ruppelt recalled:

> "About 10:00 a.m., the President's air aide, Brigadier General Landry, called intelligence at President Truman's request to find out what was going on. Somehow I got the call. I told General Landry that the radar target could have been caused by weather but we had no proof."[80]

Ruppelt later learned that Truman had been listening on a phone line while he gave his briefing to General Landry.

With this information at hand, Truman decided he would ask the CIA to have the UFO phenomenon analysed to see if these things posed a security threat to the United States.

---

79   http://www.presidentialufo.com/harrys.htm.
80   http://www.presidentialufo.com/harrys.htm.

Ruppelt continued with his work at Project Blue Book, not thinking too much about the meeting. As he looked at various cases coming into the office, he did notice (and later gave a quote about[81]) an interesting link between electromagnetism and UFOs. Again, it was too early to tell what was happening.

The Robertson Panel

The discovery of certain electromagnetic effects and Ruppelt's open-minded look at the UFO situation was enough to ruffle a few too many feathers at the Pentagon, either because not enough UFOs were being explained as familiar IFOs or because certain people in the know didn't want Ruppelt to get too close to working out what these objects were. So within a year of Project Blue Book starting its own independent UFO investigations, on the morning of Wednesday January 14, 1953, in Washington, DC, the CIA started its own investigation into UFOs behind closed doors.

The CIA's UFO panel held an impressive array of top scientists from various government and military establishments. The scientists, who had CIA security clearance at the time, were as follows:

Dr. Howard P. Robertson: a professor of mathematical physics at the California Institute of Technology, Pasadena, and a CIA employee who headed the scientific panel, and was then Director of the Weapons Systems Evaluation Group in the Office of the Secretary of Defense.

Dr. Luis W. Alvarez: a physics professor at the University of California, Berkeley. He would receive a Nobel Prize in physics fifteen years later.

Dr. Lloyd V. Berkner: geophysicist and radar specialist. He was one of the directors of Brookhaven National Laboratories. He didn't join the panel until the afternoon of January 16, 1953.

Dr. Samuel A. Goudsmit: theoretical physicist and statistician, also from Brookhaven National Laboratories.

---

81   The quote from Ruppelt on "induction fields" and other electromagnetic disturbances in
       association with UFOs is revealed later in this chapter.

Dr. Thornton L. Page: astrophysicist from the John Hopkins University; was once professor of astronomy at the University of Chicago and Deputy Director of the John Hopkins Operations Research Office.

Other participants not officially made members of the panel included an astronomer from Northwestern University's Dearborn Observatory and former scientific consultant for the USAF Project Blue Book, Dr. J. Allen Hynek; Commanding General of the Air Technical Intelligence Center, William M. Garland; Assistant Director of the Office of Scientific Intelligence (OSI) and CIA scientist, H. Marshall Chadwell; Deputy Assistant Director of OSI and CIA representative, Ralph L. Clark; Air Force Officer at Wright-Patterson Air Force Base, Colonel William A. Adams; Air Force Officers Edward Ruppelt and Dewey Founet; Navy Photo Interpretation Laboratory representatives Harry Woo and R. S. Neasham; four CIA representatives including CIA agent, Philip G. Strong; and an army ordnance test station director who was president of the American Rocket Society and of the International Astronautical Federation named Frederick C. Durant III whose job was to report on the panel's findings.

The CIA-approved Wright-Patterson and Pentagon UFO experts[82] were permitted a total of 12 hours—spread out over a period of three days—to examine the "selection of the best documented [UFO] incidents" acquired since 1947. The Air Technical Intelligence Center (ATIC) selected 75 of the best-documented UFO cases from 1951 to 1952 for the panel to analyse. A further 89 summarised cases, together with the sightings at Holloman Air Force Base, the mysterious "Green Fireballs", and intelligence reports on UFO-related information from across the Iron Curtain—in the USSR—were also presented.

After the meetings, a report classified Secret was produced titled *Report of Meetings of Scientific Advisory Panel on Unidentified Flying Objects* (Convened by Office of Scientific Intelligence, CIA, January 14–18, 1953). In the report, written by CIA agent Frederick C. Durant before the members could read it, the conclusion to be agreed on by the Robertson Panel was as follows:

---

82   Evans & Stacey 1997, p.95.

"...the evidence shows no indication...[of] a direct physical threat to national security...[and] no evidence that the phenomena indicate a need for the revision of current scientific concepts."[83]

Once agreement was reached, mainly by the official members of the panel, the entire report, originally classified Secret, was signed only by the official members of the panel. Dr. Robertson then quickly recommended the aura of mystery surrounding the UFO phenomenon be removed and would hopefully stop further reporting by the public by implementing a comprehensive program of "public training and debunking", as stated in the report.

Soon after the report was published, the government began implementing the panel's recommendation via public education programs and through changes in the Air Force's original Project Blue Book aims.

Former American astronomer, the late Dr. Josef Allen Hynek, who was invited to attend the meetings but not sign the report's conclusion, gave his view of the whole situation as follows:

"I was dissatisfied even then with what seemed to be a most cursory examination of the data and the set minds implied by the panel's lack of curiosity."[84]

He was not alone on this view. A member of the scientific panel was also not impressed. Dr. Thornton L. Page's view on the results of the study was quite clear. He said:

"The Panel underestimated the long duration of public interest in UFOs....[I]t also tended to ignore the five percent or ten percent of UFO reports that are highly reliable and have not as yet been explained."[85]

Although most other scientists in the panel were generally skeptical about UFOs, little is known of three other members—who presumably still had close CIA ties until their deaths. But after opening his mouth and making his view clear to everyone, Page is certainly not working for

83   Fawcett & Greenwood 1984, pp.126–127.
84   Spencer 1991 *The UFO Encyclopaedia*, p.325.
85   Randles 1987, p.45.

the CIA anymore. He now works for the not-so-secret organisation known as NASA[86].

The Battelle Memorial Institute Joins the UFO Debate

The world-famous Battelle Memorial Institute got involved in UFOs (yet again) by releasing its own statistical report giving full endorsement to the Robertson Panel's conclusion. When the report was released (as Special Report No.14), scientists at the institute also believed UFOs showed no evidence of a threat to national security. Well, if the institution is so adamant that UFOs are nothing out of the ordinary and no threat to the American people and the rest of the world, it should have no problems releasing all the secret reports it produced for the USAF after 1947, when at least a couple of Battelle scientists[87] were secretly contracted in May 1948 by the USAF at Wright-Patterson AFB to study the world's most powerful and toughest shape-memory alloy, known as NiTi. As we know, the aim was to develop new techniques to make super-pure titanium, and to use the purest titanium available at the time to create other titanium-based shape-memory alloys as a means of understanding how the shape-memory behavior works. Yet, for some reason, the ability to create highly pure titanium was not available to anyone in 1947, and too little in 1948 to explain the amount of shape-memory Roswell foil discovered by the witnesses in early July 1947. The USAF certainly had no knowledge how to create a super-tough Roswell foil with its shape-memory property in 1947, let alone the facilities to manufacture it in large quantities. So how did the USAF and Battelle realize the importance of highly pure titanium alloys while they were not able to manufacture the Roswell foil on their own in 1947?

To this day, the institution is unable to explain why it had to get involved in the Roswell case. Yet it chooses to get involved in the UFO situation.

---

86   Story 1980, p.311.
87   Headed by Professor John P. Nielsen (who also worked at the metallurgy department at New York University, and before then at Phillips Laboratory owned by the USAF before he quickly became professor and headed the NYU department virtually uncontested so he can perform his work on NiTi using the world's first vacuum furnace installed at NYU and developed by Battelle).

Ruppelt Retires from Project Blue Book

Following CIA efforts to convince the majority of the prestigious Robertson Panel participants of "no alien evidence", the aim of Project Blue Book was quickly changed to follow the Robertson recommendations, which was to explain away all UFOs as nothing more than mundane natural or human-made phenomena, to stifle public interest, and to slap down any belief in the extraterrestrial hypothesis. As a result of the change, Ruppelt decided to retire from the USAF in August 1953.

The new Project Blue Book team headed by Hector Quintanilla (seated).

When Ruppelt left, he made one important statement after 1957 based on the information he had been privy to during his reign as head of Project Blue Book:

"During my tenure with Project Blue Book we had reports of radiation and induction fields in connection with UFOs, however the information was sketchy and we were never able to pin it down....[The reports of electromagnetic disturbances characterised] a whole new dimension to the UFO investigation."[88]

In another statement, Ruppelt said:

"...there is a worldwide interest in flying saucers; people want to know the facts. But more often than not, these facts have been obscured by secrecy and confusion; a situation that has led to wild speculation on one end of the scale and an almost dangerously blasé attitude on the other."[89]

Ruppelt gave more details of his findings and his view in his book titled *The Report on Unidentified Flying Objects* (Doubleday & Company, Garden City, New York) in 1956. In it, he clearly stated his positive stance on UFOs and the need for proper investigations.

Then, without warning, his personal stance on UFOs would suddenly change. In 1959, allegedly under threat of losing his job because of pressure exerted by the USAF to cut contracts to the aerospace company he was working for, Ruppelt was told to revise his book. This time he went against all his previous statements, saying ostensibly that UFOs were not extraterrestrial vehicles and there was nothing to be found in the UFO phenomenon that would be of interest to the public.

People who read the revised book, including Ruppelt's close friends, decided to attack his integrity. However, Keyhoe, a friend of Ruppelt for many years, understood the situation better than most when he said:

"One of the most disturbing aspects of the cover-up operation is the heavy pressure sometimes used against individuals. Capt. Edward Ruppelt is believed by many who know the inside story to have been a tragic victim of such action.

After going on inactive duty, Ruppelt wrote a book on UFOs which jarred the censors with its verified evidence and disclosure about the secrecy. In letters to me he firmly opposed the cover-up and he swore he would never follow the Air Force 'line'. He frequently gave me valuable leads, and when I

---

88   Maney & Hall 1961, pp.70 & 186.
89   Ruppelt 1956, p.5.

became director of NICAP he steadily cooperated, praising the organisation from its struggle against secrecy.

But in 1959, this suddenly ended. Ruppelt was then working with an aerospace company with Air Force [AF] contracts. Something happened which obviously put him under pressure to stop criticising AF UFO activities. One explanation —which is unproved—is that the AF implied his company contracts and his job might suffer if he did not co-operate. Perhaps there is another answer, something powerful enough to make him give in. The main target was his revealing book. Somehow, he was forced to repudiate it completely.

Adding three chapters at the end, Ruppelt reversed all he had disclosed, rejecting all his strong evidence and ridiculing expert witnesses — some of whom had become personal friends. Now, he said, he knew that the UFO sightings were only illusions, mistakes and hoaxes—the standard AF line.

I knew Ed Ruppelt well. He had courageously revealed what he believed the public should know, and this violent change was bound to hurt him deeply. He was also a sensitive man, and when the revised book appeared there were sharp attacks on his integrity, some even by former friends who should have guessed the truth. Like several others who knew Ruppelt well, I have always believed that the enforced retraction and bitter criticism were partly the cause of the premature death from a heart attack."[90]

Furthermore, Ruppelt did admit that "maybe I was playing the front man to a big cover-up"[91] while he was in charge of Project Blue Book.

Ruppelt died in 1960.

Following his retirement, a new Project Blue Book team was assembled under a new head. The team followed the project's new aims right down to a tee.

USAF Regulation No.200-2

To further reinforce the new official military position recommended by the Robertson Panel findings, USAF Regulation No.200-2 was changed

90   Keyhoe 1973, pp.90–91.
91   Evans & Spencer 1988, p.241; Randles 1987, p.39.

in August 1953, stipulating that "the percentage of unidentifieds must be reduced to a minimum"[92]. Not exactly an iron-clad guarantee that military personnel would never report UFOs. So, to ensure that those remaining unexplained cases did not get reported to anyone except to the appropriate military channels (where the reports could disappear into the legendary black hole of the Pentagon, never to see the light of day again), a Regulation JANAP-146 (JANAP stands for Joint Army/Navy/Air Force Publication) was issued in December 1953 stating that any unauthorized release of information about UFOs (which included unidentified missiles, aircraft, or possibly more unusual objects) and unidentified objects on the ground, such as submarines, ships or ground parties, was punishable by up to 10 years' imprisonment and a $10,000 fine.

For UFOs to be described as meteors, clouds, stars, and ordinary aircraft, this is certainly an extreme effort to clamp down on the natural and man-made explanation for UFOs. Or is the Pentagon worried a genuine alien craft might get revealed to the public without authorisation?

An FBI Status Report

The new regulation may have helped to put a lid on UFOs being reported by military personnel and seeping out into the public; however, UFOs and the civilian population still seemed to act oblivious of the regulation with continued reporting of these enigmatic objects within the United States and internationally. For example, after the world's first artificial satellite, known as Sputnik 1, was launched on October 4, 1957, 701 recorded UFO sightings, or 60 percent (instead of the expected 25 percent) of the 1,178 official U.S. UFO sightings for that year, occurred between October and December. As the FBI stated in a status report dated November 12 of that year:

> "Ever since the Russians released Sputnik there has been a great increase in the number of flying saucers and other UFOs reportedly seen by people all over the U.S."[93]

---

92   Spencer 1991 *The UFO Encyclopaedia*, p.8 and Peebles 1994, p.91.
93   Randles 1993, p.58.

While the reason for the propensity to overwhelm authorities and the media with UFO reports from civilians is still not precisely known, some skeptics have suggested it may have something to do with the fact that most people had decided to look at the sky more closely and discovered natural and man-made things they did not realise existed. As Dr. Carl Sagan said:

> "Perhaps people were looking at the night sky and saw more natural phenomena they didn't understand."[94]

However, more examples of unusual and highly detailed UFO sightings at close range from credible witnesses would still emerge in the reports, including further "electromagnetic disturbances" from strange flying objects looking like giant spherical light bulbs at night and unusual symmetrical metallic discs in the daytime, making it difficult for the average rational person to explain away. Even the overseas cases were getting harder to dismiss as natural or man-made explanations with increasing reports of radiation poisoning soon after a UFO encounter, unexplained power blackouts in association with UFO sightings near power grids, extraordinarily detailed UFO abductions cases (often recounted without the need for hypnosis) and similar unexplainable occurrences.

So how could the U.S. government deal with the civilian population, including the rest of the world?

U.S. Government Interested in Gravity Research

Before a solution could be found to keep the public quiet (or at least quieten down the media in the reporting of UFOs), contracts were issued by the U.S. government and military (predominately from the USAF) in the mid-1950s to top electromagnetic experts in various U.S. research centers who could work on the problem of gravity and the unified field theories.

Why the interest in electromagnetism and gravity at this time? No one knows for sure. Surely the work of Mr Smith in Project Magnet could not have initiated this interest even though he did believe the link could be important to understanding UFOs. Something else within the United States had instigated this new line of scientific investigations for

---

94   Sagan 1996, p.71.

the benefit of the U.S. government (now who within the government would be interested in this work?). What we do know is that the interest would encompass a number of scientific fields, including Albert Einstein's Unified Field Theory, published in 1929 and linking gravity with electromagnetism.

Evidence to support this can be found in the front-page article from *The New-York Herald-Tribune* dated November 20, 1955.[95] The story revealed the names of several important scientists (including overseas experts), research centres, and aerospace companies involved in the study at the time.

On further examination of the article, we learn these contracts were part of a program set up by the General Physics Laboratory of the Aeronautical Research Laboratories (ARL) at Wright-Patterson AFB, Dayton, Ohio, in September 1956. Aha! So it really was the USAF that started this work. We wonder why? Has this got anything to do with the Roswell object recovered in New Mexico? At any rate, the aim of the program was to coordinate all research efforts into gravity and the theories surrounding the unification of the gravitational field with the electromagnetic field. Dr. Joshua N. Goldberg from Syracuse University was made head of the program.

It wasn't surprising that Dr. Goldberg had been chosen for the job. In an article by Daniel J. Kennefick titled "Controversies in the History of the Radiation Reaction Problem in General Relativity", it stated:

"In 1955…Joshua Goldberg, a student of Peter Bergmann, examined the [radiation] reaction problem….A couple of years later Goldberg was introduced to Peter Havas, a physicist with experience in the problem of radiation in special relativity, who shared his interest in developing a fast motion expansion in general relativity. Having each worked on the problem independently, they began a collaboration based on this approach."[96]

Also, Dr. Goldberg had a special interest in Einstein's general theory of relativity and eventually published a 1955 article titled "Gravitational Radiation" in the *Physical Review* while working at the Armour Research Foundation.

95   Deyo 1992, p.146 shows the U.S. gravity research article in *The New-York Herald-Tribune*, November 20, 1955, pp.1 & 36. Full text included.
96   Kennefick 1997, p.9.

For readers feeling a bit rusty with their scientific knowledge of gravitational radiation and what it is, it is nothing more than a gravitational wave in which the energy density of the gravitational field has been intensified (or amplified) for a brief period of time by the accelerating motion of uncharged matter. For example, the simple act of moving your hand up and down in an accelerating and decelerating manner will produce a gravitational wave, albeit an extremely weak one and thus beyond the detection of any scientific instrument. However, Dr. Goldberg's paper was the first to show how such a wave could be amplified to the point where it may be possible to detect this wave after discussing the effect of binary stars rotating around each other on the gravitational field of space.

Dr. Goldberg also realised a connection between electromagnetic radiation and gravitational radiation. More specifically, Einstein's general theory of relativity predicted that gravitational radiation can move at the speed of light and can act on uncharged matter over great distances. Anyone who has worked in physics will know this is remarkably similar to electromagnetic radiation in terms of its speed and influence on ordinary matter, except the reach of this field over great distances is only meant to influence charged matter. Uncharged matter, on the other hand, is supposed to be the domain of the gravitational field or gravitational radiation (also known as gravitational waves). Despite the uncanny similarities in the two types of radiation and the attempt by Dr. Goldberg to understand the reasons for this, no one knew at the time if these two radiation types were directly related. More of a reason for Dr. Goldberg to continue his study into this field as one would think.

However, the USAF had other ideas for Dr. Goldberg. This was the moment when the USAF decided it would instigate its own study into gravity and electromagnetism, and keep Goldberg busy working for the USAF—although why a pragmatic military organisation involved in the business of building flying objects for the defence of the nation would be interested in this mostly radical and theoretical work is unclear. Did the USAF think it could control gravity using the laws of electromagnetism? What gave the USAF the idea that this might be possible?

In an interview, Dr. Goldberg gave some details of how he got involved with the USAF:

"There were people, and I don't know who—this is one of those hearsay things that nobody can verify, so I will say it, but it's totally unverified—that some officer in the Air Force, thinking about the next big thing that the Air Force needed, was an anti-gravity device. And so they needed somebody to work on general relativity."[97]

No surprises as to why Dr. Goldberg was given the job[98]. But did Dr. Goldberg actually work on general relativity and gravitational radiation during his time with the USAF? We discover from Peter R. Saulson that Dr. Goldberg's work would be directed elsewhere with encouragement from the USAF. As Saulson said:

"Josh played a central role in developing our understanding of how a binary star system generates gravitational waves. Another key contribution came through his patronage of the 1957 Chapel Hill Conference, in his role as funding officer for the Air Force's support of research in gravitation."[99]

So it turns out that Dr. Goldberg's role was merely to identify key individuals involved in gravity research and to provide government funding to these individuals for the benefit of the USAF's own, possibly secret, research studies into this obscure line of work for whatever purposes. Does this mean Dr. Goldberg was able to conduct further research into gravity in his own time? Or did the USAF want him to be preoccupied with the new job as funding officer so as to avoid any new discoveries about the Unified Field Theory from coming out?

At any rate, the USAF's interest in gravity research continued for at least another couple of years, as confirmed by the 30 December 1957 issue of *Product Engineering* where it was revealed:

"A few weeks from now, at a special session of the Institute of the Aeronautical Sciences (New York City, Jan. 27–31), a group of dedicated men will discuss what some people label pure

---

97   Interview published on http://www.aip.org/history/ohilist/34461.html. as of July 2013.
98   Dr. Goldberg ended his work with the USAF in 1963.
99   Abstract of Saulson's paper found at http://link.springer.com/article/10.1007%2Fs10714-011-1237-z as of July 2013.

science-fiction, but others believe is an attainable goal. The subject: electrogravitics—the science of controlling gravity.

...David B. Witty noted in his award-winning essay for the Gravity Research Foundation, gravitational screening is crucial in all theories of gravitation.

...E.M. Gluhareff, Pres. of Gluhareff Helicopters, suggests much progress might come if gravity were considered as "push" rather than "pull"—with all matter being pushed toward the centre of the earth by a sort of "electronic rain" from outer space...

Perhaps British aeronautical engineer A. V. Cleaver (see 'Electro-Gravitics: What it is—Or Might be' by A.V. Cleaver, F.R.Ae.s., Fellow B.I.S., published in the *British Interplanetary Society Journal* for April–June of 1957 in Volume 16, Number 2, pp.84-94) is right in insisting that if any antigravity device is to be developed the first thing needed is a new principle in fundamental physics not just a new invention or application of known principles....Nevertheless, the Air Force is encouraging research in electrogravitics, and many companies and individuals are working on the problem. It could be that one of them will confound the experts."[100]

Not long after, all research into this field would suddenly go quiet. The public would never learn more about this work or what the USAF had achieved. The only further details we would learn of the USAF interest in this work were from Dr. Goldberg himself and a handful of articles written by the aerospace firms involved in gravity research with the apparent aim of developing some kind of so-called anti-gravity propulsion systems.

In 1992, Dr. Goldberg wrote an article summarising gravity research as he understood it at the time, including those supported by the USAF for the period between 1955 and 1972 inclusive[101]. It looks as if the USAF had finally figured out whatever it was that it needed to know about gravity by 1972.

Or more likely, it was closer to 1974 if the following quote from Wikipedia in the article "United States gravity control propulsion research" can be relied on:

---

100   *Product Engineering* 1957, p.12.
101   Goldberg 1992, cover title page.

"Articles about the gravity propulsion research by the aerospace firms ceased after 1974."[102]

Does this mean the USAF gave up on gravity propulsion research? Or did it find a solution and decide to go at it alone, stopping all further funding of aerospace companies that were doing the research for the USAF?

Without this admission from Dr. Goldberg, all we would know is that information to the public about this USAF involvement in gravity research and the connections to electromagnetism ended prematurely, for some reason, by the late 1950s. And yet it unashamedly continued studying the area in further secret contracts with some notable aerospace companies until 1974, when the USAF decided to go quiet. No reason is given by the USAF as to why all this work suddenly ceased.

More foreboding was the fact that there was a sinister side to the USAF secrecy on this research on gravity and Einstein's Unified Field Theory at the end of the government-funded research in the 1950s. As we shall see later in this chapter, anyone who had the slightest advanced knowledge of the electromagnetic connection to gravity and was willing to explain this connection to the general public could face dire consequences—even death.

USAF's Best Attempt at a Man-Made, Saucer-Shaped Aircraft

At the same time as this gravity research was taking place, the USAF worked on a classified study in 1956 to develop a supersonic flying saucer known as Project 1794 (the number represents a simple rearrangement of the year 1947, named for the most important time in the U.S. military calendar regarding UFOs, presumably because of what happened near Roswell). The project never made it to the final stage of building the aircraft, which is rather odd considering that the Roswell witnesses to the July 1947 second crash site claimed to have seen a dark-greyish metallic saucer-shaped object that somehow successfully flew before crashing to the ground and throwing bodies outside. If it was a man-made flying object, the USAF was acting like it wasn't

102   Wikipedia https://en.wikipedia.org/wiki/United_States_gravity_control_propulsion_research_
       %281955%E2%80%931974%29 as of July 2013.

involved in its construction and/or testing if Project 1794 was anything to go by. It is definitely looking like a foreign object.

At any rate, just prior to this project, the USAF expressed interest in the flying saucer-shaped aircraft known as Project Y-2 from the aerospace company Avro Canada. In July 1954, the first of two U.S. Air Force contracts totalling $1.9 million was awarded to Avro Canada to conduct further studies.

By 1958, the U.S. Army joined in, and by 1959 the final design, known as the VZ-9 Avrocar was built and tested. Despite all the work, the Avrocar could not lift itself higher than a couple of meters above the ground without losing stability. Furthermore, the aircraft could not reach the planned 300 miles per hour.

By the time the project was abandoned in 1959, the best available technology that the U.S. military could muster for a disc-shaped aircraft was the aluminium-based VZ-9 Avrocar of empty weight 1,400 kg, and powered by three large Continental J69-T-9 jet engines, each having their own fuel and oil tanks. Absolutely no advanced electromagnetic propulsion system would be found in this design.

The Unified Field Theory Linked to UFOs?

In 1955, a scientist and lecturer named Dr. Morris Ketchum Jessup, who taught astronomy and mathematics at Drake University in Iowa and at the University of Michigan, published a book, *The Case for the UFO*. He stated throughout his book the importance of placing pressure on politicians to demand research into Albert Einstein's Unified Field Theory, which he thought might be of great importance to understanding the propulsion systems of UFOs (assuming UFOs were alien, of course). He said:

"If the money, thought, time, and energy now being poured uselessly into the development of rocket propulsion were invested in a basic study of gravity, beginning perhaps with continued research into Dr. Einstein's Unified Field concepts, it is altogether likely that we could have effective and economical space travel, at but a small fraction of the costs we are now incurring, within the next decade."[103]

Jessup continued researching the Unified Field Theory until his life came to an abrupt end on the evening of April 20, 1959. Just as he was about to discover and reveal to the world the important secret behind the theory, he was discovered dead in his gas-filled car in Dade County Park, Florida.

Some fellow scientists suggested it was suicide, because Jessup had appeared depressed about the continued government denials and evasions on UFOs and the skepticism he received from his colleagues. It was also reported that he had sent a suicide note to a close friend just before his death.

No so, according to another friend. He claimed Jessup was probably murdered, saying "...he knew too much, they wanted him out of the way".

So what was Dr. Jessup close to discovering?

Dr. J. Manson Valentine, an oceanographer, zoologist and archaeologist, conversed with his close friend Dr. Jessup months before his friend's death. Dr. Valentine said his friend appeared to be suffering from depression and was in need of someone to talk to. Jessup confided to Valentine that as he learned more about the secret U.S. Navy experiment on invisibility, known as the Philadelphia Experiment, and had discussions with U.S. Navy officers with knowledge of the experiment and about its connection to the Unified Field Theory, Jessup believed he was on the verge of discovering how the invisibility experiment worked, and its connection to UFOs. Perhaps Dr. Jessup was privy to some interesting UFO cases where UFOs could render themselves invisible in a similar way?

Valentine recalled what Dr. Jessup was close to uncovering in the following quote:

---

103   All quotes obtained from Berlitz & Moore's *The Philadelphia Experiment*.

"In practice, it [the U.S. Navy experiment on invisibility] concerned electric and magnetic fields at right angles to the first; each of these fields represents one plane of space. But since there are three planes of space, there must be a third field, perhaps a gravitational one. By hooking up electromagnetic generators so as to produce a magnetic pulse, it might be possible to produce this third field through the principles of resonance. Jessup told me that he thought that the U.S. Navy had inadvertently stumbled on this."[104]

This remarkable quote is suggesting a pulsing magnetic field, which is another way of saying an oscillating electromagnetic field (or radiation) is creating a gravitational field of its own. This is exactly what Einstein had proposed in his Unified Field Theory. The link between the electromagnetic field and the gravitational field was already clear in Einstein's mind when he tried to understand how an oscillating electromagnetic field, or light in its most general sense, can move uncharged matter.

Regarding the claimed connection with invisibility, as allegedly took place with the secret U.S. Navy experiment (known today as the Philadelphia Experiment) some time in 1943, we should point out that witnesses have reported situations where some UFOs have been able to render themselves invisible to the naked eye when in operation (when hovering or moving through the air) and sometimes oscillate between invisibility and visibility at low frequencies of whatever energy the objects were emitting to the surroundings. It is almost as if light from and/or around the object is able to bend back on itself and/or around the object, using a strong gravitational field generated by this energy in some mysterious way. The same has been reported of UFO occupants so long as they are within range of the UFOs and usually standing inside a light beam.

In other cases, light beams from UFOs have been observed by witnesses to bend at right angles or back on themselves. Light bending? Now, how is this possible?

Well, scientists know invisibility is a reality after observing this effect in black holes, which, due to their extreme mass, create a powerful gravitational field that bends light. But could the Unified Field Theory reveal another way to imitate this extreme mass situation with the help

---

104   Berlitz & Moore 1991, pp.131–132.

of strong oscillating electromagnetic fields, and so achieve the same light-bending effect? The idea suggested by Dr. Jessup could be the answer: we merely use electric charge to generate a pulsing magnetic field, which in turn generate a gravitational field. Even if we could confirm this interesting idea through an experiment, we are still none the wiser as to why UFOs have to create a pulsing magnetic field, if this is the energy we are dealing with. What's the point of rendering UFOs invisible? Or is this just a side effect, and pulsing a magnetic field has another purpose?

As for the Philadelphia Experiment, we do not know whether this was a real event or another secret and anonymous attempt by someone in the military to explain to the public the latest knowledge uncovered about UFOs.

The Work of Thomas Townsend Brown

Another interesting development in the field of electromagnetism and in understanding the relationship it may have with UFOs comes from the work of Thomas Townsend Brown (1905–1985).

It began in his high school days. Brown, a highly ambitious 16-year-old at the time (1921), purchased a Coolidge X-ray tube from an electrical store. He already had his head in the clouds, so to speak, as he dreamed of building a machine that could go into space, but he could not figure out what sort of propulsion system would be essential to making this a reality. When he heard a Coolidge X-ray tube was available to the public, he wondered whether the X-rays emitted by the tube could provide the solution he was looking for.

Brown walked back to his home in Zanesville, Ohio, rested the tube on a weight machine and switched it on, thinking the X-rays might exert a force on the weight machine. To his amazement, he noticed the tube temporarily lost weight. But after carefully investigating the problem, he realised the loss in weight was not caused by the emission of X-rays. Something else was causing the tube to move. Not sure if what he was seeing was real, he tried again by turning the power off, then on again. The effect repeated itself.

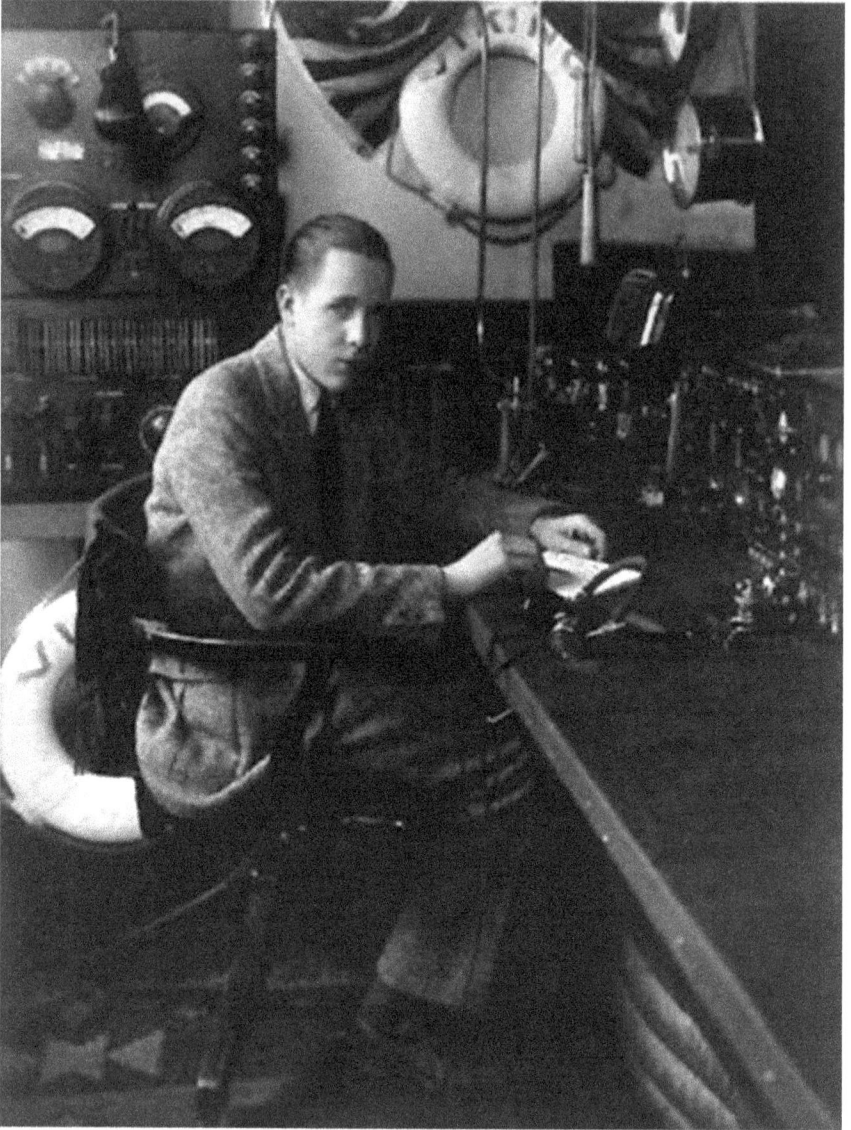

Thomas Townsend Brown working in a laboratory (taken some time during the 1920s).

In another interesting experiment, he carefully mounted the tube on one end of a thin wooden rod, counterbalanced it with a weight on the other end of the rod and suspended the entire contraption from the ceiling. The result of turning on and off the high voltage electric current astounded the budding young physicist. As author Gerry Vassilatos stated in his book, *Lost Science*:

238

"Once again, as before, he electrically impulsed the tube. The wonderstruck teenager saw a dream before his very eyes, as the tube began to rotate the entire suspension rod! Each time he impulsed the tube with a sudden jolt, the tube gained speed. The power was cumulative. Each successive jolt drove the rig around with increasing speed."[105]

The observation was interesting, for it was not like turning on an electric fan, which would be set to a certain voltage and spin at a *constant* speed, and where if you wanted to set the fan to a higher speed, you simply raised the voltage (or added more energy). What Brown saw was a fixed high voltage being turned off and on to maintain this "jolting effect" causing his tube to *increase* in speed. The tube was actually *accelerating*. It was as if no further energy was required to accelerate the tube.

At what rate was the acceleration? Was it linear or non-linear (e.g., exponential)? Brown didn't know. As a young boy, he was more fascinated by the way his tube was increasing in speed without any increase in the voltage. He was an experimental physicist, not a theoretical physicist. Here was an opportunity to exploit a potentially new type of propulsion system.

Brown looked at the tube more closely and tried to identify what was causing the accelerating effect with each jolting action. He eventually removed the tube and pinpointed the area around a pair of oppositely-charged electrodes separated from each other at a fixed distance.

By extracting the electrodes and putting an insulator between them, he effectively created what is known in the world of electronics as a *capacitor*. As any electronics enthusiast knows, a capacitor is primarily designed to temporarily store electrical energy when the electrodes are charged. However, Brown was the first person in the world to discover something else: the capacitor can move when given an oscillating high voltage. All he had to do was make sure one electrode was made different from the other before charging the electrodes.

We call this an *asymmetric capacitor*.

On further examining this curious movement of his capacitor, he noticed the direction of the mysterious force pushing the capacitor was

---

105   Vassilatos 1999, p.171.

usually toward the positive electrode, although in other designs he developed later, it seemed the movement was toward the negative electrode. Somehow, the size of the electrode played a part. Whatever direction the capacitor was moving, Brown knew it moved in a linear way.

This was exactly what Brown was looking for.

But, more importantly, the young teenager was also excited by another possibility in relation to his capacitor: he suspected that his device could somehow affect the gravitational field in its general locality at the moment it temporarily "lost weight".

Was this possible? Or was it a young man's fanciful thinking?

In support of his "controlling gravity" claim, he discounted X-rays and literally every other form of radiation, thinking the energy was too weak to contribute to the motion of his capacitor. Even though X-rays can move matter more effectively than radiation of a lower frequency, his observations suggested that X-rays were not strong enough to explain the significant movement of his capacitor. Indeed, he thought radiation would only move uncharged matter (or the mass of charged matter) and there was nothing that charging the electrode plates could do to amplify the force of the radiation when pushing against the electrode plates. He simply assumed another, more mysterious force was being created by the oscillating electric charge, leading to the movement. As confirmed by Vassilatos:

"Applying the high voltage suddenly, the wires jumped…and the tube jiggled. [So he suspended the tube and on applying the voltage, Brown noticed] the impulse tub clearly moved in a single direction.

…the tube was demonstrating real translation through a fixed distance, without ANY visible reactants. He counted X-rays out of the puzzle; they had no mass at all…

The tube was now mounted on the end of a thin wooden rod, counterbalanced, and suspended from a strong ceiling position. Once again, as before, he electrically impulsed the tube. The wonderstruck teenager saw a dream before his very eyes, as the tube began to rotate the entire suspension road! Each time he impulsed the tube with a sudden jolt, the tube gained speed. The power was cumulative. Each successive jolt drove the rig around with increasing speed.

The tube always moved in a specific direction, always with the electropositive side forward...."[106]

Perhaps it was a gravitational force?

In addition, here was a device with absolutely no moving parts. Or, to put it another way, there was nothing Brown could see that could cause air to be physically pushed away by rotating or flapping appendages. Something else was making the capacitor recoil in the opposite direction.

Of course, the other possibility was that the air was being ionised and moving from the positive electrode to the negative electrode. In other words, the air molecules were being stripped of their electrons, making them positively-charged by the positive electrode. These ionised air molecules were then repelled by the positive electrode and brushed across the second electrode to regain electrons. If these air molecules acquired excess electrons (considered highly probable given how high the voltage was), these negatively-charged air molecules would be repelled once again. It is this repelling of the ionised air molecules that Brown thought was making his capacitor move in the opposite direction (i.e., a recoiling action).

Despite what seemed to be a reasonable explanation, Brown was still not entirely satisfied with it. There was something peculiar about the movement.

Just to add to his confusion, Brown noted how the force that made the capacitor "lose weight" varied depending on the position of astronomical objects in the sky, especially the Sun and Moon—an observation he claimed was reproducible. For example, if the Moon (or Sun) was present, the force to lift his capacitor would increase for the same voltage[107]. Alternatively, he could reduce the voltage to generate the same force. Either way, it was like something from the sky is pushing down on the capacitor and this was being reduced in the presence of these astronomical objects. "How could this be?" he must have thought. Or was the air pressure being affected? Only problem was, if somehow the ionised air molecules were the cause of his capacitor's moment by being repelled and so exert a recoiling force, the

---

106   Vassilatos 1999, p.171. Also available from
        http://customers.hbci.com/~wenonah/history/brown.htm.
107   Could this be the reason why UFOs tend not to glow as much or appear more metallic during
        the daytime because the voltage needed to make them move (perhaps based on Brown's
        asymmetric capacitor design) is lowered to help reduce the heating effect and with it the glow on
        the outer metal hull?

position of the astronomical objects should not be able to change the density of the air sufficiently to cause this. Or if it did, the Moon would gravitationally attract the atmosphere slightly to reduce the air pressure, and this would mean less ionised air molecules to move the capacitor. In other words, a reduced air density should reduce the movement. Apparently this was not the case. Quite the contrary—he observed that the force lifting his capacitor had somehow increased. So how could a reduced density of the ionised air lead to an increase in the capacitor's movement with the position of the astronomical objects? Something else was affecting movement.

Brown was left with only one conclusion. The astronomical objects in the sky were affecting the Earth's gravitational field, and his invention was somehow picking up on the changes. And if that was the case, then any movement of his invention must be an example of "controlling the gravitational field".

No doubt it was a radical idea to consider, but Brown was young and ready to consider new possibilities. The only thing was, would other people see it the same way he did?

It turned out that Brown had no luck convincing academic staff at Caltech, when he entered this institute in 1922 to further his studies in physics, including the illustrious Robert A. Millikan (1868–1953)— famous for his oil drop experiment, where he measured the electric charge of an electron in 1909—of his ideas or research work. Millikan was steeped in the traditions of 19th-century physics and was unable to see a possible connection between gravity and electromagnetism through Brown's moving capacitor. Taking the rejection to heart, Brown returned to his home state of Ohio where he entered Denison University in Granville. For a while he quietly tinkered away with his capacitor, thinking no one would be interested. Then one day he opened up sufficiently about his work to Dr. Paul Alfred Biefeld (1867–1943). Fortunately, Biefeld did express an interest in Brown's work. In 1923, Biefeld helped Brown to study the interesting accelerating effect of his asymmetrical capacitor. Brown eventually completed a thesis on his work, and tried to submit his manuscript, "Tapping Cosmic Energy", dated October 28, 1927, to the *Physical Review* for publication. Again he was rejected, mainly because his ideas were too radical and he had not established a definitive explanation for the mysterious force responsible for his capacitor's movement through a series of experiments.

On November 15, 1928, after much refinement, Brown decided to go at it alone and patented his moving capacitor[108] in England. In the patent, we see that Brown maintained the idea that his invention influenced the gravitational field in some way. He wrote:

"This invention relates to a method of controlling gravitation and for deriving power therefrom, and to a method of producing linear force or motion. The method is fundamentally electrical."

A rather bold idea but not an entirely unreasonable one, considering an increasing number of scientists were (and still are to this day) seriously questioning the very nature of gravity following the publishing of Einstein's Unified Field Theory in 1929. The great man himself believed there was a link between electromagnetic and gravitational fields and thought it was time to marry them in an inseparable yet monstrously complex mathematical relationship known as the unified field equations. Of course, the problem was how to find a solution to the equations that would allow scientists to verify the link in experiments. Knowing that Einstein had been correct in his previous ground-breaking special case theories relating to relativity, a number of scientists (and some of the general public) wanted to look for evidence of any link between the two fields that might support Einstein's work.

Brown was definitely among those open-minded scientists fired up about this possibility. Realising his moving capacitor was essentially an electromagnetic device moving in an unusual way based on the position of astronomical objects in the sky, it didn't take much to convince him to consider the likelihood that his capacitor could be influencing the gravitational field. It would probably explain why Brown decided to name his capacitor a "gravitor". The only thing eluding him was how it actually worked, and hence the nature of this mysterious force moving his device. Such a critical question no doubt burned in Brown's fertile mind, sparking a lifelong interest in his invention and a desire to find the answer.

The Great Depression of the early 1930s saw Brown leave his comfortable job of four years as a staff member at the Swazey Observatory in Ohio and enter the U.S. Naval Research Laboratory (NRL) until 1933. While at NRL, he managed to find a little time to

---

108   See appendix E, British Patent No. 300,311.

join the International Gravity Expedition to the West Indies in 1932 before continuing with his Navy work. He also joined the Johnson Smithsonian Deep Sea Expedition in 1933 and later returned to the Navy Department as Officer-in-charge of Acoustic and Magnetic Minesweeping between 1940 and 1941.

In 1938, Brown set up a non-profit organisation based in Ohio under the name of the Townsend Brown Foundation to further his research into his invention.

The arrival of the UFO phenomenon in the United States in 1947 certainly didn't escape Brown's attention. He had ample opportunities to read and listen to numerous observations from UFO witnesses. Some were obviously ridiculous or had to be some form of natural phenomena, but then he noticed how other UFOs showed indications of electrical glows similar to that displayed by the negatively-charged electrode of his moving capacitor. Furthermore, talk of highly manoeuvrable UFOs seemingly able to defy gravity with ease made him wonder whether he should compile all available evidence on the nature of gravity (and later on UFOs). He went ahead with the decision to do so in 1952, when he established Project Winterhaven from an office in Washington, D.C. This was also about the time Brown, either inadvertently or deliberately, changed his moving capacitor design to look more like a disc.

In 1952, as reported in the March 23, 1956 issue of *Interavia* and in a personal interview with researcher William Moore, an Air Force major general was present to witness a demonstration of Brown's disc-shaped capacitor. At the time, Brown succeeded in sending afloat two disc-shaped electrokinetic "airfoils" measuring 60 centimetres in diameter that were tied to a pole by a wire supplying 50,000 volts, with a continuous input of 50 watts and extremely low amperage. The "airfoils" spun around with a top observable speed of 19 kilometres per hour. Even though the results were interesting, an investigator from the Office of Naval Research (ONR) wrote a report later in the year which concluded the movement was probably caused by "ion wind".

Not convinced by this argument, Brown wanted to disprove the "ion wind" theory by making his next demonstration more spectacular, using bigger discs charged to a higher voltage for greater speed and increasing the size of the course.

In late 1954, Brown certainly upped the ante when he flew a pair of 3-foot-diameter discs around a 50-foot diameter course, tethered by wires to a pole supplying 150,000 volts to charge the airfoils. The ambitious demonstration was witnessed by numerous military officials, government scientists and representatives of unnamed major aircraft companies. Everyone who saw it was impressed. When Brown revealed that the discs could attain speeds of several hundred miles per hour if using voltages up to several hundred thousand volts, the whole experiment was immediately classified[109]. Even so, the people who saw the experiment continued to remain skeptical, claiming an "electrical wind" was causing the movement[110].

Then one day in April 1955, an opportunity arose for Brown to test his capacitor in a vacuum chamber.

Jacques M. Cornillon (1910–2008), a U.S. technical representative for a French aeronautic company, Sociètè Nationale de Construction Aèroneutique du Sud-Ouest (SNCASO, or commonly known simply as Sud-Ouest), learned about Brown's work from an article published by Mason Rose, Ph.D., written by a nuclear scientist named Bradford Shank. Cornillon was in the library of the Institute of Aeronautical Science of Los Angeles when he came across the article, "The Flying Saucer: A Simplified Explanation of the Application of the Biefeld-Brown Effect to the Solution of the Problem of Space Navigation". In it, Cornillon noticed the details of Brown's capacitor and the possibility of a new force present in the device that could represent a new form of propulsion for aviation. Cornillon got in contact with Rose and, after a fruitful and lengthy discussion of Brown's work, decided to contact Brown personally.

Cornillon met with Brown in Washington on April 7, 1955. He asked if Brown was interested in working for the French company while getting support to pursue his work on his invention. Brown was probably taken aback by the kind offer but pleased to see someone showing interest in his work. It was far more than any of his American compatriots in the military, scientific and aerospace industries were willing to do for him, despite all his efforts. Brown explained his current dilemma, emphasising the fact that he didn't have the expensive vacuum chamber equipment he needed to carry out the next crucial experiment to disprove the "ionised wind" theory. Cornillon explained what his company could do for Brown. All Brown had to do was create

109   *Interavia* 1956, p.373.
110   Even to this day, NASA believes the movement is caused by ionised air molecules and electrons creating a "wind effect" on the electrodes leading to some form of "recoiling force" acting on each electrode.

a proposal and blueprints of the equipment he needed, and Cornillon would see to it that Brown got what he wanted.

Brown expressed interest in the offer, but he had already organised a trip to England and so asked Cornillon to give him a few days before making his decision. Cornillon agreed. As things turned out, it didn't take long before the beleaguered Brown took up the French offer.

As history tells us, Brown went to England to seek funding for his work from what he had hoped would be more willing investors. He tried everything he could to convince people of his invention. As usual, he gave top-notch presentations and demonstrated a small-scale version of his capacitor for all to see. Unfortunately, nothing eventuated from all this effort.

Without wasting any more time, Brown rang up Cornillon to say he was ready to take up his offer. Cornillon was pleased and notified his boss. Papers were hastily prepared, and Brown signed them on his arrival at the French company in Paris.

The main work Brown was involved in for the company was building a state-of-the-art vacuum chamber (which would be useful in other areas of research). He would later give the company shared rights to commercialise his capacitor, should his testing show great promise for the building of a practical device (e.g., if it could carry a payload). In return for his technical expertise, he would be allowed to pursue testing of his invention.

The testing was done at a facility known as B-12, located on the outskirts of Paris. It was conducted in secret under the auspicious name of Project Montgolfier, in honour of the French brothers who had taken the bold flight across the French countryside in the world's first hot-air balloon.

Soon after the vacuum chamber was built, Brown succeeded in testing his moving capacitor inside the chamber. This was the moment when he made his next important discovery: his discs flew more efficiently at a lower voltage in a vacuum than in the air.

Among other results that later emerged from Brown's testing were the following:

- Increasing the area of one electrode plate (or reducing the plate area of the other) increased the movement.
- Reducing the distance between the electrode plates increased the movement.

- Increasing the curvature of the electrode plates increased the movement.
- Dielectric materials used between the plates having high permittivity values (such as titanium dioxide) gave greater movement than those with lower permittivity values.
- Repeatedly varying the voltage supplied to the electrode plates (i.e., the rate of turning off and on the power) at higher frequencies increased the movement[111].
- When the voltage (and/or frequency) exceeded a certain value, the capacitor would accelerate in a dramatic manner, unless the voltage was reduced to stop the device from flying away from a prescribed circuit[112].

Following these interesting results, plans were underway to build a bigger vacuum chamber and supply voltages as high as 500,000 volts. Brown was literally on the verge of a major breakthrough, even if he did not know how his device flew. However, just when things were looking up for Brown, once again the project had to suddenly end, much to Brown's disappointment. The reason for the termination was because SNCASO agreed to merge with a larger company called Sociètè Nationale de Constructions Aèroneutiques du Sud-Est (SNCASE or Sud-Est), and the new president of this company[113] showed no interest in "far-out propulsion research efforts", preferring substantial interest in "air frame manufacturing". Brown, thoroughly disappointed by this latest turn of events, was forced to make some hasty final tests before returning to the United States in the summer of 1956, taking with him all of his papers, including data he had collected while working at the French company. The Final Report on Project Montgolfier was eventually completed on April 15, 1959, according to Cornillon[114].

Armed with the latest data and thinking he would finally get the support he was looking for, he contacted the U.S. Navy to present his results showing the "electrical wind" theory was wrong. However, far from the Navy welcoming Brown with open arms and wanting to learn

---

111   Childress 1990, p.108.
112   Childress 1990, p.106.
113   The merger was finalised on March 1, 1957 under the new company name Sud Aviation.
114   Jacques Cornillon's recollection of the events surrounding Project Montgolfier can be found at http://projetmontgolfier.info/INTRODUCTION.html.

more, Brown claimed that Admiral Hyman G. Rickover[115] approached him and told him in no uncertain terms that he wanted him to drop the whole matter and not to pursue his work on the moving capacitor any further. There was something about Brown's work that was embarrassing the military, or touching the raw nerves of certain people at the Pentagon. Was this because the government thought Brown's work was unscientific and had no value for the military, or because his invention was a failed technology? Apparently not. Although Brown didn't know it, behind closed doors, the Pentagon had already begun sponsoring research into Brown's device at major aerospace firms, with the study classified as Top Secret. In fact, it was the USAF that would show the greatest interest in Brown's moving capacitor.

It was clear that something about Brown's work ruffled the feathers of top defence chiefs. Brown was not privy to the reason for this interest.

Despite the terse recommendation from Admiral Rickover, Brown was not ready to give up. Without wasting time, and perhaps with a subtle hint from the Admiral that his work may have some link to UFOs, he established a reputable UFO research organisation called NICAP on October 24, 1956 with the aim of gathering information about these enigmatic flying objects. Brown made Keyhoe the head of the organisation while keeping a low profile himself, helping out where he could and getting involved in numerous meetings while listening to the discussion of various interesting UFO reports. Unfortunately, the time spent working on his invention meant he neglected his extra duty of managing the finances for NICAP. As a result, he was forced to leave NICAP within a year of its inception, but this only gave him more time to pursue further development and testing of his moving capacitor.

Within the next twelve months, Brown received sponsorship from a Delaware company he had formed called Whitehall-Rand Corporation to pursue his work. Brown then filed the first American patent application for his moving capacitor on July 3, 1957 and was eventually granted a U.S. patent on August 16, 1960 (see Appendix E, U.S. Patent 2,949,550). In this patent, Brown gave a more conservative name for his invention—an "electrokinetic device"—which did not allude to the possibility of a link to gravity. So, definitely no terms like "electrogravitic" or "gravity control". Is this because Brown was

---

115   LaViolette 2008, p.86.

encouraged by the Patent Office to provide a more plausible explanation in order to receive approval for the patent? If so, Brown may have decided to mention the "electrical wind". But after receiving the patent, he would still use "electrogravitic" as his preferred term to show this link between gravity and electromagnetism that was the underlying force he believed to be responsible for moving his devices. In other words, Brown continued to think that his devices were moving too fast to be an electrical wind. Something else was happening to them.

Brown continued with his work, this time focusing on the gravitational link with his devices.

During the period from 1957 to 1958, a new anti-gravity investigation was set up by multi-millionaire and Gravity Research Foundation trustee, Agnew Hunter Bahnson Jr., president of the Bahnson Company of Winston-Salem in North Carolina. Bahnson, a man with a strong interest in UFOs, constructed a well-equipped private laboratory and asked Brown to become chief research and development consultant. But, incredible as it may sound, just when everything was going well, Bahnson, an experienced pilot, died when his private airplane reportedly struck a high-tension wire under somewhat unusual circumstances. Bahnson's heirs expressed no interest in the project, and so it was terminated.

In 1958, Brown went on alone, renaming his corporation to Rand International Limited. During his time as president of the corporation, he filed several more applications and was granted further patents in the United States for his unique electrokinetic devices. But again, as if someone or something was working against him, all his efforts to demonstrate his devices to interested government and corporate groups came to nothing.

Just to add to his woes, Brown claimed there were repeated break-ins into his laboratory by unknown person(s). Under most circumstances, these events would have been enough to discourage most people—not so for the determined inventor. After more than 30 years of working on his invention and trying to find the elusive scientific explanation for how his capacitor moved, and after having spent nearly $250,000 of his own money in getting his invention to a level where he hoped people would take notice, he knew something was working against him. Never giving up on his work, he continued to believe that he had invented something important to society. He felt certain that his devices revealed a link to gravity due to the relationship he observed between the force on the capacitor and the positions of astronomical objects. He kept working on his devices right up to the end of his life.

Perhaps if Brown had had the scientific concept to explain how his devices moved, more people would have taken notice? Or was there someone in the U.S. government trying to thwart Brown's attempts at success?

Lacking a better explanation for why his devices moved so appreciably, Brown went into semi-retirement after 1965 and lived in California for the rest of his life.

As Brown's health slowly deteriorated, a university graduate named Mark M. wrote to him on January 9, 1982, expressing interest in his work. Brown acknowledged receipt of Mark's letter on February 9, 1982:

> "I regret to advise you that electrogravitic research has been taken over in its entirety by a Californian corporation which has imposed secrecy—at least until their investigations are completed. No further publication or release of information is permitted, possibly until next year."

Unfortunately, no further release of information would arrive, ever. The unnamed company allegedly conducting the research—perhaps Lockheed Martin or Hughes Aircraft—remains tight-lipped about its results.

Brown died in Avalon, California, on October 22, 1985, aged 80 years. As his daughter Linda Brown recalled:

> "At the end of his life, Dad organised his material, shut down his recorders, and saw to the final dispersal of his special papers. I wondered if he was accepting defeat after a lifelong struggle to obtain recognition for his work, but there was no sadness in his actions. He set about a purposeful completion of these final tasks and, with everything in order, he slipped away from us a few days later."[116]

---

116   http://www.ttownsendbrown.com/entrance.html (as of July 2013).

Before he died, Brown revealed to researcher William Moore that he believed that the Unified Field Theory was somehow important to understanding how his devices worked and could have important implications with regards to UFOs and gravity, but he could never pinpoint exactly how.

So why has the work of this American physicist gone unnoticed to this day? Brown himself suggested that the human race was not yet ready to accept such a revolutionary scientific concept nor to consider his devices as worthy of scientific study with the potential to revolutionise the transport industry as well as reveal insights into the nature of the gravitational field.

Or could it be that certain individuals in the U.S. military and government (more specifically those working within the USAF) are doing everything they can to ensure Brown's dream never becomes a reality in the public domain in case the implications regarding UFOs and how they work become fully understood? This would appear to be the case, as leaks would later emerge suggesting government scientists have secretly worked on Brown's devices since the late 1950s without telling Brown or providing him with funding. As a U.S. researcher stated:

"The government did not give him funding to develop electrogravitic propulsion, but they did manage to steal some of his best ideas, and used their own scientists to continue development and refinement of his engines in secret."[117]

Why the secret government work without Brown's knowledge? Unless the USAF or someone else discovered something about Brown's devices and how they related to certain other secret projects (e.g., UFOs, crashed discs and the need to reverse-engineer the technology, or if someone else already knew how to build a man-made electromagnetic flying object with the help of Brown's invention and needed to keep it quiet), we can only speculate. Certainly it would explain Brown's incredible run of bad luck in not being able to get his invention to a level that would convince others of a new electromagnetic propulsion system capable of replacing jet engines. Even if by some miraculous chance Brown was not subject to dubious methods by unknown persons to keeping him quiet, one cannot say his

---

117   http://www.go2emedia.com/reports/misc/R1012.htm.

invention is a failed technology. For the USAF to work on his invention in secret would suggest that there is something else we need to learn about the technology.

We can only wonder what it is—or maybe we already have the answer, as we shall see later in this book.

In 2013, President Barrack Obama, in his second term in office, appointed personnel in his administration to find ways to reinvigorate the U.S. economy. Among the initiatives sought to be implemented was the anticipated release of classified technologies based on anti-gravity propulsion systems held by military and corporate entities for more than 60 years.

As a classic example, it was revealed in the March 1992 issue of *Aviation Week & Space Technology* in an article titled, "Black World Engineers, Scientists, encourage using highly classified Technology for Civil Applications" that the USAF had used Brown's moving capacitor concept on the Northrop Grumman B-2 Spirit Stealth Bomber's leading wings and exhaust for some reason. As the article stated:

> "Take-off thrust of [each of] the [four] F118-100 [B-2 engines] at sea level is given as '19,000 lb (84.5 kN) class' by Northrop Grumman and as '17,300 lb (77.0 kN)' by the USAF. These are startlingly low figures for an aircraft whose take-off weight is said to be 336,500 lb (152,635kg) and which was until recently said to weigh 376,000lb (170,550kg). Aircraft usually get heavier over the years, not 20 tones [sic] lighter. Even at the supposed reduced weight, the ratio of thrust to weight is a mere 0.2, an extraordinarily low value for a combat aircraft."

Unless there is something unique about the aerodynamics of the B-2 wings that give it extraordinary lift for less thrust, the only way to achieve adequate lift is to increase thrust. But that would involve employing a radical new technology. Now we have learned that the B-2 does make use of Brown's moving capacitor, presumably to achieve significant thrust and/or reduce its weight. Using what little we do know about the aircraft's design and whisper-quiet take-off and flight through the air, the article claims the B-2's sharp leading edge along the length of the wing is charged to "many millions of volts" and the exhaust jets coming out of its four engines are given a negative charge. The article speculates that it may be possible that the charge present along the sharp leading edge of the wings is there to ionise the air molecules and push them away to reduce air pressure. However, to attain reasonable thrust for its weight suggests something else must be going on between the negatively-charged exhaust openings and, presumably, the positively-charged wings. Does this mean we have what is effectively a large-scale capacitor design, where the negatively-charged exhaust openings are "chasing" the positively-charged wings to create the additional thrust (or movement), as we see in Brown's invention?

And why stop there? What about adding more of these capacitors to lower its weight?

The highly-respected aviation writer Bill Gunston (1927–2013) gave support to electrogravitics as an important addition to the B-2 design when he said:

> "I have numerous documents, all published openly in the United States, which purport to explain how the B-2 is even stranger—far, far stranger—than it appears. Most are articles published in commercial magazines, some are openly published U.S. Patents, while a few are open USAF publications by Wright Aeronautical Laboratory and Air Force Systems Command's Astronautics Laboratory. They deal with such topics as electric-field propulsion, and electrogravitics (or anti-gravity), the transient alteration of not only thrust but also a body's weight. Sci-Fi has nothing on this stuff."

If a moving capacitor has been added to the B-2 design, why is it that this technology remains highly classified to this day? We all know about Brown's invention, so why the secrecy?

The person appointed to the task of releasing this kind of technology to the public is Obama's National Security Advisor and retired U.S. Marines General James L. Jones. Will he succeed in releasing the information?

As of 2019, the unnamed Californian company allegedly doing the secret work on Brown's devices has remained remarkably tight-lipped about its results, despite spending more than 25 years studying the subject. Not even the USAF wants to comment on Brown's technology, let alone the Californian company in question.

From the way things have been progressing, it looks like General Jones has his work cut out for him.

The USAF Shows Interest in the Scientific Study of Accelerated Charges Emitting Radiation

In 1959, the USAF became the first military organization to show interest in the work of scientists studying electrically-charged particles emitting radiation. Since a device like Brown's invention is essentially an electrically charged device emitting radiation from its oscillating charged surface, with higher energy density radiation coming off one end and so causing the device to recoil in the opposite direction (and looks like a charge particle emitting radiation in one direction when viewed from a great distance), it is possible its charged surface may emit large-scale radiation in a manner that might contribute to, or be the sole cause of, its movement. Somehow, the timing was right for the USAF to focus on this area.

As experiments using the radiometer and other devices have shown, we know radiation can move ordinary matter, whether charged or not. So, when an object emits radiation, it will recoil (or accelerate) in the opposite direction. However, in the case of a charged particle, scientists have noticed something interesting. When propelled by radiation from a charged surface, the rate of acceleration is highly unusual and quite dramatic. Further details about this acceleration rate will be discussed in Chapter 8, but for now let us just say that there has been, and continues to be, much debate about this "accelerated charge emitting radiation" problem among scientists. We see this in the scientific literature, including questions as to whether an accelerated charge does emit radiation.

As a result of this controversy, two scientists were partly supported to study the problem in 1959 by the National Science Foundation and the USAF through the Air Force Office of Scientific Research (OSR) based at Wright-Patterson AFB. Those two scientists were Dr. Thomas Fulton from the Department of Physics at Johns Hopkins University, Baltimore, Maryland, and a visiting professor from the State University of Iowa between 1958 and 1959.

By November 1959, a paper was submitted by the scientists to the *Annals of Physics*. It was accepted and published in Volume 9, Issue 4 in April 1960. Titled "Classical Radiation from a Uniformly Accelerated Charge", its main focus was on the problem of whether a charged particle in constant acceleration emits radiation. As the abstract for the paper stated:

"The old and much-debated question, whether a charge in uniform acceleration radiates, is discussed in detail and its implications are pointed out. Many contradictory statements in the literature are analysed and those answers which can be given on the basis of the standard classical Maxwell-Lorentz equations are presented. Although the questions that remain open are difficult and fundamental, some simple results can be proved: Contrary to claims in some standard sources (Pauli, von Laue), a charge in uniform acceleration *does* radiate. The radiation rate is finite, invariant, and constant in time in the instantaneous rest system. There is no contradiction of this fact with either the principle of conservation of energy or the principle of equivalence. Finally, the group of conformal transformations is found to be not physically meaningful."[118]

In other words, the scientists were confident that a charged particle in constant acceleration does emit radiation. At the same time, it was realised that there were enough contradictory statements in the scientific literature surrounding the emission of radiation from a charged surface to make them see that much work was needed to fully understand what is going on when a charged particle emits radiation linearly.

Nevertheless, the question remains: did the USAF gather anything interesting from this work? And what made the USAF decide to show interest in this specific area of physics?

Indeed, would the work somehow relate in any way to observations of UFOs?

The Yale Perspective

The highly respected *Yale Scientific Magazine* of Yale University, made its own independent assessment of the work of the Air Force by stating in April 1963:

"Based upon unreliable and unscientific surmises as data, the Air Force develops elaborate statistical findings which seem impressive to the uninitiated public unschooled in the fallacies

---

118   https://doi.org/10.1016/0003-4916(60)90105-6 as of August 2019.

of the statistical method. One must conclude that the highly publicised periodic Air Force pronouncements based upon unsound statistics serve merely to misrepresent the true character of UFO phenomena."[119]

A Hollywood Celebrity's Untold Secret about UFOs

Hollywood actor and socialite Marilyn Monroe[120] (1926–1962) may have had her life abruptly cut short by certain individuals after learning a bit too much about UFOs. Whether this is true or not depends on the veracity of a leaked CIA document discussing the results of a wire tap on Monroe's phone as well as the testimony of several key witnesses versus the official explanation, as accepted by the U.S. Authorities, of what happened.

If we accept the official position, it would appear that Monroe was found dead on her bed at around 3:40 a.m. on Sunday 5 August 1962. She had either accidentally overdosed on prescription sleeping pills, or she committed suicide[121]. As the official coroner's report written by Thomas Noguchi, M.D. concluded:

> "Miss Monroe has suffered from psychiatric disturbance for a long time. On more than one occasion...when disappointed or depressed, she has made a suicide attempt."

The alternative version of events is that Monroe died between 10:30 and 11:00 p.m. on Saturday 4 August 4 1962, because certain people wanted her dead to prevent her from revealing certain secrets to the public.

So, what really happened to the famous Hollywood star? Was it murder or suicide?

---

119   *Yale Scientific Magazine,* Volume XXXVII, Number 7; Flammonde 1976, p.268.
120   Born as Norma Jeane Mortenson; baptised as Norma Jeane Baker.
121   No suicide note was left behind.

What we do know is that written letters have surfaced[122] confirming a number of secret affairs between Monroe and U.S. President John F. "Jack" Kennedy. Historians are in agreement that at least one sexual encounter probably took place on March 24, 1962, at a hotel in Palm Springs. However, the letters also suggest repeated encounters starting in February 1962, only to raise some public suspicion of a possible affair by May 19, thanks to her sexually provocative rendition of the "Happy Birthday" song to the president, which happened on live television. Afterwards, the president, who was still married to his wife Jackie, decided to end the relationship with Marilyn on July 16, 1962, fearing it could affect his career and marriage. Luckily this didn't prove difficult, as Monroe had become infatuated[123] and eventually had affairs with Kennedy's younger brother, Robert Kennedy, who was the country's attorney general at the time. But even after the affairs ended with John Kennedy, Monroe continued to attend social functions with the president right up to July 26, 1962 as confirmed by a confidential FBI memo.

The only tricky part in all of these affairs with Robert Kennedy is that Monroe was falling in love with him. This meant that any end to this relationship with him could turn out to be a much harder proposition for Monroe to handle. She could easily get upset and angry, as she may have felt like she was being used by Kennedy for sexual gratification and that he never really loved her to begin with.

Perhaps a good enough reason to commit suicide?

However, the CIA document alludes to another possibility. One that is more disturbing.

The alleged CIA document first came to light in the spring of 1992 when security officer and private detective Tim Cooper allegedly received a poor-quality photocopy of the CIA memo from Thomas Cantwell, who Copper believed to be a former CIA employee. In 1994, Cooper sent a copy to another private detective Milo Speriglio, writing anonymously as "an unknown well-wisher". Speriglio passed away in 2000 at the age of 62 without commenting on the document. Mr Steven Greer reviewed Speriglio's work and his vast collection of documents. He came across the anonymous "well-wisher" note attached to what was purportedly a genuine top-secret CIA document. Further investigation eventually led to an identification of the man who

---

122   https://clickhole.com/amazing-read-these-newly-discovered-letters-between-jf-1825121023
123   Monroe Monroe was already divorced from her husband and American baseball player Joe DiMaggio.

sent the document. However, in an attempt to avoid creating undue attention, Cooper sent an email on 23 April 2009 to UFO researcher Robert Hastings claiming the document is a fraud.

The only other indication that the document could be a fake is the appearance of the word "MJ-12". Other leaked government documents mentioning MJ-12 in more recent times have been found to be fakes due to the type of typewriter used to create the documents (it was only available well after the alleged dates shown) as well as the obvious forged presidential signature (it was photocopied from another document).[124] Even so, it need not be the case that an MJ-12 group had never existed in history. Apart from Cooper choosing to change his mind and the mention of an MJ-12 group, there is virtually no other evidence to put this CIA document on Marilyn Monroe in the "fake" category at this stage.

However, the unusual chain of events and discrepancies surrounding Monroe's untimely death cannot be explained properly unless we consider the contents of this intriguing CIA document and the latest testimony from key and new witnesses who were present on the day of Monroe's death. For all intents and purposes, the document could still be genuine.

Analysis of this document dated Friday 3 August 1962 suggests that not only was Monroe's phone wire-tapped[125] for several months prior to her death following her relationship with John[126], but also Monroe told Howard Perry Rothberg, an antique dealer who also acted as a designer for the rich and famous in New York and a close associate of journalist Dorothy Kilgallen with insightful and reputable articles on UFOs to her name, that her diary contained sensitive secrets obtained from President John F. Kennedy. Unbeknownst to both women at the time, the CIA also had allegedly also wiretapped Kilgallen's phone. If the CIA document is meant to be a fake, it must be a particularly good one considering one would expect the CIA to monitor Kilgallen. Her work

---

124   More details about these documents mentioning MJ-12 are discussed later in the chapter.
125   Presumably by the CIA or a request was put through by the CIA to the FBI to have the phone tapped.
126   Author Fabulous Gabriel confirmed that the authorities were aware that Monroe's home was wired. He believes it was set up by the FBI. He claimed "four different sets of bugging tapes" have conveniently and mysteriously disappeared following her death, together with police records, Monroe's organ and tissue samples, and all incoming and outgoing calls at the Hollywood star's home. Such critical evidence disappearing so soon after her death is reminiscent of what happened to John F. Kennedy immediately after he was assassinated just over a year later. Remarkably, we find a confidential memo he had written just 10 days before he was killed requesting that UFO information be released in an attempt to end the Cold War with Russia.

on UFOs was already enlightening and showed an effort to reveal the truth. It just wouldn't be proper for the CIA not to have a tab on her whereabouts, what she was doing, and who she was talking to. What makes this document more interesting is how, in the conversation between Kilgallen and Rothberg, the R word was mentioned in relation to the most famous crashed UFO in history. Given what we know of the continuing secrecy behind the Roswell case and the USAF's lack of transparency of the USAF in explaining the work it was doing in shape-memory alloys and the bodies that were found, we can be sure the Roswell case remains of utmost importance to the U.S. government. In that case, it makes sense for any mention of Roswell to affect certain people in this government. As it turned out, the proverbial manure did hit the fan, considering Monroe's death was on either the night of Saturday 4 August, or early hours of Sunday 5 August 1962.

If this is true, Rothberg may not have necessarily known who Monroe was. But it seems like Monroe probably knew who he was and his connection to Kilgallen. She may have read further into the UFO situation, and Kilgallen's name may have cropped up in relation to the Roswell case.

If anyone would be interested in the Roswell case and could cause the most damage to the U.S. government, it would be Kilgallen.

Now, for this scenario to have traction, there must be a moment in history when the U.S. president could pass on such a secret to Monroe. This can be observed during the period when the two were cavorting with each other between February and March 1962.

So, apart from sharing his naked ambitions with the famous Hollywood actor, the president may have also been a little too relaxed and willing to share certain government secrets with her. Not an inconceivable situation. Looking back in history, it was well-known during the Cold War with Russia how the KGB in the Kremlin made full use of female spies to have sex with targeted men in order to obtain certain secrets. The tactic did prove to be highly effective on more than one occasion. In the case of the president and Monroe, it is not difficult to imagine following an intimate moment in the bed that the topic of UFOs could easily have cropped up, which was a popular topic at the time. Perhaps the president told Monroe of a secret visit he made to an air base to view the remains of the original Roswell debris and recovered bodies. It may have given her considerable confidence in the reality of UFOs because the president was the one to have seen the evidence.

It is probably worth mentioning that rumours did circulate some years later of another U.S. president who allegedly viewed the original Roswell materials and bodies. It occurred on one mysterious day in which the whereabouts of President Eisenhower were said to be the least documented in history. As the story goes, a certain time frame for the president on a particular day was allegedly spent meeting a dentist. However, the dentist in question had no recollection or written evidence to support this visit. No one else has seen the president to vouch for his whereabouts, either. Then, within days of the time frame in question, people noticed Eisenhower had visited a church with his wife to pray for which he was not known to do previously. Prior to this day, Truman was the only other president to be aware of what happened near Roswell. However, if John Kennedy was the third president to have had a secret private viewing of the UFO evidence, this may have occurred just before Kennedy suddenly made his ambitious announcement before a special joint session of Congress that he would put the first human on the Moon in his speech on 25 May 1961. From this moment on, John Kennedy felt confident in the reality of UFOs and became personally interested in the subject.

Robert Kennedy, on the other hand, was not so open and willing to divulge government secrets, even when lying in bed with a famous Hollywood celebrity.[127] If anything, the affairs Monroe allegedly had with Robert Kennedy were kept more secret to a level where he could later deny this claim. Of course, if anyone could set the record straight in this regard, it would be Monroe and her diary, as there is no reason to believe she would not have written a record of these intimate him, the diary no longer exists.

At any rate, a time came when it was alleged that Robert Kennedy had to end the relationship with Monroe.

A de-classified FBI document[128] seems to explain what happened. It claims Robert Kennedy was "deeply involved emotionally" with

---

127   Let alone someone in the Kremlin in Moscow. In terms of support for the affair, a released FBI memo dated 23 July 1973 to T. J. Smith (available at https://vault.fbi.gov/Marilyn %20Monroe/Marilyn%20Monroe%20Part%201%20of%202/view) suggest that the rumours of an affair with Robert Kennedy was known and spreading among people along the West Coast following Monroe's death and "her death was in some way related to this". Another theory mentioned by this document is that some extreme R-wing CIA and FBI agents aware of Marilyn's associates who had past Communist Party affiliations could have had the "motivation" to kill Monroe.

128   https://vault.fbi.gov/Marilyn%20Monroe/Marilyn%20Monroe%20Part%202%20of%202/view

Monroe and was contemplating on whether to divorce his wife Ethel and marry Monroe. After much thought, he decided against this. Monroe then learned of his decision, which was enough to affect her work at the studio; eventually, her contract was cancelled due to a "reliability" issue—that is, not arriving at the film set. Monroe made several phone calls from her home to Robert Kennedy at the Department of Justice in Washington for help. At first, he said he would take care of it. When nothing was done, Monroe made a final call to Robert Kennedy in which some unpleasant words were exchanged. The document claimed "She was reported to have threatened to make public their affair".

However, the CIA document indicated another secret, along with the affairs, Monroe was prepared to reveal to the press. With Monroe in love with Robert Kennedy, one can imagine how quickly she became upset over his decision to stop the relationship. She wanted to be with Kennedy—when she couldn't, she decided to do something to get back at him (and his brother). She wanted to hurt the Kennedys by doing the one thing that was most damaging to them: revealing to the press the secrets she kept in her diary.

But there was one secret that would also hurt the CIA. As the CIA document claims, Monroe contacted Rothberg to let him know she would make an announcement to the press and potentially pass on sensitive information to Kilgallen with his help in relation to this other secret.

As people would say, "Never throw the baby out with the bath water". There could be some truth to this document. It is either that or the CIA document is not genuine, and Monroe was simply having mental health issues. Her death was presumably inevitable due to her depression, and there was nothing going on in her life to cause all of this. It was entirely an accidental overdose and that was it.

Perhaps.

However, certain eyewitness accounts of what happened next have given support for the CIA document and reveal a sinister chain of events that suggest someone was not happy, that Monroe had to be "taken out"—but surely not on the grounds of revealing the secret affairs with the Kennedys, of which the Kennedys could easily deny if the diary can be taken from Monroe. Ii would be her word against the Kennedy's in a legal sense. Women in those days were not likely to be believed in the face of two powerful men. But to kill someone suggests that Monroe was aware of something else—something that required more drastic action.

In essence, murder.

But why go to such drastic action on one woman? Only the CIA document can shed light on this matter.

Indeed, there are enough discrepancies that have emerged even prior to the witnesses speaking to the police and private investigators for a reasonable person to seriously consider the "murder" theory for Monroe.

For example, when the first policeman arrived at the scene, he became immediately suspicious as if time had been spent preparing Monroe's body to make her death look like a suicide, even right down to the fact that someone later inserted a tall drinking glass on the floor next to the small table beside Monroe's bed to suggest that she had ingested 30 to 40 pills. However, John Miner, the assistant Los Angeles District Attorney, who was present at the autopsy has re-evaluated the autopsy report and has concluded that Monroe did not die by ingesting sleeping tablets. In fact, the number of pills needed to be consumed should have shown signs of crystallisation of the chemicals in the lining of Monroe's stomach walls, the upper intestinal tract, kidneys, and urine. There was none. Only her blood showed the chemicals. Therefore, the only way to administer the chemicals into the blood was through an injection of a barbiturate (pentobarbital, also called Nembutal).

Unfortunately, the report did not reveal signs of needle puncture wounds, even though there are ways to hide the mark. For example, by not killing Monroe straightaway and allowing blood flow to heal the tiny wound, the body can assist to hide the site of injection, making it difficult to detect, even by a coroner. However, even despite this knowledge, the level of concentration of the chemicals in the blood would not have killed a human as medical experts have stated—which means Monroe died by another means.

We also have certain witnesses' testimony that casts doubt over the suicide claim.

For example, friends (e.g., Henry Rosenfield, Sydney Guilaroff, Joe DiMaggio Jr. Mickey Rudin, and others) and Monroe's ex-husband and American baseball star Joe DiMaggio recall speaking to Monroe over the phone on that final day. All agree that she seemed happy enough and was looking forward to certain things she would do in her life. Not the type to contemplate suicide.

For example, Joe DiMaggio Jr. the son of the American baseball player, called Monroe at around 7:15 p.m. on 4 August. He found her voice cheerful and upbeat with a sense of purpose in life.

If there was ever a moment on the day Monroe was allegedly not quite herself was observed early in the afternoon in the presence of the house keeper Mrs Eunice Murray, and her son-in-law and handyman Mr Norman Jefferies who was working on re-tiling the kitchen in Monroe's home. Jefferies thought Monroe looked a little unwell and sounded grumpy about something, but he couldn't quite put a finger on what it was. She was wrapped in a bath towel and told Mrs Murray and Jefferies that she wasn't feeling well and would go to bed to get some rest. However, by 9:00 p.m., Mrs Murray and Jefferies were watching a movie on NBC called *The Day the Earth Stood Still*. Monroe came in to watch some of the movie. There was no evidence she had taken drugs or alcohol. Indeed, she was observed to be in high spirits, even laughing and enjoying the movie. Then she returned to her bedroom.

Jefferies is a key witness of the events that took place at Monroe's home. He had much to say about the people he saw come and go then. Surprisingly, the police never took his statement—nor did the press. In the testimony he gave many years later to investigator Anthony Summers, he claimed to have suspected foul play on that night but could not pinpoint exactly who carried out the fatal drug overdose on Monroe.

Confined to a wheelchair and terminally ill, Jefferies was not worried about getting into trouble with the U.S. government with his decision to explain his version of events. He wanted to tell the story as he recalled it and without embellishment. He said he arrived at the Los Angeles bungalow at 12305 Fifth Helena Drive owned by Monroe with his mother-in-law and Monroe's longtime housekeeper Mrs Murray at 8:00 a.m. on Saturday 4 August 1962. It was around 2:00 p.m. on Saturday when Robert Kennedy arrived. He was not alone. Actor Mr Peter Lawson drove and accompanied Robert. Lawford entered the house and told Jefferies and Mrs Murray to leave and gave them money to buy a drink. The two left and later returned to the house. The only other location Jefferies was present was in the neighbour's house when he and Mrs Murray were again asked at around 9:45 p.m., this time by Kennedy, in the presence of two unidentified men wearing dark suits and sunglasses with one carrying a small black bag to leave the house. By no later than 11:00 p.m., Jefferies saw Robert and the two men run away in a haste. This prompted Jefferies and Mrs Murray to return to the house. Since then, Jefferies only left the house the following morning at 7:30 a.m. Sunday 5 August. As he stated:

> "I was there in the living room with Eunice when Marilyn died, and after that all hell broke loose."[129]

Robert Kennedy disputes any claims of him having been at Monroe's house, saying he was in northern California during the weekend. But as the retired Los Angeles Police Chief Daryl Gates has admitted, Kennedy was in Los Angeles on the day Monroe died[130]. Mrs Murray (hired by Monroe's psychiatrist, Dr Ralph Greenson, as a reliable and trustworthy person) stated on the BBC documentary *Say Goodbye to the President* that she saw Kennedy that Saturday afternoon. And nearing the end of his life, Jefferies reaffirmed the claim, adding that Lawford[131] was with Kennedy. But they were not the only witnesses. A number of neighbours and their friends saw Kennedy arrive at Monroe's home on two occasions—the Saturday afternoon, and later that night. For example, Monroe's next-door neighbour Mary W. Goodykoontz, her guest Elizabeth Pollard, and two other ladies having an afternoon of bridge party recall seeing Kennedy leave Monroe's home at around 2:20 p.m. He ran back to a white Lincoln convertible and raced off.

As further supporting testimony, a distraught Monroe telephoned her hairstylist and makeup artist Sydney Guilaroff telling him that Kennedy was at her house and had responded, "If you threaten me, Monroe, there's more than one way to keep you quiet." In the light of the CIA document, it looks like Monroe was prepared to use something against her former lover, and Kennedy was not happy about this. He matched her threat with his own and seemed determined he would carry out the threat if he had to.

She called Sydney again between 8:00 and 9:00 p.m., telling him that she had just met with her psychiatrist. Sydney thought Monroe sounded

---

129   https://www.angelfire.com/stars/mmgoddess/JEFFERIES.html
130   An FBI document at https://vault.fbi.gov/Marilyn%20Monroe/Marilyn%20Monroe%20Part
        %202%20of%202/view claims Robert Kennedy was staying at the Beverly Hills Hotel on 4
        August 1962.
131   Peter Lawford is John F. Kennedy's brother-in-law.

fine and certainly better than during the call she made earlier. This time, however, Monroe offered an intriguing hint that she knew of a number of secrets in Washington and that she intended to do something about them after the weekend (apparently first thing on Monday). Again we can only appreciate what she meant by this when we refer back to the CIA document.

Despite these important eyewitness accounts, however, the authorities had no evidence to implicate Kennedy in Monroe's death so the matter was not pursued. The reason for Kennedy being at the house has never been satisfactorily explained. Unless, of course, we refer back to the CIA document. This is the thing. We cannot ignore this document simply because it seems too incredible, unless there is something else to put it into context. Thus far, the document makes the events surrounding Monroe's death appear perfectly sensible. The alleged event does not work if it was entirely a relationship breakup. The Kennedys could deny any affairs (assuming the diary can be found and destroyed). Furthermore, if the missing diary was taken by Robert Kennedy, any other secret would not make sense for why Monroe had to die unless it is a secret so sensitive and damaging to the U.S. Government (specifically the CIA) that having Monroe survive and mention it to anyone, especially the press, was considered too much of a security risk.

In 1975, Mr Lawford was tracked down. In a recorded interview, he told detectives that he was not at Monroe's house and never saw Robert Kennedy on the day. He claimed to have spoken to Monroe by telephone (telephone records confirm this). He said the woman sounded about the same as before, which was her usual depressed self. He had a "gut feeling" on that day that something was wrong and regretted not visiting her home to talk to her. As the detectives mentioned in their report:

> "She stated she was tired and would not be coming. Her voice became less and less audible and Lawford began to yell at her in an attempt to revive her."

The detectives said Lawford described it "as a verbal slap in the face."

"Then she stated, 'Say goodbye to Pat, say goodbye to Jack [John F. Kennedy] and say goodbye to yourself, because you're a nice guy'," the report said. The phone then went dead.

So what is going on? Why are there plenty of people willing to testify that Robert Kennedy was at the house twice and doing something to suggest that these were not social calls to the Hollywood actor. More interestingly, why was he seen running away from the house twice and claiming to police he was not near Monroe's home on the day of her death? Something isn't quite right, unless, of course, it has something to do with the CIA document.

Only the CIA document can make sense of all these events.

But does this mean Monroe was murdered? Could the contents of the CIA document reveal the true motive for having Monroe gone?

Mrs Murray is certainly another key witness. Her version of events revealed more details of a possible homicide or murder. However, she has been frightened by the events that transpired on the night and the things she was told to do and later had to say to the police by Dr Greenson. She has felt compelled in her initial statement to police to stick to one story.

Jefferies is in general agreement with much of what Mrs Murray has said except for one important detail: Mrs Murray never actually saw a light under the bedroom door and later called out to the woman to see if she was okay. Rather, the two of them returned to the house at around 10:30 p.m. on Saturday night and found Monroe in the guest cottage next to the main house. Beyond that, Jefferies followed the recommendation of Greenson to hide elsewhere in the house claiming Greenson and Murray would handle the situation when the first police officer arrived at the scene. When the police officer was dismissed and taken over by a new group of policemen handling the case, Greenson received an agreement from Monroe's personal secretary and press agent, Pat Newcombe, to support his story of what happened. Later, as a reward for Murray and Newcombe cooperating with Greenson, Murray was able to pay for a trip to Europe two weeks later, and Newcombe was put on the federal payroll "as top assistant" to George Stevens, Jr., head of the motion picture service for the United States Information Agency.

If Jefferies' testimony can be relied on, then what happened was that at around 9:40 p.m. on the evening of 4 August 1962, Robert Kennedy and two unidentified men (one of whom was carrying a small black bag) entered Monroe's home unannounced. Murray and Jefferies were in the house watching a movie when Kennedy told them in no uncertain terms to leave. The two got out of the house and walked to a neighbour's house. They waited there. Jefferies was watching through the window. They heard noises from the guest cottage as if someone was looking for something. Monroe's large filing cabinet was broken. As if they still couldn't find what they were looking for, more noises could be heard. No one saw Monroe come outside. It was unclear to Jefferies what she was doing.

Jefferies and Mrs Murray were unaware that Monroe was on the phone in her bedroom talking to her friend and occasional lover Jose Bolanos at around 9:45 p.m. Bolanos recalls her happily chatting away with him. There was nothing to indicate she was upset about anything. Indeed, there was a sense of confidence in her voice as if she knew what she wanted to do. Then, she mentioned having a secret that "will one day shock the whole world". Before she could say more, she said she heard noises coming from the guest cottage. Bolanos could hear her put down the phone without hanging up and went to investigate. She never returned.

The men eventually left the house in a haste, running back to their car. There was screeching of the tyres on the road, as Jefferies remembers it—as if the men had done something in the house, or they had found whatever they were looking for.

By the time the housekeeper and Jefferies returned to the home, they either the two found Monroe in a comatose state in the guest cottage according to Jefferies, or the housekeeper claimed she saw a light in Monroe's bedroom through the gap between the door and floor. The door was locked. She had a key herself to get in, but didn't think too much about it. She took a nap. By the time she came back later to see the light and called out to Monroe at around 3:00 a.m., something was amiss.

Whichever story is true, the two eventually realised Monroe was not responding. Mrs Murray called the ambulance and then the psychiatrist, Dr Greenson. This decision from Mrs Murray is confirmed by Jefferies. Soon Peter Lawford and close friends of the Hollywood star began arriving.

The ambulance attendant James Edwin Hall (and his driver Murray Liebowitz) arrived just before 11:30 p.m., according to Jefferies. With Liebowitz's help, they moved Monroe into the main house. One of the men accidentally dropped Monroe onto the floor, leaving a bruise on the "left side of [her] lower back". The autopsy report later showed the site of the bruising as evidence that Monroe was very much alive at the time she was carried. Hall decided to rest Monroe on the hard floor and told Liebowitz to get the resuscitator from the van. He inserted an oxygen tube into the woman to assist with her breathing. Something suggested to him that Monroe was still alive at this point. Hall claims Monroe's colour was returning to normal and believed it might be a good time to take her to hospital.

Suddenly Dr Greenson arrived and what happened next was unusual to say the least.

Dr Greenson tried to look like he was assisting Monroe. He ordered the ambulance officer to remove the breathing tube. Even though the officer disagreed with this, he had to do as he was told, as Greenson was an MD (the ambulance officer was trained to never challenge an MD). Then, the ambulance officer noticed another odd thing. Greenson then took out a syringe with a long heart needle out of his bag. He filled it with a fluid from a "pharmaceutical bottle of adrenaline" in an attempt to stimulate the heart. Hall was surprised to see Greenson had made no attempt to dilute the solution. Another strange observation noted by the ambulance officer was how the psychiatrist had to count how many ribs to go down (which an experienced doctor would not need to know). Hall remarked later that it looked like "he [Greenson] was still in premed school and had really never done this before". Then, it occurred to him that Greenson was just a psychiatrist and perhaps he may not have much experience with a needle. He had to let him continue. Greenson became aware his methods were alarming the ambulance officer. He said that his clumsiness and working things out were more of an attempt to make "a show of this" in front of others. With that said, he immediately plunged the needle only to hit a rib because of the incorrect angle he chosen. Remarkably, instead of pulling it out, he continued to push in the needle. He eventually entered Monroe's chest by cracking the rib and quickly injected the fluid in or near the heart. If Monroe was not yet dead at this point, the psychiatrist would pronounce her dead within minutes of giving her the injection.

Greenson told Hall his services were no longer required. The ambulance officers had to leave the scene.

It was about this time when some mysterious plainclothes officers in a police car arrived claiming to be from the LAPD intelligence division. They never mentioned their names.

Jefferies watched the events unfold as he observed the men move Monroe's body to the bedroom of the main house. The story of "suicide in the locked bedroom" started to take shape as the men placed pill bottles on a small table next to the bed, broke the bedroom window from the inside (later Greenson would claim to police that he was the one to break into Monroe's bedroom and discover her body even though the glass was lying on the ground outside), and the door was locked from the inside and closed.

After the men left the house, Greenson remained at the scene until just after midnight. He went into Mr Lawford's Lincoln Continental sedan. The inebriated Lawford forgot to turn on the headlights as he drove at 70 to 80 mph with Greenson in the front seat. Beverley Hills Detective Lynn Franklin observed this and pulled the car over. Astonishingly, he chose not to give a ticket because Robert Kennedy was in the back seat. Instead, he gave Lawford proper directions and let the men go. Still no mention on the police radio of the death of Monroe. It would be several more hours before news of her death reached police.

Los Angeles International Airport had records to show Robert Kennedy boarded a plane for San Francisco sometime between 12:30 and 2:00 a.m. in the morning.

By 4:25 a.m., Norman Jefferies, Pat Newcomb, Mickey Rudin, and Monroe's personal physician Dr Hyman Engelberg were at the house when Greenson, also at Monroe's home, decided to call the police (or perhaps Murray was requested to make the call). Finally, to make sure there was no evidence of someone else having rectally-administered a barbiturate enema into Monroe at the time when Robert and his men arrived at the house and had to sedate her properly as if the initial injections through the skin did not have a sufficient effect of keeping her still and quiet, Greenson told Murray to clean the soiled bedsheets in the cottage.

The first police officer to arrive was watch commander Sergeant Jack Clemmons. He was told by Greenson that Monroe committed suicide. Greenson pointed to the empty Nembutal bottle as alleged proof. Clemmons was not entirely convinced. He noticed the way the body was face-down in the soldier's position with her arms at her side and

269

legs straight. And the bed sheets were too clean. From his experience, the way the body was laid down on the bed suggested that someone was trying to disguise needle marks. Furthermore, he noticed the lack of a drinking cup with water to explain the presumed number of pills ingested[132] by the woman based on the empty pill bottles. Later someone else placed a glass of water into the bedroom without Clemmons awareness. And just to add yet another level of strange behaviour, he also noted Murray was using the washing machine to clean bedsheets.

Clemmons wasn't aware of anyone else at the house other than Murray and Greenson. He acknowledged not checking all the rooms, including the guest cottage. If he did, he would have noticed that Jefferies, Newcomb and Mickey Rudin were hiding in those rooms at the request of Greenson. He spoke to Mrs Murray for her version of events, followed by Greenson. Mrs Murray was told by Greenson to choose the story of noticing the light beneath the door of the bedroom and calling out. Mrs Murray had to say that Greenson entered the bedroom by breaking the window on the outside where he noticed for the first time Monroe slumped on her bed. Greenson added that he found Monroe dead around 3:40 a.m. There was no need to call the ambulance. With nothing else to go on, Clemmons had to accept the statements.

The Village Mortuary employees Mr Guy Hockett and his son Don arrived at 5:40 a.m. Hockett noted the state of rigor mortis taking place due to a darkening of the skin on Monroe's face. He estimated the time of death was between 9:30 and 11:30 p.m.

Yet the police did not appear concerned by all of this or by the lengthy delay in being contacted. It was assumed Monroe was in her bedroom the whole time and had simply overdosed on the pills.

By the time reporters arrived, Sergeant Marvin Iannone dismissed Clemmons from the scene, and Dr Greenson made his quick exit from the house.

Still more discrepancies would emerge, this time with the autopsy report. When the first official autopsy report was written, it accepted Greenson's time of 3:40 a.m. for Monroe's death. However, this contradicts the advanced signs of rigor mortis on Monroe's face (i.e., it

---

132   A toxicology report found 4.5mg percentage Pentobarbital and 8mg per cent chloral hydrate. That would be equivalent to swallowing 30 to 40 pentobarbital pills. Medical professionals are certain no one has died with such high concentrations in the blood. Something else had killed Monroe.

was darkening the skin) as seen in photographs taken at the morgue. The report had to be adjusted to show the time of death closer to around 11:00 p.m.

Also, a confidential source relayed to Jay Margolis claims that, in the first revision of the report (which, like the diary, first police records and many other key forms of evidence, has also gone missing), Thomas Noguchi noted the needle marks behind Monroe's knees, the jugular vein in her neck, and under her left armpit. There was also the needle mark to the heart. When the final revised version of the report was released, however, Noguchi decided to handwrite "no needle mark"— not even the one to the heart.

About the only thing that had not changed in the report was the lack of crystallisation of the chemicals in Monroe's stomach and small intestines given the number of pills allegedly ingested. But because Noguchi wrote "no needle marks", he had to assume the pills were swallowed.

The cause of death was marked as accidental suicide. Case closed.

But now, with the advent of the CIA document and latest testimony from a larger number of witnesses and those not afraid to speak out, it is looking more like foul play had occurred. Either that, or Dr Greenson killed Monroe through gross medical negligence and incompetency. This is unlikely based on the testimony and the fact that the ambulance officer was present at the scene and assisting the patient with signs that she was looking better and a decision was being made to take her to the hospital. Therefore, it raises the question, Why kill Marilyn Monroe? To be more precise, what kind of secret warranted this kind of action?

One can safely assume the diary was found for the decision to kill Monroe to proceed (depending on the nature of the secrets she had written). And it is reasonable to think that the diary had influenced someone to kill Monroe. The diary cleared contained various secrets, such as the whirlwind affairs with the Kennedys. But would this be a good enough reason to silence Monroe? In the case of the affairs, this is not likely. Rumours were already spreading of a possible affair, at least with John Kennedy, and historians have learned that Jackie became aware of this and told him to sort it out or face a divorce. Furthermore, an affair is not exactly an Earth-shattering secret that would "shock the whole world". Too many people were already thinking this secret might be true. If that is not enough, we know that if Monroe had survived to

271

tell her secret affairs, without the diary it would be extremely difficult for anyone to take Monroe's statements seriously. The Kennedys would simply deny any claim and life would go on.

The other possibility is that there could have been a secret regarding a possible sexual relationship with the psychiatrist. Unprofessional conduct on the part of Greenson with at least one high-profile Hollywood client may have sent his reputation into tatters, thereby ending his career. This is something that could have seen Greenson take matters into his own hands. However, the problem with this theory is that it makes no sense why Monroe had to be silenced. We can be confident the diary was found. Any secret about Greenson could easily be destroyed, along with secrets about the Kennedys. All that would be left is mere hearsay from an essentially unreliable and emotionally unstable woman. However, there is one more thing that makes this scenario even more unlikely. You see, the gripe Monroe had was not with Greenson. It was with the Kennedys, in particular Robert Kennedy. There was nothing Greenson had to worry about even if he did something unprofessional. Monroe was not going after him—only Robert Kennedy. The ones who had more to worry about were the Kennedys, and potentially the CIA, depending on how sensitive those secrets were.

Not even a secret decision by the president to kill the Cuban leader Fidel Castro would have warranted Monroe's death. If one could somehow show such a secret was true, the president might get a slap on the wrist and a downturn in his polls, but things would continue as usual.

Therefore, the only reason to go as far as Greenson did must be because he was told by someone (everything is now pointing to Robert Kennedy with Robert Lawford in on the plan without knowing all the details of what was contained in the diary, although we cannot completely rule out the CIA as not having some involvement in the Hollywood star's death) to silence Monroe permanently because of a more devastating and significant secret contained in her diary—one that would hurt the U.S. government right to the very core, a secret so big and sensitive that it could not be told by Monroe to anyone, even without the diary present as confirmation.

Well, what other secret could be bigger and more sensitive than what really happened in the Roswell case? Hard to imagine a secret so big that Monroe had to die for it. More of a reason why the CIA document should not be ignored completely.

Thanks to the CIA document, we may have one possible scenario to consider. And a particularly good one at that given what we now know about UFOs. The strong possibility that UFOs could be a real phenomenon and remain a highly sensitive matter to the U.S. government cannot be overlooked, and all stemming from that one key moment in July 1947.

U.S. President John F. Kennedy's Interest in UFOs

Another individual not immune to the UFO situation was former U.S. President John F. Kennedy (1917–1963). In an unusual attempt to be more open with the public and his Russian counterparts (and so end the Cold War), he openly expressed his interest in the subject, perhaps to his own detriment.[133]

On November 12, 1963, Kennedy wrote a formerly Top-Secret memorandum to a director of the CIA to review the classification of all UFO intelligence files so all unknowns could be separated from the knowns, including the high-threat cases. He also indicated in the memo that he wanted the unknown cases to be shared with NASA and the Soviets, and ultimately wanted to work with the Soviets on the UFO phenomenon, with a view to creating a joint space and lunar exploration program. He ended the memo by stating:

"I would like an interim report on the data review no later than February 1, 1964."

Unfortunately, he would never get the opportunity to find out more (and certainly no reply to his memo). At 12:31 p.m. on the hot, sunny afternoon of November 22, 1963, President Kennedy was assassinated in Dallas, Texas. The identity of the assassin has remained one of the most controversial in history. The man allegedly responsible for the shooting, Lee-Harvey Oswald[134], was killed by Jack Ruby, a nightclub

---

133   Perhaps the president had time to contemplate on what happened to her former lover, Marilyn Monroe, and realised how important it was to curtail the power of the CIA by forcing the organisation to release UFO secrets before any more American citizens got hurt by this powerful government entity.

134   Mr Oswald was vulnerably being led by two policemen from the county jail on Houston Street in Dealey Plaza through a crowd of 75 reporters and numerous policemen and civilians when somebody shouted "Do you have anything to say in your defense?" Oswald didn't have time to

owner with Mafia connections who was already going to die of lung cancer (he died in 1967), before he could reveal his knowledge to the public.

A copy of the official Top-Secret memorandum[135] released under U.S. FoI is shown below:

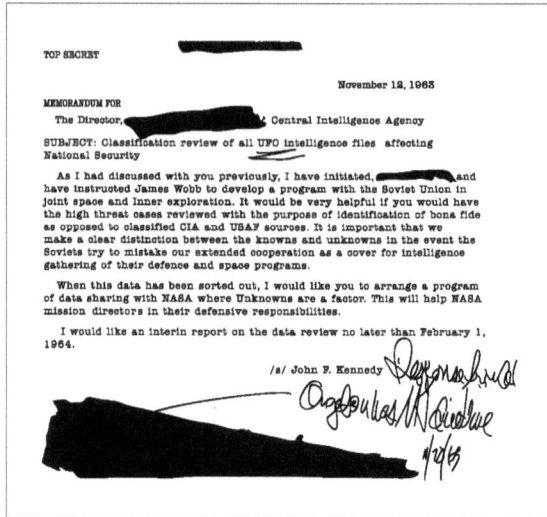

TOP SECRET

November 12, 1963

MEMORANDUM FOR

The Director, ▓▓▓▓▓▓▓ Central Intelligence Agency

SUBJECT: Classification review of all UFO intelligence files affecting National Security

As I had discussed with you previously, I have initiated, ▓▓▓▓▓▓ and have instructed James Webb to develop a program with the Soviet Union in joint space and lunar exploration. It would be very helpful if you would have the high threat cases reviewed with the purpose of identification of bona fide as opposed to classified CIA and USAF sources. It is important that we make a clear distinction between the knowns and unknowns in the event the Soviets try to mistake our extended cooperation as a cover for intelligence gathering of their defence and space programs.

When this data has been sorted out, I would like you to arrange a program of data sharing with NASA where Unknowns are a factor. This will help NASA mission directors in their defensive responsibilities.

I would like an interim report on the data review no later than February 1, 1964.

/s/ John F. Kennedy

This was not the first time the CIA has used the Mafia to do its dirty work: the Mafia was hired during a tempestuous time to kill Cuban leader Fidel Castro, unsuccessfully, during the Bay of Pigs fiasco. So any involvement by the Mafia in stopping Mr. Oswald from revealing evidence could be linked back to the CIA. Other oddities, such as a man in the crowd opening up the only umbrella on a sunny, hot day, holding it high above his head just before the assassination and then closing it after the event, together with claims by witnesses of one or more men on the ground carrying rifles, how the bullets hitting the president seemed to come from two different directions, and the missing autopsy report (including photographs and biological samples of the president's head) all add to the conspiracy theory that Mr. Oswald was not the lone gunman we were all led to believe. Someone else appeared to be involved. The CIA? Certainly there would be a

---

reply. A man later identified as Jack Ruby suddenly pushed himself through the crowd, pointed a handgun at Mr Oswald and shot him. Mr Oswald later died in the hospital.

135   Original with Robert and Ryan Wood, investigators of UFO-related U.S. government documents (available from http://www.presidentialufo.com/kennedy&1.htm).

motive for doing so, given its connection to UFOs and the Roswell case. Perhaps we may never know.[136]

Before this tragic event unfolded, it was well known that U.S. President Kennedy had a strong interest in UFOs and wanted to get to the bottom of the mystery by forcing the U.S. Department of Defense (via the USAF) and the CIA to reveal all documents and secrets it had about the phenomenon. Due to Kennedy's history of taking risks, "getting to the truth" and achieving greatness in the eyes of the public while he was president, he could have easily been considered a threat to national security by the CIA.

Was President Kennedy assassinated by the CIA to avoid highly sensitive UFO information getting through to the public? Or was it the crazy antics of a lone gunman, as the CIA and FBI-backed Warren Commission report on the assassination in September 1964 wanted to conclude?

The View of Former President Gerald Ford

Despite the death of a U.S. president and the considerable efforts of the new Project Blue Book team to turn all UFOs into IFOs in accordance with the Robertson Panel recommendations, UFOs continued to be reported by the American people, not to mention by people in other countries.

As another wave of UFO reports swept through the civilian population in the United States (and abroad) in 1965, former U.S. President Gerald Ford wrote in a letter he sent as a congressman to the Chairman of the Armed Services Committee, March 28, 1966:

---

136   In *JFK: The Smoking Gun* (2013), veteran Australian police detective Colin McLaren has examined all the available evidence and presented a compelling case that the fatal shot that killed Kennedy probably came from a secret service agent in the car traveling directly behind Kennedy's own vehicle. However, the documentary could only assume in the end that it was an accident as the evidence was only able to focus on how the president was probably killed based on available witnesses testimony, ballistic evidence etc. Closer examination of the behaviour and actions of the secret service agent in question immediately after the event and those of his colleagues who knew what happened leading to the cover up of crucial evidence in the autopsy report strongly suggests it was not an accident.

"In the firm belief that the American public deserves a better explanation than that thus far given by the [U.S.] Air Force, I strongly recommend that there be a committee investigation of the UFO phenomenon."[137]

Soon the U.S. government was put under considerable pressure once more by Congress and the general public to do something about the UFO situation, and fast.

The Condon Report

Eventually, on October 6, 1966, a team of 12 scientists from the cash-strapped University of Colorado, headed by a highly respected nuclear physicist, Dr. Edward Uhler Condon, decided to investigate the UFO controversy on behalf of the U.S. government in return for extra funding.

However, despite assurances in the USAF and Colorado University contract F44620-67-C-0035 that "the work will be conducted under conditions of strictest objectivity"[138], when the Condon committee first convened at a public meeting, Edward Condon revealed his own biases going into the project, nearly a year before the conclusion was announced.

The Elmira, New York *Star-Gazette*, January 26, 1967, published the following news item shortly after the meeting was held at the Corning Glass Works:

"MOST OF UFO SIGHTINGS EXPLAINABLE - SCIENTIST ADVISES CORNING AUDIENCE. Unidentified flying objects 'are not the business of the Air Force,' the man directing a government-sponsored study of the phenomena, Dr. Edward U. Condon, said here Wednesday night.

In an hour-long rundown on the government's interest in the field and the recollection of some baffling and spectacular claims by UFO 'observers', Dr. Condon left no doubt as to his personal sentiments on the matter: 'It is my inclination right

---

137   Statement issued in a letter written by Gerald Ford on April 3, 1966, to the House Committees on Armed Services and Science and Astronautics.
138   Keyhoe 1973, p.29 .

now to recommend that the government get out of this business. My attitude right now is that there's nothing to it.'

With a smile he added, 'but I'm not supposed to reach a conclusion for another year.'"[139]

The University of Colorado's inquiry into UFOs started off extraordinarily well when two team members, Dr. Norman E. Levine and senior Colorado University psychologist Dr. David R. Saunders, were fired on February 1, 1968, after leaking a memorandum dated August 9, 1966 written by assistant project director and the coordinator of the inquiry, Dr. Robert J. Low. It stated:

"The trick would be, I think, to describe the project so that, to the public, it would appear a totally objective study but, to the scientific community, would present the image of a group of non-believers trying their best to be objective but having an almost zero expectation of finding a saucer."[140]

Also, Condon's administrative secretary, Mary Louise Armstrong, quit on February 22, 1968, saying, "[the] attitude from the beginning has been one of negativism."[141]

The 1,465-page Condon Report, officially titled *Final Report of the Scientific Study of Unidentified Flying Objects,* was delivered to the Air Force in November 1968 and publicly released on January 8, 1969, with the blessing of a special committee of the National Academy of Sciences. The so-called final report on UFOs was impressive. Thirty-seven first-rate scientists, including Condon himself, contributed in part or entirely to the chapters of the report. The report discussed 91 cases[142] and contained detailed analysis and explanations for 59 of them[143].

But not everyone was impressed with the conclusion.

For example, Dr. Joachim P. Kuettner of the ESSA Research Laboratories in Boulder, Colorado, and chairman of the UFO subcommittee of the American Institute for Aeronautics and

---

139   Hynek & Vallee 1975, p.212; *Mysteries of the Unknown: The UFO Phenomenon* (Time-Life Books) 1987, p.116 and Peebles 1994, p.180.

140   Memorandum from Robert Low, Project Administrator of the Condon Report (Oct 1966 -Jan 1969), to Colorado University V.P. Thurston Marshall. Quote also published in *Mysteries of the Unknown: The UFO Phenomenon* 1987, p.116; Sachs 1980, p.70; Story 1980, p.78; Keyhoe 1973, p.145.

141   Blundell & Boar 1984, p.169; Peebles 1994, p.183.

142   *Mysteries of the Unknown: The UFO Phenomenon* 1987, p.119.

143   *Encyclopaedia Britannica* 1995, Volume 12, p.130.

Astronautics (AIAA), formed in 1967, explained in his article titled, "UFO: An Appraisal of the Problem"—published in the *Journal of Astronautics and Aeronautics* in November 1970—that "not all the conclusions contained in the Report itself [by other team scientists] are fully reflected in Condon's summary", and also, "Condon's chapter, Summary of the Study, contains more than its title indicates; it discloses many of his personal conclusions."

With support from the group of independent scientists from the AIAA UFO subcommittee, Kuettner added:

> "From a scientific and engineering standpoint, it is unacceptable to simply ignore substantial numbers of unexplained observations...
>
> The Committee has made a careful examination of the present state of the UFO issue and has concluded that the controversy cannot be resolved without further study in a quantitative scientific manner and that it deserves the attention of the engineering and scientific community."[144]

Another to express his concern was Peter Sturrock. He wrote an analysis of the Condon Report for the *Journal of Scientific Exploration*, where he stated:

> "The 'Condon Report', presenting the findings of the Colorado Project on a scientific study of unidentified flying objects, has been and remains the most influential public document concerning the scientific status of this problem. Hence, all current scientific work on the UFO problem must make reference to the Condon Report. For this reason, it remains important to understand the contents of this report, the work on which the report is based, and the relationship of the 'Summary of the Study' and 'Conclusions and Recommendations' to the body of the report. The present analysis of this report contains an overview, an analysis of evidence by categories, and a discussion of scientific methodology. The overview shows that most case studies were conducted by junior staff; the senior staff took little part, and the director took no part, in these investigations. The analysis

---

144   AIAA UFO Subcommittee, *Astronautics and Aeronautics*, November 1970, p. 49.

of evidence by categories shows that there are substantial and significant differences between the findings of the project staff and those that the director attributes to the project. Although both the director and the staff are cautious in stating questions, the staff tend to emphasis challenging cases and unanswered questions, whereas the director emphasises the difficulty of further study and the probability that there is no scientific knowledge to be gained.

Concerning methodology, it appears that the project was unable to identify current challenging cases that warranted truly exhaustive investigations. Nor did the project develop a uniform and systematic procedure for cataloguing the large number of older cases with which they were provided. In drawing conclusions from the study of such a problem, the nature and scope of which are fraught with so much uncertainty, it would have been prudent to avoid theory-dependent arguments."[145]

If Condon wasn't trying to ignore the results from his junior staff, surely he would have acknowledged in his conclusions that 30 percent of cases analysed in the report remained unsolved[146]. Not exactly an insignificant figure to not deserve a closer study. As the prestigious American Institute of Aeronautics and Astronautics UFO Subcommittee said:

"The opposite conclusion could have been drawn from The Condon Report's content; namely, that a phenomenon with such a high ratio of unexplained cases (about 30 percent) should arouse sufficient scientific curiosity to continue its study."[147]

And why were junior staff members doing the investigative work? Extra work experience? Or was this because the work was not considered sufficiently important enough for senior staff to perform and determine if anything could be found from it that could advance science? Or was this to protect the reputation of the university if some independent scientist outside the university suddenly made a discovery

---

145   Sturrock 1987.
146   Hynek & Vallee 1975, p.213.
147   Story 1980.

showing that UFO reports could indeed help to advance science? It could help to protect the senior staff by pointing the finger of blame at the junior staff for not doing first-rate scientific work. Luckily for Dr. Condon, no one found anything significant at the time, so his conclusion still stands.

Far from destroying any belief in the extraterrestrial hypothesis (ETH), the report still left a number of scientists wondering whether it was truly the final answer to the UFO phenomenon. As stated in the 1995 edition of the Encyclopaedia *Britannica*, Volume 12, page 130:

> "This [Report] left a wide-variety of opinions on UFOs. A large fraction of the American public, and a few scientists and engineers, continued to support ETH. A middle group of scientists felt that the possibility of extraterrestrial visitation, however slight, justified continued investigation, and still another group favoured continuing investigations on the grounds that UFO reports are useful in socio-psychological studies."

Even the late Dr. James E. McDonald (1920–1971), a senior physicist at the Institute of Atmospheric Physics and professor in the Department of Meteorology at the University of Arizona, USA, was not impressed by the conclusion. He maintained his view on UFOs until his death in 1971 and said on March 12, 1968:

> "…[L]et me stress that my own studies of the UFO problem have forced me to the conclusion that it is an international scientific problem of potentially enormous importance….
>
> If there is admitted to be even a very slim possibility that UFOs are extraterrestrial surveillance devices of some type (and I incline to that view at present, as do many other serious students of the UFO problem), then it should be obvious that a very energetic scientific investigation of that possibility ought to be launched. Instead, to date, world scientific opinion leans predominantly in the direction that UFOs constitute a 'nonsense problem', a bothersome host of reports of misidentified natural phenomena….Ridicule and official mishandling of the problem have kept the true nature of the UFO evidence well out of sight….

UFOs are, in my opinion, the greatest international scientific problem of our times."[148]

Condon's final words on the UFO controversy, which would echo through the corridors of time, were revealed in Section I of the Condon Report, where he said:

"Our general conclusion is that nothing has come from the study of UFOs in the past twenty-one years that has added to scientific knowledge. Careful consideration of the record as it is available to us leads us to conclude that further extensive study of UFOs probably cannot be justified in the expectation that science will be advanced thereby."[149]

And this was supposed to be an impartial and open-minded scientific investigation into UFOs. As former American astronomer, the late Dr. Josef Allen Hynek, once remarked:

"One does get the feeling that somehow the slate should be wiped clean and the job done over—properly."[150]

The Official End of All U.S. UFO Investigations

If a notable nuclear physicist (but, unfortunately, not a specialist in electromagnetism, much to the delight of the USAF) heading a group of scientists at an American university thought UFOs were nonsense, why would the USAF want to be seen publicly as being linked to any work that might amount to nothing more in the minds of these scientists than a bunch of fairies flying around in the skies of Earth? Either that, or top military brass saw this as an exit strategy from the UFO mess.

As a result of the Condon Report, Secretary of the Air Force, Robert C. Seamens Jr., announced the termination of Project Blue Book on 17 December 1969 and instructed General John D. Ryan, the Air Force Chief of Staff, to end all investigations into UFOs, as he felt

---

148   Phillips 1981, pp.5–7.
149   *Mysteries of the Unknown: The UFO Phenomenon* 1987, pp.118–119; Condon 1969, p.1; Story 1980, p.84; Peebles 1994, p.187.
150   Hynek 1972, p.212.

any continued examination of the evidence could not "be justified either on the grounds of national security or in the interest of science".

Soon after, the U.S. Department of Defense issued a news release[151] on 17 December 1969 stating the official end of all U.S. government-funded UFO investigations. The conclusions were as follows:

"(1) no UFO reported, investigated, and evaluated by the Air Force has ever given any indication of threat to our national security; (2) there has been no evidence submitted or discovered by the Air Force that sightings categorised as 'unidentified' represent technological developments or principles beyond the range of present-day scientific knowledge; and (3) there has been no evidence indicating that sightings categorised as 'unidentified' are extraterrestrial vehicles."[152]

New Public Opinion Polls

Despite the Condon Report and the conclusion of the U.S. Department of Defense, public opinion polls over a 30-year period from 1966 to 1996 showed that around half the American population believed UFOs were real.

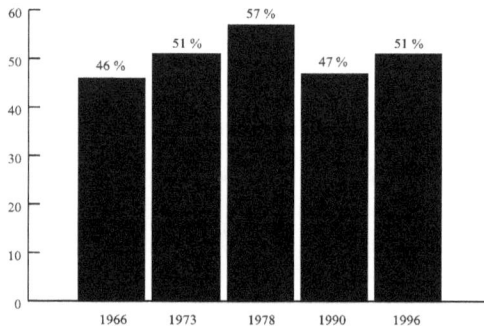

The percentage of respondents who answered "Yes" to the question "Are UFOs real?"

The highest number of respondents who answered in the affirmative was in 1978 with a total of 57 percent[153].

---

151   *Mysteries of the Unknown: The UFO Phenomenon* 1987, p.119.
152   *Mysteries of the Unknown: The UFO Phenomenon* 1987, p.119.
153   Evans & Stacey 1997, pp.238–242 (Statistical data for public opinion polls between 1966 and 1996. Opinion polls asking this question were conducted by George Gallup in 1966, 1973, 1978,

As for the trend of believing in the alien explanation for some UFOs, it had been steadily increasing over the same period. Unless some kind of more pragmatic event on Earth can successfully capture the imagination of the public, such as terrorism or some other issue, people will continue to be open-minded about the alien explanation for UFOs.

As the statistics show, the highest number of people who answered in the affirmative was in 1982.[154]

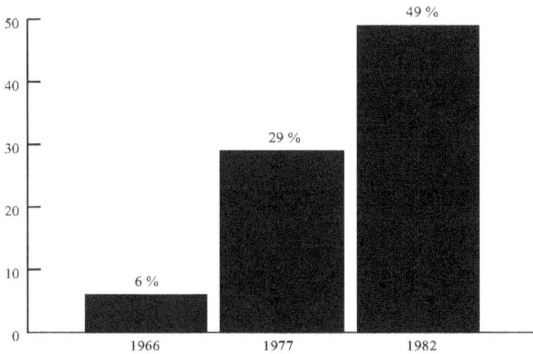

The percentage of respondents who answered "Yes" to the question "Are UFOs alien?"

The U.S. Freedom of Information Act

Before a pragmatic event could be found to keep the public busy (e.g., the Cold War with Russia and the threat of nuclear war, terrorism, and now climate change) and hopefully away from a discussion of anything extraterrestrial flying in the skies of Earth, the study of UFOs by the U.S. government went on well after 1969. This fact, together with claims of secrecy into UFOs first touched upon by Keyhoe's sensational article in *True* magazine, became more evident with the release of official documents from a number of government departments via the Freedom of Information (FoI) Act, which became available to civilian researchers and investigators on American soil after 1975.

The directors of Ground Saucer Watch—Todd Zechel and William H. Spaulding—were the first to pave the way in this field by initiating in

---

154   1987 and 1990; The Roper Organization in 1977; and several international surveys in 1995–96).
Evans & Stacey 1997, pp.238–242 (Statistical data for public opinion polls between 1966 and 1996. Opinion polls asking this question were conducted by George Gallup in 1966; The Roper Organization in 1977; and Audits & Surveys in 1982).

1975 the world's first FoI request for the release of U.S. government's UFO documents. Their attempt was partly successful, with the release of several thousand pages from government holdings in 1976. An example of a U.S. government UFO document released under FoI in 1976 is one written to the FBI director by an FBI agent in Washington, DC, dated March 22, 1950, discussing crashed flying saucers in New Mexico. It stated:

> "An investigator for the Air Forces stated that three so-called flying saucers had been recovered in New Mexico. They were described as being circular in shape with raised centres approximately 50 feet in diameter. Each one was occupied by three bodies of human shape but only 3 feet tall, dressed in metallic cloth of a very fine texture...
>
> ...the saucers were found in New Mexico due to the fact that the Government had a very high-powered radar set-up in that area and it is believed the radar interferes with the controlling mechanism of the saucers."

Further analysis of the available information has shown that ever since 1947, certain clandestine government organisations, such as the FBI and CIA, have written UFO-related memorandums and have kept surveillance on certain individuals involved in the study of UFOs without public awareness. For example, Keyhoe was one individual who was considered a particularly high security risk after his enlightening work on UFOs for *True* magazine and was put on surveillance by the FBI and the CIA[155].

We also cannot ignore other government departments showing interest in UFOs after 1969, such as the National Security Agency (NSA) and the Defense Intelligence Agency (DIA), despite the recommendations of the Condon report and the end of Project Blue Book. These are departments that had actively claimed not to be involved in UFOs.

For example, before the FoI request was put through and processed, a letter dated October 25, 1973 by former FBI director Clarence M. Kelly claimed that the FBI was not investigating UFOs at any time in its history. Then after 1976, over 1,100 pages of UFO-related information were released from the FBI files.

---

155   Randles 1987, pp.33–34.

This is just the tip of the iceberg. The U.S. government has acknowledged in the U.S. courts of law that literally tens of thousands of pages on UFOs still remain hidden away, owing to the alleged risk to national security, whatever that means. But for how long? And what could be so sensitive that the government has to keep the documents secret, if the public was meant to be convinced that UFOs are natural or man-made phenomena?

For example, how can the U.S. government say the planet Venus or aircraft, seen decades earlier, often mistaken for UFOs by the public, are a matter of "national security" and must be kept secret indefinitely? Even if we are dealing with a secret military aircraft, for how long can they continue with this argument? In the case of UFOs, too many people have seen them and a number of witnesses have suffered radiation symptoms and died. Keeping a military aircraft secret in the face of such cases is less likely to fit the "national security" argument. At the very least there is a "health" argument of telling the public what to avoid when observing a secret electromagnetic flying vehicle in the skies of Earth and near the ground. And to continue maintaining secrecy over this type of aircraft seems pointless when the USAF consistently fails to hide the work they are doing with each UFO that gets reported. These so-called UFOs keep getting observed and reported. And so far the objects are not exactly doing anything to affect the "national security" interest of the United States or any other nation. No evidence of weapons being deployed to shoot down civilians. Rather, the most people are seeing of these objects is a quick scooting around in the sky and disappear; or have a quick landing, take a few samples of plants and animal issues, and take off straightaway. Not exactly a sign that people are being invaded and having their lives threaten. Seriously, where is the "national security" argument for these UFOs? The argument quickly diminishes the further time passes and nothing is found to suggest that these objects are dangerous. To continue supporting this position will only work if the U.S. government can keep the UFOs to itself and do all its aircraft testing in secret and away from everyone else so as to avoid radiation poisoning, burns and other electromagnetic side-effects. Unfortunately, UFOs continue to be reported to this day. And people do get affected, even die from exposure to radiation. Surely at some point the USAF have to come clean to explain what it is doing. And if the USAF is not responsible, to come clean by explaining exactly who is flying these electromagnetic

UFOs. The U.S. government might as well tell the public what it has and end the UFO secrecy. Or, at the very least, the military should warn the public of what they need to do to protect themselves from the radiation effects of UFOs. Or else, explain who exactly is sending these UFOs.

But if the authorities are not doing any secret testing of a new flying technology and cannot find an explanation for all UFOs, why retain these documents for many years? What we should be seeing is an annual release of UFO documents to the public, showing that the authorities have nothing to hide and are just as perplexed as the rest of us. But they don't.

And if UFOs really do have a natural or man-made explanation, surely the release of UFO documents would help to confirm this fact. To not do so can only raise suspicion of something far more significant to be found in the UFO reports. But if not, then as an American newspaper headline stated, "If there are no UFOs—why all the secrecy?"[156]

Or is there something else the U.S. government has learned about UFOs through the U.S. military that it does not want to reveal to the public?

We wonder.

Has this anything to do with what happened near Roswell in 1947? And does this have a connection to Brown's electrokinetic devices? Or has the USAF discovered something interesting through UFO observations entirely on its own that could advance science and benefit humanity?

The Files of Project Blue Book are Released to the Public

After this embarrassing release of official UFO documents through the U.S. FoI process, the U.S. government felt it was necessary to immediately sanction the records of Project Blue Book for public scrutiny. The aim was to convince the public that all UFOs were essentially familiar man-made or natural IFOs and, therefore, it was not worth anyone's time looking into the matter. The records show that of the 12,618 sightings investigated by the USAF between 1948 and 1969, 11,917 sightings were identified, while 701 reports were not solved[157],

---

156   Randles 1987, p.169.
157   Sachs 1980, p.346; Encyclopaedia *Britannica* 1995, Volume 12, p.130 (statistical information for Project Blue Book between 1947 and 1969).

most of which were categorized as "insufficient information", whatever that means.

In the meantime, the U.S. government had an ingrained responsibility to educate the general public and media into accepting all UFOs as being IFOs (i.e., of a non-alien nature). Either it was trying to protect the public of something alien in the UFO reports that it knew about but could not say (as if there could be some truth), or it felt that it simply needed to inform the public of the more prosaic explanations for UFOs. This seems to have succeeded to a certain extent, given the reduction in worldwide media coverage on UFOs. Alas, good riddance to UFOs.

The only slight problem for the U.S. government was the dedicated efforts of some UFO researchers and investigators to continue probing the subject and uncover more insights. How could the authorities deal with these people?

Civilian UFO Researchers and Investigators

The interest in UFOs would not die down completely, much to the disappointment of those individuals in the top echelon of the Pentagon in Washington, DC and anyone else who wanted to see UFOs disappear quicker than a funny smell in the wind. The public would continue to report UFOs, and UFO researchers and investigators would take notice and continue to work on these reports privately and independently.

There were efforts by some skeptics to argue that there is nothing to study in the UFOs. All should have a natural or man-made explanation. However, the UFO researchers and investigators counter this by saying there is a build-up of an impressive list of common UFO observations that no rational and reasonable person with an open mind can ignore. While many UFOs can be explained as natural or man-made phenomena, there continues to be a growing number of highly detailed, close-range observations of UFOs from rational people that simply cannot go away. These include metallic discs with portholes, landing legs, glowing effects at specific positions on the objects, and the glows and their colours affecting the proper operation of electrical circuits inside automobiles and people's homes, not to mention humanoid occupants revealing themselves to witnesses. And how can one ignore the radiation poisoning effects as well, which lead to long-term health problems or even death for some witnesses? It looks like we are dealing with real flying objects, but they do not look like any familiar natural or man-made phenomena.

More remarkably, these genuine UFO cases are not restricted to the United States. They are a worldwide phenomenon. As former French minister of defence M. Robert Galley said to Jean-Claude Bourret, in an interview broadcast on radio station *France-Inter* on February 21, 1974:

> "There is a steady accumulation of sightings of luminous phenomena that are sometimes spherical, sometimes ovoid, and which are characterised by extraordinarily rapid movement.
> …I will indeed go so far as to say that it is irrefutable that there are things today that are inexplicable, or poorly explained….I must say that if your listeners could see for themselves the mass of reports coming in from the airborne gendarmerie, from the mobile gendarmerie, and from the gendarmerie charged with the job of conducting investigations…then they would see that it is all pretty disturbing…"[158]

No doubt these UFOs will continue to interest dedicated UFO researchers and investigators trying to get to the truth of the matter. With the way things are going, it is unlikely UFOs will disappear any time soon.

It seems the U.S. government will have to find something to keep the UFO researchers and investigators busy, or else infiltrate or become one of these people and feed these more curious people with different information.

The USAF Still Unable to Build an Electromagnetic Flying Saucer by the 1970s?

An interesting USAF report has been declassified showing the state of the USAF's technological understanding of any type of electromagnetic flying machine by the early 1970s. The report is titled *Advanced Propulsion Concepts: Project Outgrowth*, edited by Franklin B. Mead, Jr. The

---

158   Randles 1987, pp.180–181; Brookesmith 1984, pp.62–63; Blundell & Board 1984, p.184.

people who conducted the two-year study worked at the Air Force Rocket Propulsion Laboratory at Edwards AFB in California. According to the report:

"The study of advanced [propulsion] concepts was initiated and directed by Mr. Donald M. Ross, then Deputy Director of the Air Force Rocket Propulsion Laboratory."[159]

When the report was released in June 1972, it mentioned a range of new propulsion systems that could be developed in the next 30 to 40 years. Among the systems considered was "electrical propulsion", of which electromagnetic/gravitational field propulsion and photon propulsion were mentioned under Part II for non-chemical propulsion in chapters 2 and 3, respectively.

In the area of field propulsion, we learn under the section titled "Electromagnetic Spacecraft Propulsion" how Air Force researchers looked at magnetic field propulsion using Earth's magnetic field, and later interplanetary and interstellar magnetic fields when interacting with the magnetic field generated by a solenoid coil. The solenoid coil would contain a large superconducting magnet, with cryogenic cooling of the magnet easily achieved "because of reduced convection and the low blackbody temperature of deep space".[160]

The superconducting magnet would also act as an efficient energy storage system and not just as a generator of a powerful magnetic field. Of course, all energy has to be initially generated from some source. The Air Force researchers recommended "solar energy or a small isotope reactor" to create the energy. Once the magnetic field is generated, the coil effectively becomes a magnetic dipole (just like a bar magnet) with a north magnetic pole at one end and a south magnetic pole at the other. The result of the interaction of the coil's magnetic field with Earth's magnetic field will be like having two bar magnets with like poles facing each other, resulting in the coil experiencing a repulsive magnetic force.

In another field propulsion concept, Air Force researchers looked at the possibility of using magnetic fields ("the engine") to accelerate superconducting and charged particles ("the fuel") for propulsion. This

---

159   Mead 1972, p.v/vi.
160   Mead 1972, p.II-117.

is described in the report under "Superconducting Particle Accelerator" on page II–130.

The 1972 USAF report on advanced new propulsion concepts

The Air Force researchers also looked at a more advanced concept known as an anti-gravity propulsion. In this system, the researchers suggested that there could be a way to control the gravitational field through two methods:

"One utilises a new physical concept of gravitational absorption (gravity screens), and the second is based on the concept of a unified field theory (using electromagnetic analogies to gravity control)."

Of course, the biggest problem researchers had with the latter method was working out precisely what a gravitational field was, and the relationship between the gravitational field and the electromagnetic field. For if there is a relationship, it may be possible to control the gravitational field through an electromagnetic approach and hence lead to the ability to make a flying object levitate, and later move through the air or in space with minimal energy requirements while using another form of propulsion. But without a full understanding of Einstein's Unified Field Theory, researchers could only speculate as to how this might work.

On the gravitational screening method, researchers noted an experiment where a lead sphere was introduced into a hollow sphere of mercury (also a metal) and somehow the weight of the lead sphere reduced slightly. As the researchers put it:

"An experiment was conducted to determine whether a barrier could effectively reduce the gravitational attraction of the earth on some body. During this experiment an effort was made to determine the gravity absorption coefficient. In the test a lead sphere was introduced into a hollow sphere of mercury and the weight of the lead sphere decreased by 10^{-6} grams, which is equivalent of liberating twenty million calories of gravitational energy....Many attempts were made to refute this test, but none have been completely successful."[161]

Perhaps this interesting result depends on the perfect symmetry of the mercury sphere for the gravitational field screening effect to be effective. But until we can discover the inner workings of the

---

161   Mead 1972, p.II–138.

gravitational field and which materials act as effective gravitational field barriers (it seems like metals are important), there is still much work to be done here.

In chapter 3 of the report, we learn of the Air Force researchers' attempts to look at various photon propulsion systems. One involves creating antimatter by some means and letting it combine with ordinary matter, resulting in a massive release of photon energy that is then directed out of the exhaust (looking like a powerful "light beam"). The photons effectively cause the spacecraft to recoil in the opposite direction.

The other photon propulsion concept discussed uses solar energy to move a spacecraft with the help of a massive metallic sail attached to the spacecraft. This then grabs enough photons from the sun to create sufficient electromagnetic force to propel the spacecraft at a reasonable speed.

Beyond that, no other electromagnetic propulsion concepts were mentioned in the report, suggesting the USAF had nothing on the table to implement straightaway in a real-life electromagnetic flying object, which would have shown how it could have been responsible for the UFOs reported by witnesses, especially where electromagnetic observations were mentioned, such as light beams and oscillating electromagnetic fields. Not even a tiniest mention of the work by USAF researchers to study the radiation emission from a charged surface at the end of the 1950s. Why was this overlooked in this latest study?

The report is really a snapshot of where the USAF was at in its understanding of electromagnetic propulsion systems at the time the report was released. Essentially, the USAF had nothing available by the beginning of the 1970s to show it could have been responsible for the UFOs.[162]

The Emergence of Fake U.S. UFO Government Documents

During the 1980s, interest in the Roswell case among UFO researchers and investigators grew thanks to the release of Charles Berlitz and

---

162   One should bear in mind that the people who wrote this report were unaware of super-secret studies taking place in the USAF by those who have access to the Roswell materials and recovered disc. Like Project Blue Book, one can imagine this declassified study into innovative and upcoming propulsion systems was a cover to a much more secret study into UFOs taking place at another location.

William Moore's *The Roswell Incident*. This book would bring to light the secret USAF recovery operation, the existence of bodies, the unusually lightweight and tough nature of the materials, and the first official publishing of statements revealing a shape-memory effect in the Roswell foil.

As if to counter this unexpected and apparently sensitive leak of UFO information to the public and to deal with the civilian UFO investigators and researchers involved in the Roswell case, two mysterious government documents purporting to be in support of the Roswell case were suddenly released to the public.

On 11 December 1984, TV producer Jaime Shandera received the documents from an anonymous source allegedly working within the U.S. government. Known as the Majestic-12 (MJ-12) documents[163], they included an alleged top-level summary of the Roswell UFO crash during the early stages of recovery, with presumed involvement by U.S. Presidents Truman and Eisenhower. These documents would keep researchers and investigators busy for quite some time until eventually someone noticed something wasn't right and figured out they were, in fact, elaborate fakes.

The documents have been established as fakes. It turns out that someone had gone to the considerable trouble of preparing the documents, but wanted to make sure people could see they were typed using a Smith Corona typewriter, which was not manufactured until 1963, while these documents were dated September 24, 1947 and November 18, 1952. Furthermore, someone was careless enough to copy a president's signature from another source without putting in enough differences. Whoever created the documents had wanted to make sure someone would discover the mistakes and, therefore, claim the documents are a fake.

If someone from the U.S. government was making a special effort to reduce the credibility of crashed-disc cases and, by implication, UFO research as a whole in the United States and abroad—and to keep scientists and the media away from the subject—they were doing a particularly good job.

---

163   Majestic 12 may also be named as Majic 12, MJ-12, M12, MJ XII, Majority 12, or even the more obscure Mars–Jupiter 12. It should be noted that even though certain leaked MJ-12 documents are considered fakes, it does not necessarily mean MJ-12 never existed or could not represent a genuine group of people who were aware of the UFO situation and possibly have had knowledge of the events near Roswell in early July 1947.

At any rate, when one reviews these documents, it is interesting to see how the idea of a crashed UFO keeps cropping up in them[164]. Seriously, why bother to put in the effort to provide sophisticated fake documents on the crashed-disc scene if the whole thing is meant to be bogus? If UFOs are IFOs and crashed discs are figments of our imagination, it is a waste of time and taxpayers' money trying to put researchers and investigators off the track with documents designed to have flaws and hence be discovered as fakes. Let UFO researchers and investigators study as they feel is necessary and leave it as that for all time.

Whoever is sending the documents through to the public is only succeeding in making some people wonder if there might be some truth to the claims.

Fake Alien Photographs and Films

Things got more interesting when fake footage of alien autopsies and fake alien photographs from crashed discs started to appear in the late 1980s and 1990s. Before then, two fake alien photographs were quietly circulated in the UFO community. They eventually turned out to be an April Fools' joke.

  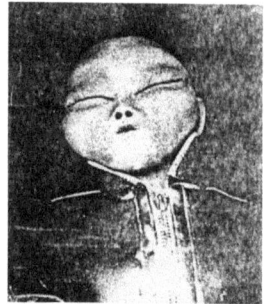

A   B   C

---

164   Even the idea of a secret UFO study group naming itself as MJ–12 should not be totally dismissed. Some government scientists have testified in support of the existence of such a group. Furthermore we find the group was led by a top electromagnetic engineer named Dr Vannevar Bush, which would make sense if UFOs do represent some kind of a new electromagnetic technology. If one wanted to fake the MJ–12 documents, it is incredible the name of the scientist and his expertise had not been changed to someone who was an expert in nuclear technology to really put people off the track. But incredibly the censors within the U.S. government didn't do this. Thus the only problem we have is the considerable effort by someone within the U.S. government to discredit the existence of the MJ–12 group as well as crashed discs and UFOs in general through a couple of carefully incorporated changes to the MJ–12 documents. The question is, why?

The photograph shown in A was an April Fools' Day joke created when William Sprunkel, editor of the German newspaper *Wiesbadener Tagblatt,* asked photographer Hans Scheffler to set up the scene. The photographer used his 5-year-old son to stand between two "MPs" as a template and later painted over the picture of his son to reveal what looks like an alien. The photograph was later photocopied and kept in FBI files since May 24, 1950, until it was released under U.S. FoI in the late 1970s.

Another fake alien photograph that was circulated at the time can be seen in B, showing what is believed to be a small monkey from around the same time period as the previous example.

Yet the fake alien photographs would make their comeback in the 1980s, and later as a sophisticated film in the late 1990s all presumably inspired by the events near Roswell.

Shown in C is a slightly more sophisticated photo of a dead alien that was allegedly retrieved by the U.S. military. The photo was circulated in the 1980s among UFO researchers. It has now been discovered to be a fake.

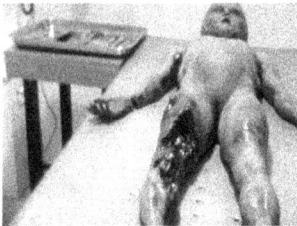

Not to be outdone, the fake alien photographers really got creative when, yet again, another anonymous U.S. military source decided to release the infamous alien autopsy film (see left). In over an hour and a half, the film allegedly showed the autopsy performed on an alien. Ray Santilli of Merlin Films, a London rock video entrepreneur, claims to have bought the film off a retired U.S. military cameraman. Considerable media attention was given to the film when it first appeared publicly before it was determined to be a fake.

Yet it never ceases to amaze how the same ideas keep cropping up: the U.S. military allegedly recovers a crashed disc, and the dead aliens get examined or photographed.

While all these photographs and movies are fakes, could the ideas themselves be possible and even probable in reality? Perhaps someone in the U.S. military is "testing the waters", so to speak, just to see the public's reaction and to decide when to release the real evidence? Or, more likely, someone is overdoing it too much by releasing fake evidence in the hope of putting scientists off the track and keeping them away from UFOs.

A Fake U.S. Government Employee at Area 51?

Still, UFO researchers and investigators remained fervid in their search for the truth, and the public was just as curious about the subject as ever before. As recent Gallup Polls show, more and more people are becoming convinced that UFOs are real and likely to be extraterrestrial in origin.

Then after 1989, the area of interest shifted to an isolated military facility in the Nevada desert known as Area 51, where testing of highly secret and unusual-looking symmetrical aircraft, including strange glowing objects flying over the area at night, was seen by American civilian observers at a distance.

In what seems to have been yet another fascinating attempt to keep the curious civilians, media and scientists away from Area 51, a man by the name of Robert Scott "Bob" Lazar approached a UFO researcher and revealed something he had allegedly witnessed while working in Area 51.[165]

According to Lazar, he worked from 1988 to 1989 as a physicist at a hidden U.S. military location in Sector 4 (S-4), allegedly located near Papoose Lake southwest of Area 51. He understood Area 51 was the place where the military would secretly test unusual aircraft, whereas S-4 served as the place to reverse-engineer flying discs recovered from previous crashed disc cases.

Leaving aside S-4 for a moment, testing of unusual aircraft is not entirely a bombshell statement considering the USAF is in the business of flying all sorts of machines. The more important questions we should be asking are whether the USAF is responsible for the menagerie of unusual UFOs sighted over the decades in many countries, and if so, how long it has been involved? If not for very long, when did the USAF find out something interesting about UFOs, and how did they figure it out? And who was sending up these UFOs for people to observe prior to the USAF discovering something about them? On the other hand, if the USAF had already known about UFOs for a long time (presumably before 1947) and is, in fact, responsible for

---

165   Area 51 is also known as Nellis AFB. Declassified documents released in August 2013 (which can be obtained from the National Security Archive at the George Washington University via http://www2.gwu.edu/~nsarchiv/NSAEBB/NSAEBB434/) prove the U.S. government is aware of the existence of the air base and the name Area 51. Originally built at the end of World War II, it was used to test the high-altitude spy plane known as the U-2. Today, the isolation of the air base makes it ideal for the USAF to conduct further secret flights of new types of flying objects, a number of which have similar characteristics to UFOs, including symmetrical designs that glow at night.

sending these objects into the air, what is the USAF testing at the present time? Clearly the UFOs are flying around perfectly fine, so what more is there to learn?

Whatever the USAF is trying to figure out (and still cannot get right to this day), it seems reasonable to assume that some tests on symmetrical aircraft designs, probably based on electromagnetic concepts, could have been taking place at Area 51 in the late 1980s and early 1990s.

Because of these tests, Lazar initially and quite logically assumed that any symmetrical flying objects of the type perceived as flying saucers by the media and public were probably inventions of the U.S. military. This seems reasonable enough so far. But after he was allegedly shown nine different types of symmetrical flying objects called discs and multiple briefing documents at S-4, together with an opportunity to have a closer examination of the discs (why he would be given access to such information and the discs themselves is beyond anyone's comprehension), he concluded that the flying saucers are extraterrestrial and that the U.S. military had captured at least one and was trying to reverse-engineer the technology (it seems to have been largely unsuccessful, given how long the work has been going on— unless, of course, someone has figured it out and the witnesses' claims of flying lights in the vicinity of this secret location is evidence of successful testing, in which case everything Lazar might say now should be taken with a grain of salt).

Flying discs? There might be some truth to this, considering we are trying to find out whether those close-up reports of symmetrical flying objects are part of some man-made experiment.

And given the sudden change in the military's behaviour toward UFOs after 1947, Lazar might have been right in assuming these discs had some kind of extraterrestrial connection. It would change anyone's attitude in a jiffy if one discovered the evidence to support the alien presence.

However, Lazar then explained the propulsion system of the discs. It is here we find that he overly complicated what was presumed to be the science of these discs and made it virtually untestable. More surprisingly, he would rely on nuclear concepts to explain the essence of the technology. It sounds like things have not progressed since the days when General Twining allegedly wrote his July 16, 1947 document on the construction of one of these discs; it makes you wonder why the USAF even bother with hiding the evidence, if it cannot, to this day, see how UFOs work.

According to Lazar, the propulsion system is based on the use of a reactor (the exact term used in Twining's document that examined the recovered disc) containing an extremely rare and difficult-to-produce atomic element. Lying at 115 on the periodic table, the element called *ununpentium* is believed to be important to the propulsion system and for running the instruments on board. How? He talks about anti-matter being created when the element is bombarded with tiny positively-charged particles known as protons, which when combined with ordinary matter release a tremendous amount of electromagnetic energy. This is then converted into electricity for use in powering the instruments aboard the discs.

Not the propulsion system? How odd.

Well, yes, scientists do know that a considerable amount of energy is produced when anti-matter and matter are combined. However, why is so much energy needed to power the instrumentation panel? Couldn't this energy be derived from the environment by using a metal antenna as it passes through the natural magnetic fields of space? Better still, why not just use an ordinary battery to power the instruments? Nowadays we have rechargeable lithium ion batteries that can easily power a compact instrumentation panel. Surely something so advanced would not need a lot of power from an impossible-to-manufacture element just to power the instruments.

But how do the discs lift off the ground and move?

Lazar says there is another purpose to this reactor. He claims that the bombardment of the element with protons also creates and amplifies a nuclear force field to the point of being able to distort and bend the gravitational field around it. Once the field is bent, the disc can levitate, and it becomes relatively easy to accelerate the disc to very high speeds.

So how does it actually move? We don't know. Lazar hasn't explained, so we are none the wiser. The only thing Lazar added, and presumably obtained from the documents he read, is that the object can allegedly achieve infinite speeds (or certainly well beyond the speed of light). Yes, that might explain how UFO occupants might be seen in association with these flying discs. Fly fast enough, and it is scientifically feasible for a biological entity to participate in the flight and reach the stars.

Already we see two problems in what Lazar has said.

The first problem is that element 115 cannot be created by any known means (i.e., it is too unstable), making Lazar's claims about creating a nuclear force field untestable.

According to the scientific literature, it is understood that element 115 has never been produced by scientists. Efforts in 1999 by scientists at the Lawrence Berkeley National Laboratory to create heavy atomic elements, despite the extremely short-lived nature of these elements, were met with limited success. According to BBC News online science editor Dr. David Whitehouse in his article "Superheavy Elements Created" (published on June 9, 1999), it is possible to create a few atoms of element 118, although they quickly decay into 116, and after a slightly longer period of time these decay to element 114 and other lighter atomic elements. But never in scientific history have scientists been able to create element 115. Either Lazar is not correct in his claim, or it makes Lazar's claim of alien discs sound plausible. But even if it is alien, there is the question of whether Lazar is choosing an element that can never be produced by humans as a way to convince independent researchers it is too hard to figure out the technology, so why pursue UFO research? As they say, anything alien must be impossible to re-create or build by humans. This is the familiar motto for many UFO investigators and Hollywood producers of science fiction films, and as we shall see later, the view of the USAF too. How can mere human mortals ever work out an alien technology? Our brain is too feeble and small to work it out. Just accept the technology is too difficult to understand, let alone for us to create element 115 because it requires an alien technology to make it. No need to further investigate UFOs. As a counter-argument, if something is so beyond our comprehension, and virtually indistinguishable from magic, as one science fiction writer has said, wouldn't this be more of a reason for the USAF to stop hiding the evidence? The USAF cannot work it out. No one can. Surely, it would make no difference if the evidence is given to the scientific community and the public. Might as well admire the technology, or else increase the probability of someone in the world learning something from it by getting people to notice the alien technology. The more people who become aware of it, the better the chances of someone figuring it out, right?

Unfortunately, the USAF does not see it this way. By keeping the evidence secret to this day, the military is effectively saying that humans are too stupid to figure it out. Either that, or the USAF really has discovered something and is growing in confidence after learning about the real technology behind UFOs, but believes it is better to remain tight-lipped on the work done.

Of course, we will have a closer look at the statement of just how incapable we are as humans in understanding the potential alien technology in greater detail later in this book, just to see how true this is.

However, a more serious problem concerns the fact that no solid matter (i.e., containing a multitude of atoms arranged in a crystalline structure), no matter how clever the supposed technology might be for the discs or how much energy the atoms can release, can ever achieve speeds exceeding the speed of light unless there is a way to reduce the energy density of space. Maybe this is what Lazar means by bending the gravitational field. But according to the Unified Field Theory, a region devoid of a gravitational field means there is absolutely no electromagnetic field in that region as well. In other words, to bend the gravitational field means to bend the electromagnetic field. In essence, we are talking about a perfect vacuum devoid of radiation. In this hypothetical "mathematically predicted" vacuum, any object can levitate perfectly without any energy input, and it can be accelerated to infinite speeds with ease[166]. In fact, you would not feel inertial forces when you accelerated, because there would be no radiation and hence no gravitational field to penetrate your body. Unfortunately, the real universe we live in will not allow a region of space to become a perfect vacuum, no matter how hard we, or any alien civilisation, can try. It is an impossibility. No amount of energy in the universe can ever create a perfect vacuum. Not even the most advanced alien civilisations in the grander Universe (going beyond the visible universe) will have the technology to achieve this.

You can also include in that picture the biggest black hole in the universe. No black hole can ever create a perfect vacuum around itself. The presence of any kind of perfect vacuum would see the black hole reverse its sucking and be broken apart into pure electromagnetic energy. There is a minimum energy density of space required and that

---

166   In the article titled "Light hits near infinite speed in silver-coated glass" in *New Scientist*, published on January 9, 2013 (downloadable from http://www.newscientist.com/article/dn23050-light-hits-near-infinite-speed-in-silvercoated-glass.html), light and any solid matter can potentially be accelerated to infinite speed in a zero energy density environment.

is the real vacuum we see above the Earth's atmosphere. A zero-energy density region is truly in the realms of fantasy, and certainly not science fiction (where some ideas might possibly become a reality, given enough time).

Thus, to suggest that perfect levitation and infinite speeds is possible by bending the gravitational field to create a perfect vacuum, as mentioned by Lazar, is mathematically correct, but fundamentally flawed in a real-life sense. The real universe simply does not allow this to happen.

However, there is another odd thing about Lazar's claims. UFOs have been seen since 1947 and for a lot longer than that. These UFOs have been reported as revealing electromagnetic effects, such as light beams and electrical disruptions to man-made technology. Knowing these observations have existed for such a long time, Lazar is still unable to explain all the common observations of genuine UFOs seen at close range by witnesses, despite his reading of many briefing documents about how these discs work (and one would think he may have supplemented his knowledge with a read of some good UFO books to see if he could come up with more clues as to how the objects might work). He offers no point-by-point, detailed explanation for simple things like the glowing surfaces of UFOs (surely there must be some kind of electric charge on the surface similar to what you find in an incandescent light bulb, so what is this charge meant to do?); the light beams beneath the UFOs (surely they cannot be spotlights since they are shot vertically downward as the object levitates or moves through the air); the design of UFOs as symmetrical metallic boxes (what's wrong with the traditional aircraft shape with wings, or any shape for that matter? Or is there a specific reason for this symmetry in the central body of the UFO?); the importance of lightweight materials used to construct the discs, if we are to rely on crashed disc reports (why hasn't Lazar picked up on this lightweight construction of crashed disc cases from the documents he allegedly inspected, as this would reveal how fast the discs really do travel—is it at infinite speeds, or close to the speed of light?); and so on. And why manufacture this nuclear element for the engine? Surely there must be other easier ways to generate electrical power and achieve propulsion. Even if one could travel at nearly the speed of light, the ones participating in the flight can enjoy very short journey times. So why is it necessary to have a perfect vacuum?

Furthermore, with all the work done on the Unified Field Theory by the U.S. government in the 1950s, right up to 1974, one would think that, by now, various other solutions would have been found to bend the gravitational field and make an object levitate. Or how about adding more gravitational (or electromagnetic) fields between the ground and the UFO, as we see with the light beams? As this invisible mass gets emitted by the UFO, it would effectively help to recoil the UFO upward (and hence defy gravity) just like we see with the exhaust gases coming out the back end of a rocket.

This is the problem. Lazar is unable to talk about these developments in the Unified Field Theory and explain the UFO observations, such as light beams, based on his understanding of UFO technology (man-made or otherwise) after looking at the secret documents.

In fact, Lazar doesn't even provide unequivocal examples of how science can be advanced from his inspection of the alleged discs and documents at his secret location near Area 51, other than claiming that the bombardment of the nucleus of element 115 would produce vast amounts of energy and amplify a nuclear force field around the element and eventually the disc to help bend the gravitational field and so achieve levitation. As the claim is virtually untestable, scientists cannot do anything to check it.

But there is something else that puts the final nail in the coffin of Lazar's bizarre claims. If the discs could be made to completely nullify their own mass by bending the gravitational field around them, then technically it shouldn't matter how heavy these discs are. Even the occupants themselves could be incredibly overweight or extremely large and still be able to travel in one of these discs to any destination in the universe. However, the strange thing is, UFO occupants have always been observed to be very thin and usually quite small. No researcher or investigator has yet seen a UFO occupant that one could describe as "fat". Furthermore, the materials used to construct the crashed discs are described as incredibly lightweight (and strong). Why lightweight? Surely it shouldn't matter how lightweight these materials are if Lazar's claims are true, but whoever constructed these discs and those operating them seems to have been greatly concerned about mass and wanted to keep it to a minimum. There is no point in reducing the mass unless the aim is to get as close to the speed of light as possible for the occupants inside the discs to benefit from short journey times. But this would imply that the discs cannot exceed the speed of light and, consequently, no bending of the gravitational field to create a perfect vacuum in space.

Just to rub it in a little more, we also have noticed in history that people who were close to discovering the truth suddenly ended up dead. If Lazar had leaked highly sensitive information to the public about UFOs, surely he would have been taken out by now, and the authorities could make up an excuse claiming his death was due to natural causes (or suicide). If not, Lazar should have been, at the very least, immediately charged with a federal crime of treason for leaking highly classified information to the public. For some reason, he hasn't. Very strange.

This suggests that Lazar either was set up by the people at Area 51 to present erroneous claims to the public, was never there at all, or is part of the group at Area 51 wanting to provide disinformation to the public to keep people away.

The Mysterious Men-in-Black (MIBs)

If all these fakes still did not have the desired effect for the USAF or whoever is responsible to convince the public UFOs are IFOs as well as to keep researchers and scientists away from the subject, there is, of course, always the option of sending in secret and rather sinister-looking government or military agents known as Men-in-Black (or MIBs).

In the late 1950s and early 1960s, MIBs were known to harass UFO witnesses and researchers to keep quiet or convince them that UFOs have a natural explanation. In the 1990s, the MIBs made way for phantom social workers. And since 2008, certain executive search consultants or contract officers working at Battelle or with links to U.S. companies having contracts with Battelle or the USAF may be used to put certain targeted UFO researchers into contract work not related to UFOs (i.e., more "pragmatic" areas).[167]

Seriously, with so many fake alien photographs and footage designed to keep journalists, scientists, researchers and investigators wondering for a while until they are revealed to be fakes, and with MIBs trying to convince people there isn't anything to worry about, it is almost as if

---

167   SUNRISE has noticed this method with a search consultant based in Boston with alleged interests in UFOs and the Roswell case, and a MUFON director who currently works at Battelle.

there *is* something to worry about and someone in the know is trying to keep the curious public and scientists away from UFOs and crashed disc cases. We can only ask—why? And by whom? Indeed, how would this serve the interests of U.S. authorities if UFOs are meant to be familiar IFOs?

Or are these efforts more of an admission that UFOs are real and represent something new to science?

If not, wouldn't it be easier for whoever is doing this to let people believe whatever they wish to believe by doing nothing? OK—UFOs are meant to be IFOs. Fine. End of story. Just let the public believe whatever they wish. Clearly no harm done, and it certainly keeps people out of mischief. Also, you cannot say that people studying UFOs are terrorists, or are interfering with the functioning of the current economic system. The system continues unabated, and some people can continue to get rich and powerful. No problem there. People believing in UFOs as alien shouldn't bother anyone. Let them believe what they like and be done with it. There is nothing to learn about UFOs, so why use disinformation and go to the considerable effort to send mysterious individuals to convince others of this fact as the USAF wants the public to think?

There is no reason, unless, of course, the effort is meant specifically to stop people from getting closer to the truth. Except that would imply that UFOs are real, possibly alien and that one of these UFOs has crashed to earth and is now in the possession of the U.S. military. Or can UFOs actually advance science? It is one or the other. So which is it?

The British Government Releases Its Own UFO Documents

If the disinformation campaign in the United States cannot stop UFO researchers overseas studying the subject, why not keep them busy with an avalanche of simple and ordinary UFO reports from certain other nations? This appears to have been the next logical step, with the British counterparts deciding it was time to release their own large collection of UFO reports to the public after 2003. The aim of the release was to support their U.S. counterparts with the perception that UFOs are IFOs.

On May 15, 2006, the UK Ministry of Defence released a 400-page report titled *Unidentified Aerial Phenomena in the UK,* supporting an official announcement on May 7, 2006 that UFOs have a natural or man-made explanation. The four-year study was formerly conducted in secrecy and originally stamped "Secret: UK Eyes Only" when completed in December 2000, until Sheffield Hallam University academic Dr. David Clarke applied the FoI laws in 2005 to release the work, asking for details of the MoD's plans for "dealing with the arrival of extraterrestrials".

The author of the report remains protected. However, an MoD spokesman said:

> "While we remain open-minded, to date the MoD knows of no evidence which substantiates the existence of these alleged phenomena and therefore has no plans for dealing with such a situation. Both this study and the original 'Flying Saucer Working Party' concluded that there is insufficient evidence to indicate the presence of any genuine unidentified aerial phenomena.
>
> It is unlikely that we would carry out any future studies unless such evidence were to emerge."[168]

As the report concluded:

> "No evidence exists to suggest that the phenomena seen are hostile or under any type of control, other than that of natural physical forces. There is no evidence that 'solid' objects exist which could cause a collision hazard.
>
> Evidence suggests that meteors and their well-known effects, and possibly some other less-known effects, are responsible for some unidentified aerial phenomena.
>
> Considerable evidence exists to support the thesis that the events are almost certainly attributable to physical, electrical and magnetic phenomena in the atmosphere, mesosphere and ionosphere.
>
> They appear to originate due to more than one set of weather and electrically-charged conditions and are observed so infrequently as to make them unique to the majority of observers."

---

168   http://www.mod.uk/ (Quote from official UK MoD report released after May 15, 2006.)

As for electromagnetic effects, such as the electrical stalling of vehicles, glowing effects and radiation hazards—not to mention cases of abductions in the presence of UFOs—the report gave a possible explanation:

> "The close proximity of plasma related fields can adversely affect a vehicle or person.
>
> Local fields of this type have been medically proven to cause responses in the temporal lobes of the human brain. These result in the observer sustaining (and later describing and retaining) his or her own vivid, but mainly incorrect, description of what is experienced."

To put it simply, the UK spokesman tried to mirror the results of the Condon report in 1969 and the preferred position of the USAF in Project Blue Book by getting the public to believe UFOs are familiar IFOs, and nothing in the reports would advance science if further study was conducted, especially in the field of electromagnetism as others have attempted.

If this is true, why the continued secrecy? It does not make sense. It is pointless for the U.S. military and intelligence agencies to continue to retain behind closed doors tens of thousands of pages on UFOs to this day. Surely there is no need to if the position of the UK authorities on their UFO reports is anything to go by and UFOs are meant to be IFOs. However, the American officials do not seem willing to release the information.

Something does not add up.

The U.S. Military Tests a New Electromagnetic Vehicle

If UFOs are meant to be IFOs, this is not borne out by certain important observations reported in the U.S.

A classic example is the Cash-Landrum UFO incident in Texas in 1980. Here we have excellent physical evidence for the existence of UFOs, including electromagnetic effects, such as induced electrical currents in the body of the car to create intense heat and the emission of ionising X-rays causing radiation poisoning for the witnesses involved. More importantly, the case shows that someone in the U.S. military was involved in the testing of a glowing and symmetrical flying object, including silent operation and emitting radiation—all the classic signs of a UFO. The only problem is, the USAF has not claimed responsibility for this object. Does this mean we genuinely do have a UFO worthy of closer investigation and study for the scientific community? And is this UFO alien? Well, let us put it this way. If no other nation has come forth to claim responsibility for the UFO that left radiation effects on three American witnesses in 1980, and the USAF wishes to deny involvement, then who else can one point the finger at? Father Christmas?

However, we are told UFOs are IFOs, right? Well, at least that is the impression we are getting from the U.S. (and recently the UK) authorities. It is an IFO that somehow delivers radiation poisoning to witnesses.

Makes one wonder what kind of IFO this is.

Or is Santa's sleigh a nuclear-powered flying machine?

Alright, let us assume we do have an IFO in our midst. Clearly the Cash-Landrum case tells us it has to be a man-made and new experimental military aircraft that utilises, for some reason, electric charge on its surface to help explain the electron emissions and subsequent radiation exposure of the witnesses standing outside the car. Furthermore, the USAF must be involved in this experiment, as flying objects are right up its alley. We know the USAF was interested in electromagnetism after 1947. However, this object is not a familiar IFO that the public can readily recognise (unless they read UFO books, but still remain at a loss as to what type of aircraft this is). Certainly the public does not know what to do in the event of encountering one of these flying objects. Yet here we have the USAF wanting the public to believe it is an IFO and definitely not something alien. Okay, so it is an IFO because the USAF knows what this object is. It knows because it has built it. Maybe not the one in 1980 that causes witnesses to suffer radiation exposure. That one must be alien and was escorted out of the area by another military group (the Army?). But whatever the USAF is doing at Area 51 is considered an IFO because they are making and testing them.

Fair enough.

Well, if that is the case, how about the military explains what this object is so that the whole UFO controversy can end immediately? If nothing else, do it for the sake of protecting the health of the public. See it as a health campaign and show exactly what the public needs to do to protecting itself from the radiation effects. That way the public can be prepared and know what to do to avoid the electromagnetic hazards associated with this so-called IFO. Seriously, the USAF cannot shrug off all responsibility for the UFO in question or hide behind a veil of secrecy. People continue to report UFOs. If all those UFOs are IFOs build by the USAF, the military is already failing to keep it secret. And if people's health is being affected, you might as well explain the hazards and what to look out for. After all, the USAF should be there to protect the citizens of the United States unless it doesn't care because it has complete autonomy to do whatever it likes and not be affected by any legislation. It wants the public to believe UFOs are IFOs, in which case we must have a military experiment on a new flying object. We know it is based on an electromagnetic principle that requires electric charge to accumulate on the surface of the flying object. And the medical evidence is overwhelming to say the least. Furthermore, we have historically seen the USAF's interest in learning everything it could about some advanced electromagnetic concepts soon after the Roswell event. The USAF must know precisely what this IFO is and how it works. Clearly the USAF has manufactured an electromagnetic flying vehicle. It is a vehicle that produces electromagnetic effects on the witnesses and the environment. While witnesses get affected by radiation poisoning, this makes for a more compelling reason why the USAF needs to explain what this new type of secret military IFO actually is. If the U.S. government refuses to release the remaining UFO documents under FoI, then you might as well provide a public announcement on the technology the USAF has developed and explain the hazards. Then give clear advice on how people can protect themselves.

It is either one or the other.

No matter how secret these experiments might be (or whether the U.S. government uses the veil of "national security" to keep them hidden, or pretends to not know about or have any involvement in these experiments), eventually they must come out and be known to the public if enough people keep seeing them over many decades and are affected by them. And they don't stop seeing them just because Project Blue Book ended with the view that all UFOs are IFOs. Not even the USAF can stop the UFOs—and certainly not while it lets the public see its research and testing get seen by the public enough times and the health of certain witnesses is affected, if the Cash-Landrum UFO case is anything to go by. And what about those electromagnetic flying objects seen worldwide? Is this all part of the USAF testing? If so, why test the object in front of so many witnesses? Can the USAF not keep the technology to itself?

With all this effort to test a new flying object and claim it is an IFO, the USAF is not helped by the fact that the government still wants to keep UFO documents to itself for as long as possible. Add to this the campaign of disinformation provided to the UFO community in a strange attempt to keep scientists away from the UFO scene, and it seems odd for U.S. authorities to maintain the IFO stance for all UFOs. If these are IFOs, just prove it by explaining what the tests are about and the nature of this new electromagnetic flying vehicle developed by the USAF. Or release all the remaining UFO documents held by the U.S. government.

What Have We Learned?

We have reached a point in this chapter where we must ask the following questions:

1. How likely is it that someone on Earth could have developed an electromagnetic technology to mimic the UFO observations given how long UFOs with electromagnetic side effects have been observed?
1. Who, or which organisation or establishment, in the world might be doing this kind of electromagnetic work if the UFOs are man-made?
2. Can we figure out how this technology works and build it ourselves?

Could the electromagnetic UFOs be man-made?

For UFOs to be man-made, it depends on the time frame we are talking about. Is it either before 1980, or after 1980?

This chapter shows that after 1980, it is possible for some symmetrical UFOs with electromagnetic side-effects to be man-made. We can say this because we know the USAF, whether or not it wishes to admit it, has built and tested one of these UFOs in the state of Texas in December 1980 in front of three civilian witnesses. It does not matter that the USAF will deny its involvement. A flying object of this nature and the fact that unmarked black military helicopters were surrounding and escorting the glowing UFO and with enough evidence left behind on the witnesses to support their observations means that the military has successfully achieved the objective of building one of these UFOs.

However, if we are referring to the period prior to December 1980, we have a problem. The only organisation we know of to be specifically interested in UFOs and to study certain electromagnetic concepts and inventions of relevance to the UFO situation is the USAF. Furthermore, we know from a USAF report released in 1974 that no one was considering building an electromagnetic vehicle. Even if there is a more secret branch of the USAF that is doing this kind of work, it would appear that no one knew how to build an electromagnetic UFO prior to 1960. Not even the USAF as it was still gathering information from various research projects and funding research to study Albert Einstein's Unified field Theory, and just starting to look at the radiation reaction force on charged particles emitting radiation in 1959-60. During which time, numerous civilian and military witnesses reported seeing UFOs of which a number were revealing electromagnetic side-effects. Clearly someone else has already built the electromagnetic UFO. But since neither the USAF nor any other organisation on Earth has come forth to claim responsibility for these observed UFOs, we know it is impossible for anyone to have built an electromagnetic UFO prior to 1960, and highly unlikely to be man-made between 1961 and December 1980.

Who created this technology?

If UFOs are meant to originate from Earth as some kind of secret experiment, only the USAF could have created a new electromagnetic technology to explain UFO observations. Forget the Russians as being responsible because as Lee Katchen, a NASA atmospheric physicist, stated on June 7, 1968:

"UFO sightings are now so common, the military doesn't have time to worry about them. When a UFO appears, they simply ignore it....Unconventional targets are ignored because apparently we are only interested in Russian targets, possibly enemy targets. Something that hovers in the air, then shoots off at 5,000 miles per hour, doesn't interest us, because it can't be the enemy. UFOs are picked up by ground and air radar, and they have been photographed by gun camera[s] all along. There are so many UFOs in the sky that the Air Force has had to employ special radar networks to screen them out."[169]

By observing a distinction between a UFO that suddenly "shoots off at 5,000 miles per hour" and the more conventional Russian targets (or any target that the USAF has identified as an enemy), the USAF has effectively ruled out the Russians as being responsible for the UFOs. Equally telling from this quote is what the USAF did not say: neither the United States nor its allied nations were involved in building or testing an electromagnetic flying machine. Otherwise, the USAF would have explained very clearly where the UFOs came from and what they were.

There are a few reasons this must be true. First, the USAF was not around prior to the 20th century, even though UFOs were reported in that period. And second, the USAF expressed a surprising lack of knowledge of what UFOs were prior to the seminal event near Roswell in mid-1947. It was only afterwards that we learned that the USAF had a sudden change in attitude toward UFOs as observed by Keyhoe and others followed by a considerable effort on the part of the USAF to figure out something in the world of electromagnetism as if this area of science was somehow connected to UFOs. Does this mean the USAF really did find something of scientific significance in the New Mexico desert?

Just to re-cap on the history of UFOs from an electromagnetic perspective, we know that the USAF had an interest in electromagnetism and its connection to UFOs. It began within a couple of weeks after a mysterious object crashed to Earth northwest of Roswell, as evidenced by a memo written by General Nathan Twining on July 16, 1947. In that memo, we learn of some of the internal construction and components used in the crashed disc. Of particular interest in this historical analysis is the allegation that the disc was a giant battery to store electrical energy in order to later produce high voltages for some purpose, which resulted in the ability to ionise the air as one of the disc's own natural electromagnetic side-effects. Speaking

---

169   https://rense.com/ufo/famousquotes.htm

of ionised air, this just so happens to be one of the common UFO observations identified in Chapter 5.

By 1948, a report from Project Sign acknowledged the existence of electromagnetic observations related to UFOs. Words to the effect of "light beams" and the magnetic effects observed on compasses and watches held by witnesses were some of the examples mentioned in the report.

Less than two years later, an anonymous magnetic science expert, who allegedly worked for the U.S. government, gave details to two men about not just the USAF's involvement in the recovery of more than one crashed disc in the southwestern region of the United States, but also, notably, the fact that the discs operated on electromagnetic principles. In particular, the scientist noted the unusual construction techniques employed by the original makers of the discs, such as the lack of any sharp points, especially on any electrically conducting metal surface, as if they had to be avoided when the discs were in operation, as well as the use of extremely lightweight materials to ensure mass minimisation was achieved as the discs approached the speed of light (in accordance with Einstein's theory of special relativity). No signs of a nuclear reactor or use of any fossil fuel propellant were observed. Upon receiving this information, the men passed it on to Frank Scully who in turn published it in his bestselling 1950 book, *Behind the Flying Saucer*.

In the same year, the Canadian government got a whiff of this link between electromagnetism and UFOs. The government decided that it was worth funding a senior radio engineer to study and measure any electromagnetic effects that were associated with UFOs. All of this came after secret discussions with an American scientific counterpart who had apparently learned of interesting observations concerning these crashed discs and their occupants. Unfortunately, as soon as a clear and unmistakable set of readings were picked up one foggy afternoon on magnetometers and gravimeters loaned to the engineer, suggesting that something had definitely flew over or very near to the wooden hut that housed the instruments, something was said soon after by the Americans to encourage the Canadian government to abandon any further work in the field. It did not matter how promising the engineer's data and reports' interim conclusions were for finding an electromagnetic technology in the UFO, all work into UFOs by the Canadian engineer had to stop.

Happy to talk the walk to the Canadians, the U.S. government would do the opposite. During the 1950s, a significant shift clearly occurred in the U.S. government's interest in understanding certain advanced concepts in relation to electromagnetism became evident. The areas of greatest interest in the world of electromagnetism related to Albert Einstein's Unified Field Theory, which linked electromagnetism with gravity (1955–56), and the scientific controversy surrounding radiation-emitting charged objects (1960), the latter of which received considerable USAF support.

As all of this was happening, Captain Edward Ruppelt was tasked with examining UFO reports of which he picked up on electromagnetic disturbances or side-effects emerging from a growing number of reports during his time as the head of Project Blue Book. However, he was later forced to retire due to changes in the aims of the project by the Robertson Panel to have all UFOs debunked at every opportunity.

Whether these events were a coincidence or something significant, things suddenly went quiet after 1960. There was, of course, the unexpected death of an American scientist who was on the verge of explaining the secret behind Einstein's Unified Field Theory in 1959, and various attempts by unknown persons after 1957 to stop the work of another American physicist and inventor aiming to developing a new electromagnetic technology that could fly and mimic the observed behaviour of UFOs. Apart from that, it would take the USAF another 20 years to finally develop an electromagnetic flying machine with many of the classic characteristics one has come to expect of a genuine UFO, including a glowing metal surface and a symmetrical design (see the Cash-Landrum UFO case).

As if concerned by the radiation hazards for civilian witnesses during the testing in December 1980, the USAF decided to move to a place called Area 51 to conduct its secret testing of similar glowing flying objects without causing interference with the American civilian population. Unfortunately for the military, not only were civilians worldwide still observing UFOs (of which some continue to receive electromagnetic hazards and injuries) as if the USAF still could not control the testing and where it ended up, but enough civilians observed from a distance the appearance of highly manoeuvrable, glowing objects over the test facility, thus raising suspicions about whether alien technology was being reversed-engineered at this location. As a result, the USAF escalated attempts to keep scientists away, preoccupy UFO researchers and investigators with the latest claims from witnesses, and convinced the media that it was all probably nonsense with the help of one American physicist named Bob Lazar. His role was to state to the media what he claimed to have witnessed at Area 51—at least one alien crashed disc and how difficult it was for the USAF to reverse-engineer the technology—when he was invited to look at some papers and presumably the original alien hardware. Meanwhile, the USAF focussed on getting UFO researchers and investigators to believe the technology is too alien to be reversed-engineered properly, so why study UFOs? It is pointless as alien technology is too advanced, so just be pragmatic about it and focus on more earthly activities, such as dealing with climate change or some other problem.

Then the USAF has the audacity to claim to the public that all UFOs are IFOs without explaining what kind of military experiment it is doing and how the public can be protected from the electromagnetic side-effects of whatever electromagnetic flying object it is testing at Area 51 and everywhere else where people around the world continue to report UFOs to the authorities and so risk their own health in the process. If the USAF still cannot get its act together and explain today what it is doing, then the USAF should not be granted continued secrecy on its new flying aircraft on the grounds of "national security" because it cannot be the inventor of the electromagnetic flying technology. Someone else is. In which case, this is more of a reason to release all UFO documents from the U.S. government not just because it is in the public interest to know, since they are the ones at risk of suffering health problems from this technology, but also because the USAF cannot and will not explain what is going on or who is responsible.

From our highly distilled look at the history, the timeframe in which the USAF appears to have started looking at an electromagnetic flying vehicle was from July 1947 to 1960. Between 1961 to December 1980, the USAF eventually built a prototype and tested it in front of three witnesses in the state of Texas. Since then, it is possible the USAF could be responsible for at least some of the UFOs reported worldwide. But as time passes and more people receive health effects associated with this technology, eventually a time will come when the USAF will have no choice but to explain what it is doing. The only other option is for the USAF to point the finger of blame at someone else. But who could that be? The tooth fairy? Not likely. How about the aliens? Well, why not? If the USAF does not want to be seen as responsible for harming certain witnesses around the world, and it doesn't matter if the USAF was not flying the military vehicle in a certain area and timeframe, or else explain the hazards associated with this technology to the public and ultimate what this technology is, then acknowledge whether there is another entity at play that could be responsible.

It is time the USAF comes clean on what it knows. We know it is impossible for UFOs to be man-made prior to 1960, and highly unlikely prior to 1980. After 1980, the USAF might be able to claim some UFOs are man-made. However, it also needs to prepare a public announcement on the nature of its military experiment involving a new electromagnetic flying vehicle so everyone can be prepared and know what to do.

This is a rather disturbing conclusion in anyone's language, but let us face the facts. If UFOs are supposedly man-made, how can we explain the considerable length of time that UFOs with common features have been observed? And why the non-curious attitude by the USAF towards UFOs before 1947 only to suddenly change its attitude after 1947 upon realising the importance of looking into electromagnetism for answers?

And why electromagnetism? Should we see it as a coincidence considering UFOs are electromagnetic in the range of interesting electromagnetic effects they display to witnesses? Everything is pushing UFOs into the realm of being electromagnetic flying objects of some sort. Either we are dealing with ball lightning, or there is something technological at its very heart. However, it would be hard to imagine the USAF wanting to spend 33 years studying electromagnetism because UFOs are ball lightning. Furthermore, the test that was carried out in December 1980 is strongly leading people to believe the humble UFO is technological in nature. Does this mean we are dealing with an electromagnetic technology in the UFO reports and the USAF had stumbled on it?

Even if we could consider all of this a coincidence, there is an even more concerning issue to discuss here: the length of time that the USAF has maintained secrecy about UFOs, despite having learned about and successfully built the electromagnetic technology before running off to hide in a secret facility at Area 51 to do some more testing.

Considering that the American public has paid considerable taxes to support the USAF's research (which is presumably intended to protect the public from whatever might threaten it), one would think that the military would have given the public some inkling of the technology it is working on. Why is it taking so long to release the information to the public regarding these electromagnetic flying vehicles built and tested by the USAF? Does the military not recognise how important it is to inform people of this new technology? A number of witnesses have already received doses of radiation from this technology, and it is likely that more will be exposed while people continue to observe UFOs. At some point, the public needs to be told and made aware of the side-effects associated with this kind of technology and how to minimise their impact.

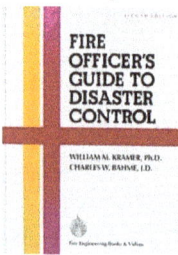

FIRE
OFFICER'S
GUIDE TO
DISASTER
CONTROL

WILLIAM M. KRAMER, Ph.D.
CHARLES W. BAHME, J.D.

For example, in both the 1952 (first) and 1992 (second) edition of *A Fire Officers Guide to Disaster Control* by William M. Kramer and Charles W. Bahme, there is a section in "Chapter 13 - Disaster Control and UFOs" titled "Part 1, ENEMY ATTACK AND UFO POTENTIAL". The following statement appears on page 439:

WARNING

Near approaches of UFOs can be harmful to human beings. Do not stand under a UFO that is hovering at low altitude. Do not touch or attempt to touch a UFO that has landed. In either case the safe thing to do is to get away from there very quickly and let the military take over. There is a possibility of radiation danger and there are known cases where persons have been burned by rays emanating from UFOs. Don't take chances with UFOs!

Wise words. This statement came at a time when Mr Bahme was conducting his own research into UFOs in the early 1950s. Probably comes across as a bit extreme in its recommendations, but at least the American firefighter fraternity is taking the UFO situation seriously by warning each other of the electromagnetic hazards associated with genuine UFOs. Perhaps the USAF could take a page out of this book for inspiration on what to say to the public when its comes to its own man-made UFOs it is testing behind the scenes? As Michel M. Deschamps said:

> "Who would have thought that a training manual on Disaster Control used by fire fighters would contain a chapter about UFOs?? This is a pretty good indicator of how important and serious the subject actually is in the minds of some people."[170]

Richard Smith adds:

> "This book is renowned as THE collector's item of the 20th century. It's reputation precedes it due to the fact that when the authors decided to publish an updated second edition, they felt it extremely necessary to include the now infamous Chapter 13 [on UFOs]."[171]

Today, however, armed with extra information about UFOs, we can now be more precise in our warning to the public. Specifically, a more appropriate warning to the public nowadays would be to explain when it is safe to approach a UFO. For example, say that it is safe to approach a UFO only when: (1) the UFO has physically touched the ground with its metal landing legs and discharged its electric charge; (2) the sound of the UFO's electrical generator has died down; and (3) any warning lights have turned green.

For example, a large glowing effect would obviously be a sign of a high electric charge flowing on the surface of the UFO to help heat up the metal hull. Naturally, care must be taken to avoid touching the UFO, as this could provoke a significant electric shock or even a serious burn. And while we are on the subject of electromagnetic hazards, the military should give some indications as to why an electric charge is used. Such information would undoubtedly go down well with the public, help us to understand the nature of this man-made technology, and give us food for thought on what to do when encountering one of these flying objects.

This is the kind of public service we should be able to expect of the military, assuming, of course, that UFOs are secret military tests of a

---

170   https://www.amazon.com/Officers-Guide-Disaster-Control-Second/dp/0912212268/
171   Ibid.

new electromagnetic flying machine. Well, some of them has to be today. More of a reason for the USAF to explain what it is doing and prepare the public on this new technology.

More importantly, warnings of this nature should not be embedded deep within a 641-page handbook used exclusively by American firefighters. The people who are on the front line of observing UFOs and getting the full brunt of any electromagnetic side effects are members of the public. They are the ones to receive more burns, radiation poisoning, and injuries from forceful light beams associated with UFOs than most.

Honesty, how hard could it be for the USAF to provide an adequate explanation and warning about the type of electromagnetic technology it has developed? Surely, the health of the public is paramount in such circumstances. Secrecy should never override this priority. Yet incredibly, the USAF does not seem to give a rat's a*se about the health of the public when it comes to UFOs and whatever testing the military is allegedly doing. Clearly, there must be another reason for the continuing secrecy.

One possible explanation for the enduring nature of this secrecy is that it could be a reflection of the amount of experimentation the USAF must perform to get things right with the technology. However, if this is true, then the available UFO reports indicate that the technology is working perfectly fine. In fact, this machine can move so well, it can out run and out manoeuvre any traditional fossil-fuel based man-made aircraft. Not even the fastest and nimblest of fighter jets available in the USAF fleet today can do anything to match the capabilities of this new electromagnetic flying machine. And apparently this has been happening prior to 1960. This electromagnetic flying machine is working exceptional nicely thank you very much, apart from that silly lightning strike in July 1947. One must assume the USAF has already solved that problem: just do go underneath a thunderstorm and hover near the ground. Simple advice. You certainly don't need to spend the next 70 years or more refining the technology just to solve this simple problem. UFOs are performing admirably and doing a fantastic job of evading military jet aircrafts. It was good enough in the 1950s, and still is today. No need to keep tinkering with it at the sides. It is time to give the public announcement of what the USAF is doing and be done with it.

And if the USAF is genuinely concerned about the health of the public when operating one of these electromagnetic flying machines, why not show a great big red light when in operation as the way to tell the public to "Don't come close to the technology as it can create radiation effects on nearby metallic objects or you could be zapped by a big enough electrical spark", and green is for "You can approach the object safely and touch it if you like". This idea of using red and green lights is heavily used in the UFO observed by Antonio Villas Boas in Brazil. It would hardly cost much to implement and it wouldn't take years or decades to figure out. Oh well. There goes the idea that the continued testing must be to make the machine safe to use and operate near other people.

But if the testing is in another area, it seems to be nothing too critical. The fact that it flies very well suggests that the testing is in an irrelevant area, such as determining where to put the fridge to hold the drinks so that the people who participate in the flight can enjoy the ride. But to say that the object cannot fly or does not have enough energy to lift itself into the air as the fundamental reason for the continued testing and secrecy simply makes no sense. We know the technology works, and rather well if the UFO reports are evidence for this.

Something is clearly not right unless there is something else about the nature of the technology and how it was acquired that is worrying the USAF.

Therefore, we are left with only four reasons as to why the secrecy behind the technology remains intact to this day:

Reason 1: The USAF does not want to admit that it did not invent and does not own the intellectual property of the electromagnetic technology.

Reason 2: The USAF wants to maintain its relevance in defending a nation against these UFOs thinking that they might be a menace or a threat in order to get paid for the services it provides (i.e., the Defense personnel need to feel wanted, have a purpose, and ultimately to have money to survive in our current economic system).

Reason 3: The USAF fears the social implications of releasing the technology in an era in which terrorists are willing to affect Western nations in negative ways.

Reason 4: Some people in the know want to enjoy the benefits of slowly releasing and selling parts of the electromagnetic technology so that they can become rich and powerful in society.

Are any of these reasons justified?

Before we examine each of these viable reasons, let's consider why it can be beneficial for everyone to keep something secret for a short period.

For example, we learn from history of an apparent discussion between military leaders and President Truman about what to do with the crashed disc in New Mexico (i.e., whether to tell the public about it or keep it to themselves). Among the things spoken about was the claim that no weapons were identified in the disc. Whether or not this is true, we can reasonably assume from this discussion that there was no national security concern that the Earth might be invaded by aliens, or by whoever built the electromagnetic vehicle. Still not one hundred per cent proof that aliens are friendly, but if the discussions did broaden to include how long UFOs were observed, there would be no reason not to conclude that aliens are friendly. Despite this possibility, a UFO secrecy would ensue with the support of President Truman. We learn why this was necessary following the radio broadcast *War of the Worlds*. At the time the broadcast was delivered, it was presented in a manner that led people to believe it was real. The subsequent social chaos seen in certain pockets of American society might have convinced the decision-makers in that secret discussion to keep the technology a secret. It was considered at the time the best way to protect simple-minded folks and make them feel safe, at least in the initial stage of establishing the secrecy.

Perfectly reasonable.

However, secrecy must come in two stages: the first stage is to keep something secret until more is understood about the thing being kept secret, and the second stage is to plan the release of the thing that is being kept secret, especially if there is a likelihood that people will find out and are being affected in some way by the thing and there is no evidence to support a threat to the public in the thing that is being kept secret.

When it comes to UFOs, one can understand the reason for keeping them secret in the initial stages. The public had some fear of anything alien, and the USAF needed to know whether UFOs were a threat and determine the best way to handle them should it be necessary to defend the nation against UFOs.

It should be noted that all of this social chaos surrounding aliens occurred more than 60 years ago, and things have changed since then. More and more people are becoming familiar with the existence of aliens in the Universe through the cinema. The only slight problem with this approach is that most aliens are not portrayed in a positive light. There is a tendency to show them as hellbent on invading Earth or hunting down humans to inject an embryo into them like wasps to insects, or to eat the humans as food, or to take advantage of Earth's other resources. Only a few movies try to change this view, such as *Close Encounters of the Third Kind* and *ET: The Extraterrestrial*. Despite the heavy handedness of some Hollywood producers in making the public feel frightened of most things alien, the UFO reports do not indicate that whoever is flying them should be feared. If UFOs wanted to cause us harm, they would have done it a long time ago. And yet here we are, still alive and going strong.

To emphasis this point, it seems rather odd that our planet has remained relatively "alien-free" in the sense that we do not see aliens walking among us or are freely walking on their own while herding humans into confined areas for some alien-driven purpose. UFOs have been around for an incredibly long time; so long, in fact, that we can effectively disprove the notion of "bad aliens" visiting and wanting to take over the Earth. If anything, as we have seen in the Biblical story of Moses and the recorded account from an ancient Greek historian, the aim of UFOs seems more to protect some humans and guide them to safety, or to stop wars from taking place between humans—not exactly the kind of observations that would be in keeping with the view of many Hollywood producers. But then, we have to keep in mind that what we see in the movies can have absolutely no bearing on reality. Whoever is flying the UFOs seems to have a different priority and want to make clear to humans what it is: to show love by protecting life and avoiding interference with it as much as possible, unless such interference is absolutely critical. And even then, the kind of interference we see is more a case of wanting to study us, grab some biological samples, and let us go. In some UFO abduction cases, the abductees may receive warnings and advice of what humans should be looking out for. Things like protecting our natural environment, be careful in the use of nuclear technology, and learn to love our fellow human beings.

Still not sure about this? That's fine. There is another important point to consider. As this book goes to press, there is growing indirect scientific evidence to support the existence of alien life on other planets beyond our solar system. Not only are planets in plentiful supply around other stars, but careful observations using more sensitive instruments have revealed Earth-like worlds of an appropriate distances from the parent star to enable liquid water to exist over a long period of time. There is nothing so far to suggest that it is not possible for alien life to exist.

As this information emerges from scientific journals to educate the masses, people are becoming prepared for the eventual realisation that aliens probably do exist. Today, more people are starting to feel comfortable about the idea of aliens compared to the sentiments of the 1950s. Noting the quick progression of things in this area of scientific study, it will not be long before more direct evidence is found to settle the debate, and people can go about their usual activities without a second thought as to who or what might be living out there among the stars.

Once the existence of aliens is finally settled, the only remaining question for scientists to solve will be whether there is extant technology that could allow other civilisations to make the journey to our planet. For the USAF, it seems the answer might be a resounding, "Yes!". Three things seem to lend strong support to this possibility. First is the USAF's rather heavy focus on electromagnetism in the 1950s (apparently commencing at the time of the Roswell case in 1947, and was also coincidentally the time when the USAF decided to change its attitude toward UFOs). Second we have claims of the recovery of crashed discs that operate on electromagnetic principles according to Frank Scully's 1950 book, *Behind the Flying Saucers*. Finally, the third is the eventual culmination of all this electromagnetic work in the development and testing of an electromagnetic vehicle in December 1980 with all the classic features of a typical UFO, right down to symptoms of radiation poisoning.

Not an unheard of observation from other UFO cases.

With the USAF now able to match the technology with those genuine UFOs (well, there is certainly plenty of UFO reports with electromagnetic side-effects from the witnesses to show what we must presume is the brilliance of the USAF's allegedly independent work in this area, assuming it is responsible for all the UFOs reported throughout history), what further reason is there to maintain the secrecy? UFOs are clearly not posing a threat to humanity. If anything, UFOs seem more concerned about our welfare and want to preserve life on Earth. Not exactly sure how in working out the technology the USAF could use the technology to defend a nation. There appears to be nothing to defend against. The UFOs can come and go at will, and after many thousands of years they are still not showing signs of wanting to harm life on Earth.

Sounds like there is no point in the military spending taxpayers money on this electromagnetic work in the naïve expectation that it would help to defend itself against something that has no need to harm us in any way.

Surely there must be another reason for the continued secrecy.

Well, there has been one other idea floating around recently from USAF personnel about why the secrecy still persists. For example, the latest claims to emerge from some military personnel allegedly involved in the Roswell case (mainly as former employees who worked at Wright-Patterson AFB) is that the USAF is presumably unable to reverse-engineer the recovered disc as needed to figure out how it flies[172]. However, this argument falls apart when we recall the Cash-Landrum UFO case of December 1980. This was when the U.S. military, in black helicopters, was observed surrounding and guiding a diamond-shaped glowing UFO in Texas. The incident received international attention partly because the three witnesses, who received significant radiation exposure, subsequently attempted to sue the U.S. government for health damages and accused the government of not informing the public of the health hazards associated with this secret military experiment. While the witnesses were unsuccessful at the time in proving the UFO did originate from the U.S. military (because the USAF was able to claim no responsibility and with no evidence to

---

172   Claims made by some USAF personnel about the Roswell disc being unable to be reverse-engineered have been released in August 2019 by Mr Anthony Bragalia, an American investigator of the Roswell UFO crash who works closely with authors Thomas J. Carey and Donald R. Schmidt.

prove otherwise), what this case tells us today is that if the USAF thinks it cannot fly a UFO, this UFO case is certainly not supporting it. We have every reason to believe the USAF had already carried out the secret test based on the work done by the USAF in the area of electromagnetism for building a new aircraft of this type in the 1950s, culminating at some point in the building and testing of a prototype. It makes sense this testing had occurred in December 1980 no matter how the USAF wishes to see it. Otherwise, the USAF had better admit that the UFOs did not originate from any military experiment, in which case we definitely do have something of scientific significance and the whole reason for having a UFO secrecy by the USAF must relate to protecting citizens from the reality that aliens are visiting the Earth. But as we have seen, the aliens appear to have no need to harm us. So where is the need for protecting humans? In that case, there cannot be a need to maintain secrecy. Time to be up front and honest and say the USAF was afraid before to tell the truth, but now things are different today and we are ready to learn the next major revelation about what is happening out there in the universe.

Not true? Alright then. So what's the holdup in releasing the evidence? The USAF is alleging it has found something in 1947 but cannot figure it out, so what is the point of keeping it secret? Still think it is not a good idea? Unfortunately for the USAF, a growing number of people outside the military will find out. It is just a question of time before the secret is known.

Speaking of figuring things out, the public and scientific community are already noticing some kind of hidden electromagnetic technology in the UFO reports. Add to this the increasing numbers of people who have gathered enough information from their education to know what is possible and, ultimately, how probable it all is, and it does make one wonder how likely is it that a secret can be maintained indefinitely. It would appear that any secrets hidden from the public will eventually be revealed, with or without the help of the USAF. When curious, the civilian population will learn things and make its own discoveries. The risk for those trying to maintain secrecy is that certain secrets will be exposed given enough time.

This brings us to a major argument that skeptics have used to explain why they believe the USAF cannot maintain a secret for very long. Even if the civilian population does not uncover certain secrets, more USAF personnel will become aware of the technology, and some may choose to leak details to the public as a way to stop the secrecy. No matter the range of mass education campaigns to convince ordinary people that UFOs are not real or alien, or if dubious or draconian methods were employed to successfully frighten military personnel into keeping quiet, there are curious people who are willing to study UFOs. It will not be long before someone will learn the truth. When the truth is known, the USAF will eventually be seen as having some involvement due to its interest in developing some kind of electromagnetic flying vehicle. Once this is understood, the USAF cannot keep an electromagnetic experiment on a new flying aircraft secret any longer.

Recounting the numerous UFO reports of a large triangular UFO observed over Belgium between 1989 and 1990, Major General Wilfried de Brouwer said:

> "Hundreds of people saw a majestic triangular craft with a span of approximately a hundred and twenty feet and powerful beaming spotlights, moving very slowly without making any significant noise but, in several cases, accelerating to very high speeds."[173]

As we can conclude, whoever was flying this UFO wanted to ensure that many people could observe it. It was not a case of trying to avoid detection where it was accidentally viewed by so many people. Quite the opposite. Whoever flew the UFO wanted to be observed by as many people as possible, to be chased by military aircraft, to make return visits, and to be viewed again and again at close range. There is enough information from this UFO case to speculate on the intention of the pilot. Someone in the UFO had the idea that the more people who can observe the UFO, the better the chances of leading some curious individuals to study the subject and eventually find a solution. It seems like curious people are the key to piercing through the secrecy surrounding UFOs.

Later, investigative journalist Leslie Kean, author of *UFOs: Generals, Pilots, and Government Officials Go On the Record,* published in 2010, commented:

---

173   Shermer, Michael. "UFOs, UAPs and CRAPs: Unidentified aerial phenomena offer a lesson on the residue problem in science": *Scientific American.* March 28, 2011. Available from https://www.scientificamerican.com/article.cfm?id=ufos-uaps-and-craps.

"Common sense tells us that if a government had developed huge craft that can hover motionless only a few hundred feet up, and then speed off in the blink of an eye—all without making a sound—such technology would have revolutionised both air travel and modern warfare, and probably physics as well."[174]

In other words, some people believe that it is just a matter of time before we do find out. If it has been built, someone will know about it, and soon the public will be informed. At that point, any secrets behind the technology will have to come to an end. Well, guess what? We are now at a stage within this book of knowing how likely there is an electromagnetic technology hidden in the UFO reports (see Chapter 8). Should we determine a new technology does exist, then the secrecy will have to end.

And then the secrecy will have to end.

Yet despite how unlikely it is to maintain a secret for so long, it is remarkable to see how well the USAF's work on an electromagnetic technology has remained hidden to this day. How is it possible for so many scientists and nearly the entire civilian population to not be curious enough to question the nature of UFOs and find out what they are in a fundamental sense, especially the most detailed unexplained cases on record?

One possibility is that, when a rational person is told something that sounds too amazing to be true, such that UFOs could be alien and may provide the solution to interstellar travel, they usually assume that it is. Well, let's face it. Has anyone come up with a solution to the interstellar travel problem? And has any scientist actually physically studied a real alien and given a scientific report to confirm this very fact? The answer is clearly no. When faced with this situation, the typically rational individual has no qualm about assuming it is impossible—even to the point of ignoring all UFO reports, no matter how detailed, that could provide an answer.

Another problem to contend with is how the few who are curious about UFOs and wanting to get to the truth of the matter have been thwarted many times by those in the know, including the USAF. Getting any answers regarding UFOs, especially truthful ones, is next to impossible when the USAF has a reason to keep UFOs secret. For

---

174   Shermer, March 28, 2011.

example, even though some UFO documents were released under U.S. FoI in 1976, the U.S. government has gone to considerable lengths to keep the remaining documents hidden on the grounds of national security. Why? What is so sensitive about a bunch of UFOs that seem more interested in protecting life on Earth and showing curiosity about what happens here, that the whole matter must be hidden from the public? And if UFOs are really misidentification of familiar man-made and natural objects, why all the secrecy? There is nothing in the UFO reports of say, the planet Venus, that indicates a threat to the United States or the human species, nor to the rest of life on Earth as a whole. Even if some UFOs turn out to be alien, the reports are not indicating they are a threat to anyone. Quite the contrary. Aliens have not cared one bit about taking over our planet given how long UFOs have been seen by humans. But never mind the possibility of friendly UFOs visiting our planet or something else; if one can create the perception of UFOs as potentially dangerous, or the notion that UFOs are too incredible to be true to make our rational minds think UFOs can be ignored and, therefore, assume they are unlikely to advance science or technology, the combination of all these factors can form an almost perfect trifecta for the USAF to hide the truth for a long time.

No wonder the secrecy has been going on for as long as it has.

However, this isn't the only way to maintain secrecy. The success of maintaining any secrecy for a considerable length of time requires a multi-pronged approach.

First, such an approach requires keeping the knowledge and technology limited to a few privileged and highly patriotic individuals who are paid well to keep it quiet. These are the type of people you will probably find in Area 51 or in other parts of the U.S. Defense and Intelligence organisations. These individuals hold the extreme view, bordering on obsession, that the U.S. is number one and the only great country in the world. This view is often reinforced with a sufficient financial reward (together with possible monitoring and threats of imprisonment or something much worse) to make these individuals see the value of maintaining the secrecy regardless of whether it is right or wrong. If someone says to keep something quiet, well-paid, patriotic individuals such as these will do so unquestioningly.

Second, there must be legislation and enforceable consequences implemented for any U.S. Defense personnel who tries to reveal information about UFOs to the public, even if the UFO sighting was a personal experience. All UFOs are seen as implicating the U.S. government in some way (i.e., will be interpreted as either a military experiment that must be kept secret at all costs, or as some kind of action by the enemy and so must be handled internally to determine the appropriate way to respond). The consequences would often involve a combination of a heavy financial penalty and long periods of imprisonment. Historically, we see this in Regulation JANAP-146, introduced in 1953.

Third, reduced media coverage on UFOs can help to reduce interest in UFOs within the civilian population, as we saw after the 1980s.

However, it doesn't stop people seeing UFOs. Furthermore, the civilian population can never be one hundred per cent controlled and made to do exactly as the authorities want, as some curious people will still want to research and investigate UFOs. So how does one deal with this situation?

Why not identify and target those individuals in society who are working on the UFO problem and try to put them into other areas that are more mundane and less likely to cause problems to those wanting to maintain UFO secrecy? Once these individuals are identified (usually a name, a mobile number, a general geographical location obtained from a mobile device, and an address in the White Pages is enough), methods will be applied to stop their work, either by affecting the individuals in certain ways or destroying the work.

As an extreme example, in 1959, we have seen in the history of UFOs a scientist named Dr. Jessup who was on the verge of working out the secret to Einstein's Unified Field Theory. Such knowledge would have probably revealed much regarding how UFO technology works by realising the the importance of radiation in affecting the gravitational field and ultimately how it can move solid matter. Then it could be a short step to working out how UFOs move and the technology that is hidden in the reports. Was this the reason his life ended abruptly? Some have suggested suicide because a note was found (although anyone could have forced him to write it), while others were not too sure. He did not seem to be the type to end his life because of his work or where he worked, nor did he have any issues in his personal life. He had a reasonably well-paying job with freedom to study different areas. From the way it looks from history and what we know about this man, he not only had much to look forward to, but also secrets to reveal from his interesting work.

Of course, the above example might seem rather extreme. But then again, we have learned of a possible reason for the assassination of U.S. President John F. Kennedy. The fact that we know from history that he was interested in UFOs and had sent a secret memo (which was never acted on) just prior to his death requesting certain government departments, including the CIA, to release UFO documents as part of a goodwill gesture to Russia in an attempt to end the Cold War means that anything is possible.

Nonetheless, care must still be taken by whoever is maintaining UFO secrecy that this tactic of eliminating people is not used too often. Otherwise it may raise suspicions of foul play by the U.S. government and make people think UFOs are genuine and represent something important to science. Therefore, it is essential to find other ways to stop people from pursuing certain areas of UFO research in order to maintaining the secrecy for as long as possible.

For example, another technique that can be used is to quietly apply stress to the targeted individual by either direct or indirect means. We see this in history with Captain Edward J. Ruppelt, the man who first headed Project Blue Book. In 1959, he wrote a book on UFOs detailing what he knew about UFOs and the situation with Project Blue Book that embarrassed the US government. He revealed important insights not readily available to the public, such as the electromagnetic effects described in UFO reports and his suspicions that Project Blue Book was just a front to a bigger cover-up. Because of this, Ruppelt was pressured, through his contract with another company, to retract everything he said. Due to financial stresses and likely threats to his own life, he had to re-write his book and go against everything he said and believed was true. As a result of this, and the verbal abuse he had received and endured from angry people and his circle of friends (who should have known better), as well as the loss of his reputation, he eventually died of a heart attack in 1960.

Again, such pressure applied to certain individuals has to be done in a judicious and limited manner to avoid attracting attention. So, another method that can be used is to convince curious individuals to devote their energy in other areas not related to the UFO matter, such as by offering contract work with the USAF or another associated company or research institution. This approach works best only for those who can be identified. This tactic was noticed by SUNRISE when it made some of its research work on shape-memory alloys and NiTi in relation to the Roswell foil known to three American Roswell UFO investigators while seeking information to assist with the research. These two principal investigators SUNRISE approached were Thomas J. Carey, a former USAF employee and an anti-crypto expert, and Donald R. Schmidt, a man with interests in UFOs and the Roswell case and whom Carey has befriended (Schmidt and Carey are currently co-authoring books written about the Roswell case). The third person to learn about the SUNRISE research was Anthony Bragalia. He was the one who first approached and communicated with SUNRISE (the other two investigators have not communicated directly with SUNRISE). Without stating his connection with the other investigators, Mr Bragalia's primary goal was to identify the SUNRISE researcher responsible for carrying out the work on NiTi and the Roswell case. He did not seem too keen on the work itself or on assisting with the SUNRISE request for information unless he knew who the researcher was.

Eventually, when Bragalia had to mention his connection to the other investigators, his aim was to work quietly in the background and provide relevant information to Carey before notifying Schmidt and then collectively deciding whether to publish the information. During the email correspondence between SUNRISE and Bragalia after the former learned of his efforts to find out the name of the researcher, there was talk of contract work being offered to this researcher, most likely with the USAF or someone with links to the USAF, such as the Battelle Memorial Institute or an American university, so long as the person involved in the research work was identified. SUNRISE was not at liberty at the time to provide that information for the researcher's safety, and certainly not before publishing the research work being carried out at the time (including this book).

As a consequence of this, the American investigators decided to publish some of the work by SUNRISE and any additional documents and quotes they could find to support the shape-memory alloy claims before SUNRISE could publish its own book. Upon hearing that the research work would be published, SUNRISE requested that the investigators provide some acknowledgement of its work, either as a footnote or elsewhere in the book. However, the American investigators appeared more interested in obtaining the name of the researcher before any such request would be granted. SUNRISE repeated the request, asking that the organisation name and a link to the research work online be included if SUNRISE was not permitted to publish the work first. When the American investigators' book *Witness to Roswell* was released in 2009, SUNRISE discovered the investigators did not comply, and no credit was given.[175]

This is the price independent researchers involved in studying sensitive areas of the UFO phenomenon must pay when the USAF and anyone else involved in UFO secrecy cannot get the information they want.

The last remaining approach to controlling the curious civilian population, especially for those individuals whom the USAF finds a little more difficult to identify as doing potentially sensitive work on UFOs (probably because they keep quiet in the same way that the USAF remains silent about its own electromagnetic technology work), the approach has to be more psychological. We have seen some examples of this approach being employed by the U.S. government in the past through the release of false documents and photographs of dead aliens retrieved from one or more crashed discs in the U.S. as a means of keeping the unidentified people busy and focused on this information until it is discovered to be a hoax. Then, their UFO work can be made to look less credible to rational people, including scientists, and the reputation of the people involved in studying UFOs can be left in tatters. Repeating the technique a few more times will hopefully be enough to reduce the number of people involved in UFO research, so the secrecy can be maintained for longer.

As can be seen, the range of methods that can be employed to maintain a secrecy for a considerable length of time is broad and far-reaching, and there is no shortage of ideas about how to implement them. When the secrecy operation is applied in its many forms over a wide enough population, its success is obvious and laid bare for all to see when we ask the one simple question: how many are aware of the electromagnetic technology in the UFO reports? The answer is hardly anyone.

For example, too many USAF personnel remain ignorant of the electromagnetic technology behind UFOs, as can be seen in the 1972 USAF report looking at advanced propulsion concepts. The report showed the state of knowledge of most USAF personnel about known

---

175   Since then, questions have been raised within SUNRISE as to whether these individuals in the US are closely working with the USAF to identify people involved in UFO research so as to assist the U.S. government in curtailing any kind of independent and sensitive UFO work by any means possible.

and workable electromagnetic technologies used for propulsion at the time the work was being carried out—essentially non-existent. The report could only focus on identifying upcoming new electromagnetic technologies that could be employed for propulsion, but nothing about what the USAF had already done in the past to create an electromagnetic vehicle to account for the large number of UFO reports.

If that is the case, imagine how many in the civilian population prior to reading this book would know about electromagnetic technology in the UFO reports. Likely no one, not even those who are supposedly "curious individuals"—that is, scientists working for civilian organisations—are aware of the electromagnetic technology. This fact demonstrates how effective the provision of false UFO information and other techniques have been in keeping the scientific community away from the subject. Unfortunately, the decision by scientists to not quietly spend even 5 minutes a day looking at the UFO problem to ensure nothing has been overlooked, has inadvertently helped to maintain the secrecy about the matter, much to the delight of the USAF.

The idea that certain secrets cannot be maintained for a long time, even indefinitely, is untrue. It can be done using the right techniques to control a population. But on the same token, we have to remember that it is also true for secrets not to last the distance and eventually get revealed. The weapon of choice for making this happen has to be a strong sense of curiosity and a willingness to learn and get to the truth —and do it quietly. In this situation, there is nothing the authority can do against such individuals in society when it comes to secrecy of any duration.

It is called the *secrecy paradox*. On the one hand you may think you have got everyone under control and the secret is maintained. But on the other hand there is a big risk. Things can suddenly fall apart when just one curious person or a group of curious people decides to blow away the secret. And no one knows when it will happen. That is why maintaining secrecy indefinitely is never a good idea. There has to be a second stage in the secrecy in which plans must be drawn up and implemented to release what is being kept secret. It has to be done at some point, especially if people are being affected by whatever is being kept secret and there are others who are becoming aware of the technology.

Let us not fool ourselves for a moment. We know that the USAF has worked out the technology and the secret behind UFOs. Everything we have seen in history indicates that the military has achieved something of significance. However, the way that secrecy is being maintained for such a long time reveals a far more concerning issue as we shall discuss. It is not a technological problem as some people might think, nor is it a question of whether the UFO is a man-made military experiment (even though this is highly unlikely, given the duration of UFO reports throughout history). There is another issue at play here. In fact, each time the USAF tries to do something to reinforce the USAF secrecy on UFOs and convince the public that UFOs are not alien, the more evident this issue becomes to those who study UFOs.

This brings us to the four main reasons for the continued secrecy.

The first reason relates to how the technology was developed in the first place. More specifically, the USAF may have trouble claiming ownership of the technology if people know where it came from. Someone else could have developed the technology, and the military's effort to reverse-engineer it and later claim ownership could be undone by someone "uncovering the secret" and telling the world what really happened, causing major problems for the military and anyone else involved in the work. Loss of significant profit from selling parts of the technology could affect not only the livelihood of the people involved, but also the American stock market. And what about the reputation of the United States as a producer of original ideas and the manufacturer of allegedly "original" technologies? That reputation would get take a serious beating if people knew the truth.

Or perhaps it wasn't so much about stealing someone else's intellectual property, but more a case of "finders keepers". With no means of contacting the original owners to ask if the USAF could use the technology, the military thought it was okay to reverse-engineer the technology and use it. However, the decision to not share this intellectual property, and all its benefits, with the rest of the world could lead to serious resentment from other nations. It could also make the United States look like a selfish child keeping an exotic toy to itself, all for the sake of a few people wanting to get rich and powerful. In that case, the reputation of the American people as an equal partner with other countries could be tarnished for a long time. Could this be a concern for the USAF?

This is the thing. The longer the secrecy goes on, the harder it will be for the USAF and anyone else involved to explain how the ideas were developed and whether any patents applied to those ideas can be upheld if the USAF or some people linked to the USAF are not the true inventors. The help obviously did not come from the Russians or anyone else on Earth (a reasonable claim to make since no one has taken responsibility for creating the electromagnetic vehicle that was found by the USAF). And based on the way the USAF behaved when it discovered something and decided to change its attitude after 1947, it does not look like the USAF had figured the technology out on its own. It would appear that the USAF had help from an entity not associated with any nation on Earth.

An extraordinary statement to make if this is true.

However, if this is all untrue and the USAF really was the inventor, why is it that the secrecy behind the technology remains intact to this day?

Perhaps the USAF is afraid that it cannot do anything to defend its nation against high-speed electromagnetic flying objects that can enter and leave nations at will and with incredible ease. As the UFO reports are indicating, these objects can land and occasionally, the occupants emerge and have plenty of time to jump up and down and over logs on the ground as a form of exercise or to play, or to take samples of plant and animal tissue (including reproductive cells) from terrestrial creatures, including humans. Should the military learn its airspace had been invaded, it could not do anything quickly enough to stop it or to investigate and see what had landed. The UFO occupants would leave just as quickly as they had arrived. If the aim of the military is to protect citizens from external threats, it looks like we have the lousiest and most expensive man-made defence system in the known universe, for all the money that has been spent on it. Indeed, people will seriously question the relevance of a military force in any nation if it cannot stop a single UFO from landing on Earth and prevent the occupants from doing whatever they like.

Could this be the problem? Certain people working for the Department of Defense and intelligence organisations could find themselves suddenly out of a job if people knew there is nothing humans can do to stop these UFOs from entering our airspace. So, in order to feel loved and wanted, and receive money from taxpayers, it is better to create the perception that humans can defend the nation. So long as UFOs are kept secret and no one knows the technology behind it, people have to rely on the defence forces to protect them and pay them accordingly.

Might as well do anything to give civilians the false impression that the military can protect the world from anything, including the aliens. Slim chance of that happening while UFOs can land and take off with impunity.

But is there another reason for the continued secrecy?

What about the social implications of releasing this technology at the wrong time in history? In today's world, terrorists lurk in just about every corner of the globe and are ready to find new ways to deliver deadly cargo, such as bombs, to Western democracies at great speed and without detection. Thus, electromagnetic technology falling into the wrong hands could potentially be worse than, say, a country like Iran or North Korea developing nuclear bombs as they only have basic rocket technology to deliver them. Even without nuclear bombs, ordinary high-grade explosives could still cause untold damage in Western nations with the help of electromagnetic technology, and no one would know for sure where these objects were coming from or who sent them. Should one rogue nation decide to use the UFO technology, the military will be unable to defend its citizens. It will be like what happened when the Americans developed the nuclear bomb and ended World War II by using it on Japan. With no military defence in Japan capable of handling the technology or stopping the Americans, there would be absolutely no point in having a military force. It might explain why Japanese people have become the most peaceful in the world. It is the only way to survive and not be a military target for another nation.

Could this be the reason for the U.S. government in keeping UFOs under wraps?

Well, if the USAF is truly worried about terrorists, there has to be a moment in human history to create a new world order. This would be a new chapter in which nations and their citizens are treated respectfully and as equals deserving of the same access to the essential resources and technologies as everyone else and the same opportunities to achieve great things. For a truly secure, long-term future for everyone, we must focus on what we need to survive and be happy and make sure everything we use must be recycled so that the things we need are always available at a very low cost, or freely, to everyone. The more we deal with this survival issue of helping everyone have what they need and be happy, the less likely people will want to do things to harm other people. Once people have what they need, they can do what they want, but in a way that repays the love we have given them to survive and be happy. That means access to free education and opportunities to let people solve different problems. Let the curiosity of the people solve the problems. There is no need to force people to work. What is work? People who enjoy solving problems in areas that interest them do not see the "work" behind it. It is something they enjoy doing. And so long as they remain curious and wanting to return the love to society, they will eventually find a solution. Great things will naturally be achieved by those who are happy and free to solve any problem set before them. Curious people will always want to learn and discover new things, and often they will do the work without financial reward (indeed, why would they need money when they have everything they need?). Everything would be done out of love for learning and helping people and all living things on this planet (the latter of which is absolutely necessary to ensure the recycling systems we implement are working optimally).

Of course, the only slight problem with implementing this new world order would be the costs of various things that we need today—they would decrease dramatically over time as we implement more recycling measures. As the things we need get cheaper, and eventually free, this could affect present-day economies in a way that would not benefit the existing rich and powerful people in the long run. In fact, it could mean the collapse of the current economic system in which people are paid to work. Or perhaps, it could be split into two systems: one to support the recycling systems and the environment and to ensure everything we need costs very little (or is free, especially for those who participate in the system[176]) and so allow them the freedom to be curious and solve problems, and another that kind of continues the current economic approach but with a different purpose. Perhaps the focus will be on implementing new ideas and solutions and building

---

176   For a glimpse into the life of those living and working in the system, consider the Bruderhof communities. However, the system used in the new world order to grow food and manage the environment need not have to have a religious element to it. Some people might naturally develop a spiritual connection and appreciate the holistic approaches to solving problems and how things are interrelated. But the thing that is in common is the growing of food, returning the environment to its pristine state, and the feeling of love that comes from sharing the harvest with each other and the social connections people will have. For other people who have higher goals and wish to get the broadest education, it is important to let them be free to try the alternative system.

the new technologies of the future. But whatever we do, it will not be for-profit motivation, especially for the needy things and the things we enjoy doing. We do it because we want to do it. No need to use the "stick" method by denying people access to money to survive just to force people to do "work" or else take the risk that people will fight in negative ways in society to get what they need and later what they want. Whether money will ever exist in the future is another issue, but with no unemployment or poor people living in the economic system to worry about (as they can support the other system and receive freely what they need and be happy), there would be no need for a police force to protect the rich and powerful, and no military force to protect a nation. No more need to defend borders between neighbours. Instead, the new economic system would have its own citizens and anyone else who might be willing to join from the other system to develop and build new technologies. But the most important thing of all in the new world order is that people will have a choice of which system they want to go into.

Humanity must face reality. Our current economic system is not exactly working properly for everyone. When the new electromagnetic technology comes out, people will have access to a new solution. And it is a solution that will force changes on the current economic system. We have no choice. If we do not change, then there will be negative consequences for everyone by those who are treated badly in the system through the wrongful application of the technology. However, if we want a solution to be positive for everyone, something will have to give from the old system. People must be given what they need to survive as the minimum standard we must reach for when treating people equally and with respect. Anything above that is a bonus and a gift to humanity. This is humanity's guarantee that the application of new technology is always peaceful and positive for everyone. Otherwise, the work we do is mainly to maintain the recycling systems for creating what we need. And once the systems are in place, and people are focused on getting the environment to be food productive and in a more pristine condition with adequate trees and a sufficient freshwater supply for as long as they are needed (which for the things we need will be indefinitely), the work we do to maintain the systems will be minimal.

As people say, "You should work to live, not live to work". It will not be like the current dysfunctional economic system that tries to switch this around through exploitative employers seeking to make high profits and forcing employees to work for very long hours. In the new world order, people will not have to work more than is required to have what they need. Finding ingenious ways to reduce the workload for the same productive output means people will actually have more time to do what they want.

Unfortunately, not good if the USAF wants to maintain secrecy. The military does not want to see people have free time and question things in case certain secrets come out[177]. That's the problem the U.S. military has when it comes to UFOs. It simply does not want people to find out the truth under any circumstances.

Do not like the idea of a new world order? Prefer to maintain the secrecy? Unfortunately for the USAF, it cannot expect the secrecy to last forever. The time is fast approaching when more and more people will discover the truth. Humans are inherently curious creatures, and the longer the secrecy is maintained and people learn about the electromagnetic technology in the UFO reports, it will not be long before the electromagnetic technology is fully revealed and prototypes are built. People will find out. It would be better for the USAF to prepare for the eventually release of this technology and start to look to a new and brighter future.

And while there are growing global problems to deal with, would it not be prudent for the USAF to release the technology if there is the slightest chance that someone could use it to solve our dependency on fossil fuels? As members of the military have claimed, there is no obvious sign of an engine in the Roswell disc, at least of a type that we are familiar with on Earth. Clearly it cannot be powered by traditional fossil fuel or nuclear technology. As we have discussed earlier in the chapter, we are almost certainly dealing with some kind of an electromagnetic technology generating intense electromagnetic fields and using them to perform work.

Yet the USAF seems unable to fathom what will happen and how the future can be better for everyone. It does not seem compelled, let alone eager, to release the evidence it has found, nor any technology it has worked out for the benefit of the world in these troubled times.

---

177   This may be seen as going a bit off-topic, but in mentioning the need for a new world order to handle social problems such as terrorism, it is important that people become aware of an alternative system that can solve world problems and bring positive benefits to everyone, even for the rich and powerful so long as they do not try to force everyone to accept the current economic system and all its flaws.

Instead, the USAF thinks it is better to hide the evidence like it is a museum piece in some secret location in the U.S., accessible by an elite patriotic American audience to walk by and be left in awe at the sight. Or else keep the technology it has built to itself. Meanwhile, the world can be left like it was "Rome burning" with dangerous climate change and all the rest when the technology could quite easily solve so many problems for humanity, quite literally overnight.

It is extraordinary.

The U.S. military and those who support it are showing every signs of having lost the plot as it tries desperately to maintain secrecy, thinking it can continue for as long as it likes. Wake up! This will not work. Curious people are discovering there is something going on. There is an electromagnetic technology emerging from the UFO reports. So why the nefarious tactics to control the population? It will not work anymore. Curiosity is the thing that will stop the secrecy in its tracks and bring out the truth.

Still not keen to start a new world order? Then curious people are left with the only viable explanation for the continued secrecy. In the current economic system, in which the USAF wants to maintain secrecy about UFOs, some people in the know do want an unfair economic advantage in the system, as well as a means of controlling everyone else to do the things to maintain the current economy and not figure out anything else, including any secrets that the USAF has learned about UFOs.

Or are there really just that many military men in the United States who would feel incredibly insecure about the future if the technology was to be released? Okay. Well, if the military and the rich and powerful are genuinely worried about their own security in the new world order, then they should ask themselves what is better: being rich and insecure, knowing that there are a growing number of terrorists willing to change the system and wreak havoc on Western democracies and the economic systems, until, eventually, we destroy the planet in the next world war (or else climate change will put an end to our way of life)? Or being happy and secure, knowing that we all have what we need at little to no cost and letting others achieve great things with this technology, such as traveling to the stars and bringing back new ideas that would be of unimaginable benefit to humanity? The answer should be clear. We must treat everyone as equal partners sharing in our knowledge and technologies while providing the things we need to survive in the cheapest way possible for all. That is the key to a genuinely secure future.

While the USAF refuses to release what it knows because of the considerable benefits some people in the know can gain from the current economic system, it is time that we, as genuinely curious people, figure out for ourselves exactly what it is that the USAF has discovered and build a prototype. We know that it is an electromagnetic technology; that's a given at this stage. There is something in electromagnetism that will explain UFOs right down to their very core. Therefore, the next logical aim in this research is to determine what kind of electromagnetic technology has been discovered in the UFO reports.

Let's see if we can figure this one out.

CHAPTER 8

Revealing the New Electromagnetic Technology

The deepest truths are the simplest and most common.

—English clergyman Frederick William Robertson
(1816–53)

NOW THAT WE have acknowledged the existence of detailed and genuine UFO cases in Chapters 1 and 4, identified a set of common UFO observations in Chapter 5, eliminated ball lightning as a possible natural explanation for those electromagnetic UFOs in Chapter 6, and have realised UFOs must be artificial, but not likely to be man-made in Chapter 7, we must now ask the following two questions:

1. Do the UFO observations indicate the existence of a new electromagnetic technology?
2. Can the UFO observations be used to advance scientific knowledge?

The Glowing Effect of UFOs

To answer these questions, we need to begin by looking at the one notable feature of UFOs observed by most witnesses: the glowing effect of UFOs.

Here we have an undeniable observation. No sensible person can overlook the glowing effect of UFOs reported at night and, to a lesser extent, during the day. We are not talking about a tiny spot that suggests some form of landing lights, search lights or a porthole. Rather, we are talking about a large area, usually covering the ends of protrusions around a central symmetrical body or at the apexes of triangular UFOs, or even the entire underside of the object. Sometimes the glow may expand to cover the entire object (see the Levelland UFO case in Chapter 1). On closer examination of some UFO reports, we learn that some witnesses who have seen this glowing region have been reminded of the way a metal might glow when heated. Furthermore, the glow can appear brighter

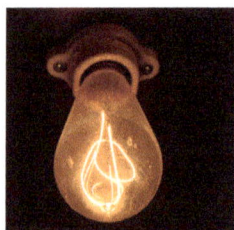

around the circumferential edge and on the bottom side when the object accelerates away from the ground. In fact, it takes astonishingly little imagination to realise that this "glowing effect" looks very much like a common household item we have all seen before: the incandescent light bulb. With its tiny filament composed of a high-temperature-resistant tungsten metal receiving a high oscillating voltage or charge, we see a rather remarkable and obvious glowing effect generated by the metal. Noting the fact that most UFOs seen during the daytime appear metallic in nature, it is not unreasonable to consider investigating the relationship between the ways in which both the metal in light bulbs and the outer surfaces of UFOs glows. Moreover, the brightly glowing filament has the ability to give off heat (another common feature of UFOs), and we know that electromagnetic fields generated around the metal by the charge can affect compasses at close range (again, not an unheard-of observation for UFOs). Such a palpable connection between light bulbs and UFOs seems too coincidental to ignore. Could this mean that the light bulb might provide insight as to how UFOs move through the air?

To be certain about this, let us take a closer look at the incandescent light bulb and see where this will take us.

What Do We Know About the Electric Light Bulb?

One thing we know about the light bulb is that it is designed to emit considerable amounts of radiation in the form of heat and light. No surprises here, considering that this has always been the purpose of an electric bulb: to bring light to dark areas. With a bit of luck, the light bulb will also illuminate another area of our lives: the UFO phenomenon.

Speaking of radiation, let us consider an interesting fact about this energy. As any physics textbook will tell us and confirmed time-and-time again through experimentation, radiation has the inherent ability to move solid matter. In the same way that we might throw a piece of solid matter, such as a tennis ball, at another object and can observe the second object move after the collision, or even the way a rocket takes off from the ground after shooting hot gases from its back end, the same effect can occur when radiation interacts with solid matter. Regardless of whether the matter is charged or uncharged, radiation will move it.

You may ask, "Why?"

Good question.

One would have to say radiation acts as though it has mass. The only problem is, what is this *mass*?

We know a typical solid object, such as a chair, is composed of innumerable tiny objects with mass called atoms. Within each atom are smaller particles with mass known as electrons, protons, and neutrons[1]. These particles come together principally through the action of electrostatic forces between negatively charged electrons and positively charged protons[2] to form the neutron, the atom, and ultimately the

---

1   There is a bit of a Russian "Matryoshka doll" game being played out in particle physics of whether these fundamental particles are made up of smaller particles called quarks, which in turn are made up of even smaller particles or strings of pulsed energy (with each pulse considered a particle in its own right). One would imagine that if we do finally get all these particles right down to their fundamental level, they are all forms of pure energy, perhaps entirely electromagnetic in nature. But let us not worry about this.

2   The neutron is said to be uncharged and, therefore, is a little more troublesome for the physicists to explain, considering that it is there to prevent the positively charged protons from flying away from the nucleus. How the neutron is able to do this is unclear, so physicists have

chemical bonds that hold atoms together in a solid crystalline structure, which in turn gives the matter the sense of volume and the feeling of rigidity and solidness to the touch that we attribute as a form of mass. With radiation, however, it is not accurate to say it is composed of the same particles as ordinary matter. The best way scientists can explain radiation at the present time is to say it is composed of a thing called *electromagnetic energy*.

What is electromagnetic energy?

Another excellent question.

Unfortunately, no one knows precisely what electromagnetic energy is. All we know is that when the energy interacts with solid matter, there is movement. More troublesome is the fact that this energy can move uncharged matter.

In the eyes of arguably the greatest physicist of the 20th century, Albert Einstein, there was a problem here. Unlike other physicists of his time who simply accepted this fact and chose not to pursue the problem to its logical conclusion (and still do to this day), Einstein wanted to find out why.

In fact, Einstein's biggest gripe was in the direction physics was taking after the 1920s. He noticed that the theoretical underpinnings of quantum mechanics via the Heisenberg Principle were designed to stop scientists from using their imaginations to go beyond what they could see at the diminutive scale known as the quantum world. Sure, scientists had trouble using any kind of instrument to observe particles in the quantum world because it required radiation to be injected into the system causing the particles to be disturbed, and that would prevent physicists from working out what the particles would do next or where they had been with any kind of precision. As a result, many physicists assumed there was nothing more to learn about radiation or anything in the quantum world. We should merely accept the concept of a mass as the thing that is causing radiation to move solid matter. If there are other practical applications of quantum theory, the mathematics will tell them based on the statistical group behaviour of quantum particles, but not their imagination.

---

decided to call the force of attraction between neutrons and protons the *strong nuclear force*. At any rate, chemical bonds and the atom as a whole are essentially formed by the electrostatic attraction of negatively charged electrons and positively charged protons because unlike charges attract. However, the particles also emit radiation as they spin and perform other accelerating motions. As a result, the particles are pushed apart slightly by the radiation. So the most the electron can do is orbit around the nucleus or to move back and forth across the bridge called the "chemical bond" to bind various atoms to each other.

Einstein was not happy.

Consequently, Einstein decided to pursue a separate line of scientific inquiry and to find out where this would lead him.

The Age of the Unified Field Theory

At the present time, and the way the standard university physics textbooks are written, it is assumed by scientists that radiation somehow generates an invisible mass of its own and that this mass interacts with the mass of solid objects. To support this claim, you might find a section in a standard university physics textbook showing in a mathematical sense how radiation moves so-called uncharged matter, mainly because physicists assume that anything that is uncharged must remain that way at all times. And when you ask scientists what, exactly, the radiation is moving, they will always state that radiation moves the mass of uncharged objects. Luckily for the scientists, charged objects just so happens to have mass as well. Indeed you can never separate mass from charge. The two go hand-in-hand. So the scientists also assume radiation is moving the mass of charged objects.

Einstein personally detested this kind of assumption. He could not understand why radiation should move uncharged objects. In particular, why the mass component of solid matter? After all, radiation is a purely electromagnetic phenomenon. Anything having electromagnetic properties should essentially affect the charges, not the mass. This naturally made Einstein wonder, what makes radiation so special that it should affect uncharged matter, and not just charged matter through its mass?

This was the fundamental problem for Einstein. After completing his General Theory of Relativity; learning of the results of an experiment conducted in 1919 during a solar eclipse on the ability of radiation (in the form of light) to bend its path near the Sun's domain; and discovering the fact that an instrument called a radiometer can have its uncharged metal plates spin on the tip of a fine needle by the radiation, Einstein became perplexed by how radiation could move uncharged matter. What was in the radiation to achieve this stunning effect?

After much consternation and many hours of careful thinking on the problem of light, he made the first important step of theoretically incorporating the gravitational field into the electromagnetic field. In other words, light (or radiation) must generate a gravitational field of its own. He thought, at the very least, this gravitational field of light must be interacting with the gravitational fields of solid matter, causing both light and other solid matter to bend and become "attracted" to each other.

In essence, he saw light as behaving like any other ordinary solid matter, and believed it should be treated as such.

From this first milestone he created his Unified Field Theory, which was completed and published in 1929. But that was not the end. The question still remained: what is the gravitational field? The core problem that had led Einstein on the quest to write his General Theory of Relativity had only given way to a new problem when he realised the electromagnetic field was somehow playing an important role in the generation of a gravitational field.

To solve this new problem, Einstein had to find ways to separate the two fields in his mind through careful thought experiments so he could show what else might be generating the gravitational field. He knew the electromagnetic field was one source. But could the gravitational field exist without an electromagnetic field? And if so, is there an exotic particle, or something familiar that Einstein could identify in solid matter, to help him pinpoint as being responsible for creating the gravitational field?

Years passed as Einstein thought about the problem further. Then, suddenly, Einstein gathered up some advanced ideas on his Unified Field Theory and decided to destroy them. This was also the time when he began to quieten down, quite possibly in response to seeing the destructive capacity of humans through yet another world war and later when the Cold War began with Russia, keeping the world on edge with no end in sight. After witnessing the destructive application of his famous equation linking mass and energy in the development of the atomic bomb, he may have felt obliged to keep quiet his subsequent ideas.

Around the early 1940s, the U.S. Navy may have accidentally stumbled upon the initial concept that led Einstein to write his Unified Field Theory. Rumours abound of the Navy having used degaussing equipment in wartime ships for an alternative purpose rather than to detonate sea mines at a safe distance. By combining several of these devices and get them to create in resonance an oscillating or pulsing

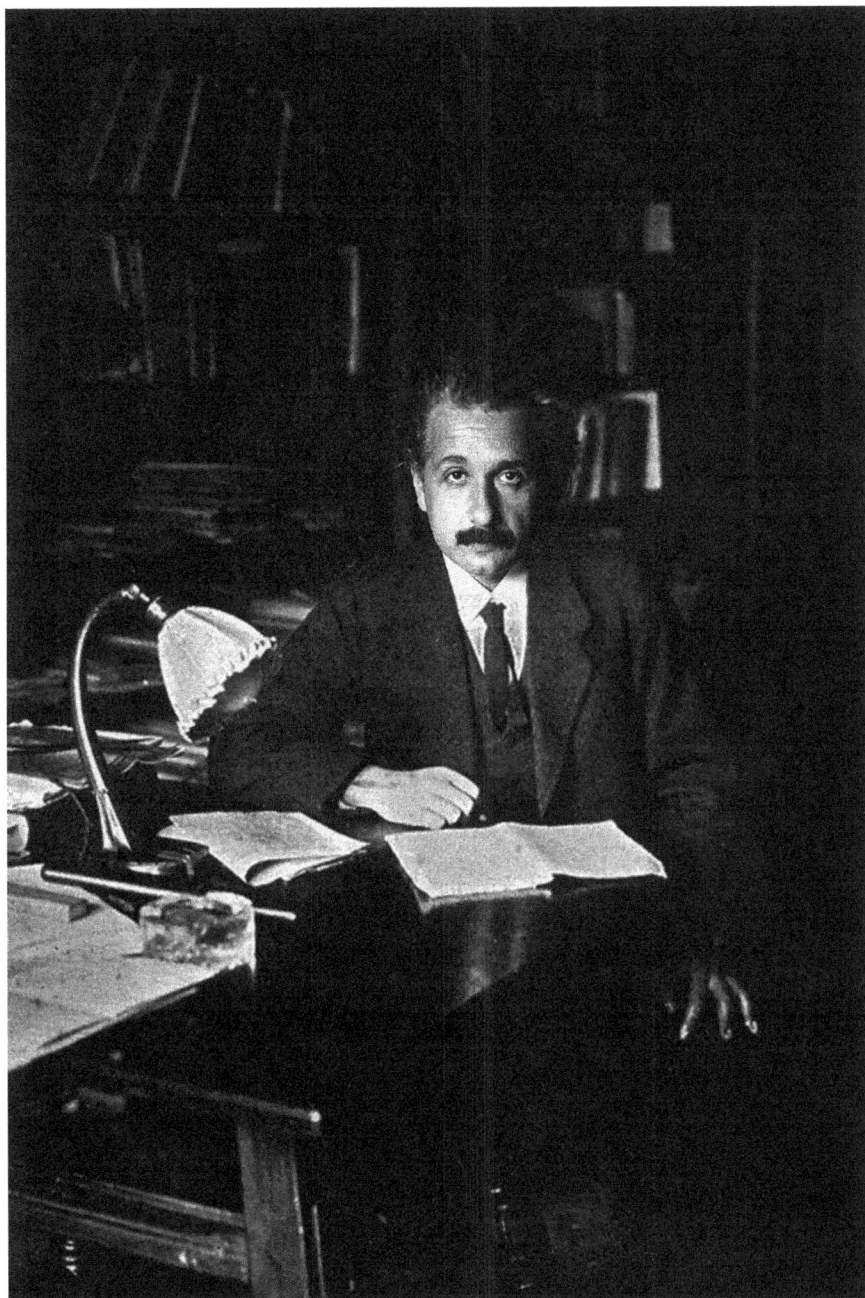

Albert Einstein in his office at the University of Berlin in 1920. It was at around this time that he began the work on the Unified Field Theory.

347

magnetic field of such intensity that light, the Navy felt rather confident that it could bend light back on itself and any light from behind a ship could be made to bend around it. In essence, the military appeared immensely interested in achieving electronic camouflage in a process known as *invisibility*. It is unclear whether the Navy did officially carry out such an experiment, but the idea is scientific enough and seems quite feasible.

The Navy did employ Einstein on some contract work for a short while. Later, two photographs were taken of Navy officers in Einstein's office conferring with the great scientist.

Afterwards, very little was known of Einstein's work. Apart from trying to solve his equations for a special case involving a non-static field problem (of which radiation is an example), which required considerable effort to solve mathematically, he stayed relatively quiet. Then, on the day of his death, Einstein asked his secretary, Helen Dukas, to bring in his most recent calculations on the Unified Field Theory. After working on his papers for a few hours, he fell asleep in his hospital bed and never woke up. The next morning his papers disappeared, probably taken as a souvenir by an unknown person(s). His office was later inspected and his work reviewed and nothing was found to indicate that his work on the Unified Field Theory was wrong. If anything, it would appear Einstein was quietly telling the world that he was on the right track but didn't want to spill the beans on precisely what he knew towards the end of his life.

So did he find a solution to the problem of the gravitational field? And can we explain how electromagnetic energy moves solid matter in a fundamental sense?

The Radiation Paradox

Presently, one of the few things scientists can add to scientific knowledge about the nature of radiation is that electromagnetic energy can be "quantified". A somewhat sophisticated term that basically means the energy comes in packets, or bundles like the beads of a necklace. While scientists have coined a term for these bundles of pulsed energy—the *photon*—we are none the wiser in our understanding of what this energy is or how it performs its magic of moving uncharged matter.

To this day, our understanding of how this mysterious energy acquires mass and exerts a force on solid matter is woefully inadequate. All scientists can do at the moment is accept that there is a thing called electromagnetic energy and that this mysterious energy can move solid matter. No need to bust our brains on the problem. Radiation must be moving uncharged matter through its own mysterious mass and nothing else.

Thus, in a simplistic sense, you can imagine radiation as an oscillating piece of electromagnetic energy creating what is called negative and positive mass (or energy) over time. If we took a snapshot of radiation over a half cycle in time (as radiation is described mathematically as a wave), we could calculate one of its negative or positive mass using Einstein's famous equation that relates mass to energy:

$$E = mc^2$$

Hence, if we know how much energy E is contained in radiation (by pretending that it is not moving), you can imagine this fuzzy ball of energy as having a mass equivalent to the following:

$$m = \frac{E}{c^2}$$

where c is the speed of light (a constant that, in the vacuum of space above the Earth's atmosphere, is very close to 3×10^8 m/s, or 300,000km/s).

This approach is okay if we apply the Newtonian picture of solid matter to radiation. However, there is another reality we have to accept with regard to radiation: the quantum picture of light. According to this mathematical picture, physicists claim that as soon as you take a snapshot of radiation (or you bring the energy to a complete rest), the oscillating electric and magnetic fields are not actually oscillating. From a mathematical standpoint, if one field is not oscillating, then the other field is not oscillating either and, in fact, must drop to zero. This means that in quantum theory, the radiation does not actually exist. Until we physically observe the radiation interact with other solid matter, radiation must always be assumed to have no mass and/or technically does not exist. Radiation can only exist when the energy is moving and oscillating. Furthermore, the only definitive way we can be sure of its existence and later measure its mass is when it interacts with solid matter.

Now when radiation does interact with solid matter, there are additional observations one can make of this mysterious energy. In particular, how much control radiation appears to exert on matter depends entirely on its energy density. As any university physics textbook will tell us:

"…the pressure exerted by collimated electromagnetic radiation is equal to its average energy density."[3]

Don't get intimidated by complicated scientific terms like "collimated" and "average". *Collimated* simply means the individual packets of radiation are moving parallel to each other, whereas inserting "average" into the above statement is just another way of saying that the scientists want to keep things simple when creating formulas for calculating the radiation pressure. Scientists know the radiation density changes over time because of the oscillating nature of radiation. But when calculating the radiation pressure, it is usually easier to get an average of the density and use this figure to calculate the total radiation pressure.

In other words, physicists are confident that the higher the density of the energy making up radiation, the more solid the mass generated by the radiation. This in turn exerts a greater force on other solid matter and, with it, the radiation pressure on solid matter increases. It is as simple as that.

With this in mind, we must use the following correct version of Einstein's total energy equation, which links mass, energy and the speed of the mass (or energy):

$$E = \sqrt{(pc)^2 + (mc^2)^2}$$

where p is momentum.

Since mass m in this equation is described as the rest mass, and we are told it is only suitable for ordinary solid matter (the stuff that is made of atoms), no such thing exists in the case of radiation. If we

3   Link between energy density and electromagnetic momentum, or radiation pressure, is revealed in this quote and can be found at
     http://farside.ph.utexas.edu/teaching/em/lectures/node90.html.

were to remove radiation from the environment and stop it from interacting with and hence contributing its energy to solid matter, we would set the momentum p of the radiation to zero. What we have left is the energy locked inside the atoms making up solid matter; and in that case, we have the familiar equation:

$$E = mc^2$$

However, if we want to look at the radiation and nothing else, physicists must set the rest mass to zero because it is assumed that this mass is not related to the mass generated by the radiation when it interacts with solid matter. Even though nuclear explosions can reveal a considerable amount of electromagnetic energy locked away in the nuclear bonds holding the nucleus of an atom together, and there is a view that electrons and protons themselves are probably composed of pure electromagnetic energy, we are told that this energy (and hence the mass) is different. If this is true, the removal of all solid matter except for the moving radiation means we get a compact equation that relates the momentum of this moving energy in the radiation to energy E, as shown below:

$$E = pc$$

Of course, if the radiation is not moving, quantum theory demands that this equation collapses and so too the radiation. That is, the energy E must be zero if the momentum p of non-moving radiation is non-existent. Radiation must always be moving in order for it to exert momentum. And detecting this momentum is the only way we can register the very existence of radiation as a form of energy and hence capable of displaying mass by the very act of interacting with solid matter.

Now, by its very definition, momentum is defined as how much of an impetus something is likely to give to something else when it interacts (e.g., collides or emits), which, in the world of Newtonian physics, is controlled by mass and speed. Thus, in order to calculate the momentum of anything having mass (whether or not it is radiation), we simply multiply the mass and speed. For radiation, the momentum is calculated as follows:

$$p = mc$$

which again brings us back to the famous equation relating mass and energy.

As this can be a little confusing, physicists like to be precise about the kind of mass we are discussing here. One is the *rest mass* (or just plain ordinary "mass" consisting of a multitude of atoms just sitting around presumably doing nothing[4]), and the other is *moving mass*. Moving mass may be refined further with another term. Should the moving mass approach the speed of light, it is known as *relativistic mass*. However, when we talk about the mass created by radiation at the speed of light, scientists have yet another term to describe it. Since radiation is generated by charge, the kinetic energy of a moving charge used to generate the radiation is also a form of mass (when it hits solid matter). When scientists detect this mass, it is called *electromagnetic mass*.

It is important to remember here that there are different terms used by physicists to describe mass of radiation and ordinary mass that is moving or at rest.

Confusing, isn't it?

Never mind. What matters most is when all these forms of mass start interacting with each other. Here, physicists realise they cannot tell the difference between types of mass when they are colliding. Not even Einstein could see a difference between, say, electromagnetic mass interacting with solid matter and another solid matter doing the same thing. Likewise, the difference between relativistic mass when it collides with solid matter is virtually indistinguishable from any other ordinary matter. When mass collides with mass, there are no terms in physics to show the differences. In the case of radiation moving something, it does not matter whether we use the Newtonian picture of light or the quantum picture of light. In fact, the equations merge into the simple Newtonian case at the moment radiation interacts with solid matter. Thus, we can imagine radiation as like a tennis ball colliding with other matter. Any specialised terms scientists use to describe mass at different speeds or in radiation are irrelevant as soon as radiation moves other mass.

---

4   In reality, atoms are constantly accelerating through its vibrational, spinning and translational motions. When the atoms are locked into a crystalline structure, the atoms still move. They simply vibrate due to the collisions of radiation in the environment and the way atoms emit the radiation back out to the surroundings to keep to a minimum energy state. Therefore, *rest mass* should more accurately be called *moving mass*.

To put it simply, we have to view radiation as acting like it has ordinary solid mass while it moves and interacts with other matter.

However, just when you were starting to settle into the idea of radiation having some mass, you may also hear scientists say radiation has no mass. As *Wikipedia* states:

> "In particle physics, a massless particle is an elementary particle whose invariant mass is zero. The two known massless particles are both gauge bosons: the photon (carrier of electromagnetism) and the gluon (carrier of the strong force)."[5]

Why is radiation considered to be massless?

It is to do with quantum theory. Basically, quantum theory does not attempt to resolve this radiation paradox. It simply accepts both views of radiation as perfectly reasonable. Thus, from this view, radiation can behave like it has mass, and it can also have no mass. It seems the only time radiation displays evidence of its mass is when it interacts with ordinary matter to cause movement, and this is when we apply classical Newtonian physics to the problem. At all other times, quantum theory tells us to assume radiation has no mass (and potentially may not exist until we know it has interacted with solid matter).

If you want a way to resolve this complicated paradox, it may be more correct to say that radiation is constantly flowing through space (so we assume it exists) and, therefore, has mass, but it is so minuscule and moves so incredibly fast through the ocean of all radiation that it is effectively "floating" in it. And anything that floats may not show evidence of mass.

You see, what makes it hard to detect this mass in the radiation is the fact that we can't physically grab a piece of the radiation, place it on a scale, and measure its mass through the use of our planet's gravitational pull. It is moving incredibly fast. Sure, scientists do know that gravity can affect radiation; this is a veritable fact after a 1919 solar eclipse allowed scientists to observe how light from distant stars can bend near the Sun. Funny that this light-bending effect should be mentioned, as ordinary matter does the exact same thing in the presence of a gravitational field. Just as gravity can pull on a tennis ball thrown through the air, so too can gravity bend light toward the

---

5   https://en.wikipedia.org/wiki/Massless_particle

ground, although to an incredibly minuscule degree. To get a sense of the scale required to see the bending effect here on Earth, consider that every second radiation moves through space, it covers a whopping 300,000km. Now that is fast. Even if you could observe as far as the eye can see, on Earth the light would still appear to travel straight. To see the bending would require a much greater distance, more than the diameter of the Earth, given the weakness of the Earth's gravitational field. However, give it enough distance and you will see it bend, ever so slightly; more so closer to the Earth's surface than further away. However, to make the radiation bend even more, you can increase the frequency of the radiation to help increase its mass-like property and so become more susceptible to bending from the gravitational field. Similarly, you can apply a stronger gravitational field to influence the radiation. This is what happens when we look at those objects described as black holes[6]. Here we find a much more significant bending effect. So incredibly powerful is this gravitational field that it forces light (or radiation) to bend back on itself. On the surface of a black hole, it is technically possible to weigh the full mass of radiation on any position by using the right instrument.

So technically speaking, you can imagine radiation as always having mass from a Newtonian physics perspective. Not a lot of mass—just enough to allow it to travel fast. And in the ocean of other radiation[7], we could describe it as floating in space to give the impression of no mass. Of course, we know it must have mass. It is just that we can only detect this mass when the radiation exerts a force on solid matter, depending on the direction in which the radiation is traveling. If it is moving parallel to the surface of a solid object (for example, a weighing machine), radiation will appear to have no mass. But if the radiation bends or travels perpendicular to the surface of the solid object, which it can hit and cause to move, then we would have no choice but to say that radiation has mass.

---

6   Black holes are fast spinning stellar remnants created from a supernova explosion. They are invisible due to the intense gravitational field generated by the rapidly rotating mass that bends the light of the star back on itself and other light behind it can bend around it to permit an observe to see what is behind the star. The best way to detect black holes is observing the superheated accretion disc that forms around the black hole. Interestingly, black holes and UFOs do have something in common, as we shall see later in this chapter.

7   The energy density of this ocean of radiation does restrict the speed of radiation to the familiar 300,000km/s. Lower the energy density and the speed of light increases. Raise the energy density and the speed of light slows down (as occurs as light bends through the gravitational field of a planet or star).

Perhaps the best analogy would be this: imagine a blue whale swimming through the waters of an ocean. We know that a whale has a lot of mass. You can easily see it with your own eyes because it is large enough (and slow enough) to detect. But if the whale moved horizontally in relation to a weighing machine, you would swear that the whale had no mass (assuming you could see it moving past). However, if you could find a way to stop the whale moving (which, in quantum theory, would imply no whale exists until it interacts with another solid object) and let its body be pulled towards the machine by the Earth's gravitational field, the weight (and thus the mass) would be registered on the machine.

As you can see, we can understand radiation using a Newtonian physics perspective. Just imagine radiation as always having mass and it exists even if we cannot see the radiation. It only gets complicated when we apply quantum theory to the problem because it cannot see everything. It must rely on mathematics for answers. And when we apply the mathematics, quantum theory essentially claims that radiation does not exist until it collides with solid matter mainly because we cannot observe the radiation. We have to assume it has disappeared into thin air. So it is like saying the whale has disappeared. In order for the whale to exist and to have mass, it must always be moving and able to interact with other solid matter for us to "see" (or more accurately in quantum theory, "know") the whale was there. It is just the way the mathematics of quantum theory work.

How Does Radiation Move Solid Matter?

Now just because we know radiation has this mysterious mass, or something that allows it to move solid matter, that does not necessarily mean physicists have all the answers and everything in the world of physics is fine. The biggest question to consider next is whether radiation is moving the mass of solid matter, or if it is moving the very charges that comprise the solid matter.

This is the thing. Physicists have naturally assumed that radiation only moves the mass. Even if solid matter is charged, it always has mass. In fact, you cannot separate charge from the mass. For example, the electron is charged, and it has its own mass. The same goes for the proton. The neutron is a little more complicated. Here, the measurements seem to suggest that there is no charge. It only has mass.

This gives physicists reason to believe that mass (and hence the gravitational field) can exist without charge and, therefore, without an electromagnetic field.

At any rate, what matters here is irrespective of whether something is charged or not, we must always accept the idea that radiation is moving the mass.

But is this true?

Well, this was the very question that preoccupied Einstein after developing his Unified Field Theory. Apart from trying to solve the equations for simple and complex scenarios to see what else was possible, and in order to understand the nature of the gravitational field, he realised the electromagnetic field was inextricably linked to the gravitational field. The whole purpose of doing this work on the Unified Field Theory was to explain the gravitational field. When he developed the Unified Field Theory, he came to a decision to accept the picture that radiation not only generates mass when it interacts with solid matter, but also generates a gravitational field of its own. Radiation is ordinary matter. Like any other ordinary matter, it always comes with a gravitational field. We must include the gravitational field in the picture for light.

Hence the Unified Field Theory was born.

But this wasn't the end. The question remains: what is the gravitational field?

Now Einstein realised that the electromagnetic field was complicating the situation for him in finding answers to the gravitational field question. Why should there be an electromagnetic field involved in the creation of the gravitational field? Sure, radiation is ordinary matter. But what is this mysterious thing called the gravitational field associated with all matter in the universe, and why does it exist?

In addition, it may be totally coincidental that anything charged should always carry mass. We observe this to be the case in both a positively charged proton and a negatively charged electron. That is rather convenient for the physicists because now it is possible to conclude that, if there is mass, the radiation can move it because it generates its own *mass*. Well, that is the theory at least.

But is it true?

To really answer this question, it is important to challenge these assumptions and ask something far more fundamental. That is, what is the nature of so-called uncharged solid matter? In other words, does uncharged matter really exist in the universe? Or is it a perception that we have created because of how our instruments have attempted to interpret the universe when measuring the charges of ordinary solid matter?

Do We Live in an Electromagnetic Universe?

In the late 19th century, Michael Faraday was the first to notice that a static magnetic field or a static electric field influenced the motion of charged matter, and apparently *only* charged matter. Forget uncharged matter as it is claimed that this does not move in the fields unless it contains a charge. But since physicists assume uncharged matter is always uncharged, we have to say that the magnetic or electric field is not affecting uncharged matter. Even when the two fields are created independently and combined to create a static electromagnetic field, only charged matter appears to be influenced. Fair enough so far. However, things start to get a bit weird when the electromagnetic field is oscillated (to create *radiation*). It is then that we have a situation where the field somehow influences not only charged matter but also uncharged matter. Well, at least that is what we are told based on what scientists have seen in experiments and how the physics textbooks are written.

But how can radiation do this?

Einstein was not unfamiliar with this problem. For many years, he grappled with this mystery. His first confident step was to develop his Unified Field Theory. He did this to mathematically cement the gravitational and electromagnetic fields together into one grand unified field. Only problem is, the gravitational field has not yet been explained. Mathematics may show an equal sign to reveal the link between the two fields, but we effectively have a "chicken and the egg" situation all served up by the world of mathematics in which no one knows which one started first? Was it the electromagnetic field, or the gravitational field? In other words, which is the source?

Indeed, how do we separate the two fields?

This was an important next step. If there was a way to show that the gravitational field is independent of the electromagnetic field, it might be possible to identify the source of the gravitational field. And to do that, Einstein needed to separate all the charges in the universe and put them into their own electromagnetic field tensor in his own unified field equations.

To put it another way, Einstein had to imagine a universe free of all charges responsible for creating the electromagnetic field. He had to identify all the sources for the electromagnetic field and to imagine eliminating them from this hypothetical universe. Then he could ask, What would be left in the universe to look at? For if there is anything left, one must conclude the gravitational field can exist without the electromagnetic field, and whatever remains must be the source of the gravitational field.

But first, Einstein had to know all the places where charges could be lurking. To do that, he had to ask, is uncharged matter truly uncharged?

Let us imagine a supposedly uncharged object sitting in an otherwise empty universe. If we were to use the most powerful microscope imaginable to look at the surface of this object, we would see a countless number of vibrating atoms sitting on the surface, held together in a rigid crystalline structure by the electrons in the outermost orbits of each atom as they constantly move between the atoms to form a chemical bond. Conversely, the rest of the electrons in the innermost orbits would be there to help neutralise the positive charge of the protons in the nucleus.

Now imagine an electrode with a very fine tip—one so small that it could measure the electric charge of one atom. We connect this electrode to a super-sensitive meter designed to measure an electric charge at any instance in time. The electrode is then placed over a tiny area directly above a single atom. What do you think would happen next?

As the electrons accelerated rapidly in their various discrete orbits around the atom and in the chemical bonds holding the atoms together, we would notice that, at certain times, some extra electrons would appear for a fleeting moment on the surface of the object closest to the tip of the electrode. Because electrons are negatively charged particles, these extra electrons would have to register a negative charge on the meter.

We would wait for a fraction of a second and measure again.

As the electrons move around to another position, there would be moments when the nucleus of the atom, consisting of positively charged protons and the uncharged neutrons, was marginally exposed.

When this happens, we would register a slightly positive charge over the same area because of the extra protons.

If we kept measuring the charge in this tiny area over time, we would notice how the electrode registered on the meter a constant variation in the electric charge as it swung from negative to positive and back again.

Hold on. According to the classical laws of electromagnetism, an electric charge constantly changing over time implies the presence of an electromagnetic field. Unless the laws as we know them are wrong, we would have to say that the atom is charged and generating time-variable electromagnetic fields. Add all the other atoms together, and we must conclude that there is no such thing as an uncharged object. All supposedly uncharged matter is charged and emitting radiation all the time.

What the problem physicists have are their oversized electrodes used to measure the charge of solid matter. No matter how sensitive the instrument for measuring the charge is, an oversized electrode must give an average result across many different atoms to the instrument resulting in a number on the instrument panel showing solid matter has a zero charge. Little do the physicists realise that radiation is already being emitted all around the solid object. Einstein could not assume uncharged matter is uncharged. It must be constantly charged all the time.

If you need support for this claim, ask yourself why do you see objects at all? It is because they emit radiation. That can only happen if there are charges present and doing things to generate the radiation, such as accelerating or changing constantly between negative and positive.

To put it simply, uncharged matter is *always* charged.

Thus, if we are going to find the source of the gravitational field, we must eliminate the charges in the atoms responsible for generating the electromagnetic field.

What Is Controlling the Gravitational Field?

So what is left?

After removing the free electrons in orbit around the nuclei, the chemical bonds making up the crystalline structure, and the protons in the nuclei, we are left with the neutrons. And by definition, a neutron is uncharged, right? Therefore, the source of the gravitational field must be from the neutron.

Well, not quite.

English physicist James Chadwick certainly threw a spanner in the works for poor old Einstein in his attempt to understand the gravitational field. In 1932, Chadwick announced the discovery of a new particle called the neutron—a strange allegedly uncharged particle that hangs around in the nucleus performing some mysterious trick of preventing the protons from flying apart and destroying the nucleus. But how do we know for sure that the neutron is uncharged at all times? Given what we now know about uncharged matter, it may be wise to reconsider this notion. Why? Because it could make the difference between whether we do have a new particle, or a particle pretending to be a new particle. By understanding what is contained inside a neutron, we may discover another picture we have not seen before.

For example, it is a fact that a neutron can decay into an electron, a proton, and some remaining energy designed originally to keep the two fundamental particles close together inside the neutron. Sure, sometimes there is a delay in the decay process in which the electron does not appear once the proton emerges. In other words, the electron can be bonded to some additional energy to form a W-boson, but quickly decays into the electron and anti-neutrino. The latter particle is just the remaining energy to come from pulling apart the electron and proton from the neutron. But have the two particles actually merged into one particle inside the neutron? There is no evidence in any scientific literature to state the electron and proton have merged completely. For all we know, it is possible that the electron has never merged with the proton. And if that is true, the picture we have for the atom changes drastically.

Physicists must consider this. Otherwise, they could be making yet more undue assumptions.

The consensus is that the neutron appears to have no charge based on measurements. Without a charge, scientists are forced to build a model that requires the electron, the proton, and some additional energy to transform into two up quarks and one down quark. Support for this is based on smashing up neutrons inside particle accelerators and noting the appearance of the three resulting structures, which we must presume are the quarks in question. However, what if this view of the neutron is inaccurate?

According to the Unified Field Theory, we cannot have an uncharged neutron, just as we cannot have uncharged matter. The neutron must be charged and done in a way that gives the impression it is uncharged. For this to be possible, there have to be two opposite charges of equal magnitude being maintained and doing something to fool scientific instruments into showing an uncharged result. Luckily for the scientists, the electron and proton are already there to provide the equal and opposite charges. No neutron has ever produced a different or more exotic particle when it decays. It is always the same electron and proton that emerges from the neutron. Furthermore, it is interesting to find that the mass of the neutron is only slightly greater than the combined total mass of the electron and proton, as if suggesting that the electron and proton may not have merged and transformed themselves into a "brand new" particle with the potential to reduce its total energy (and mass) to well below the combined masses of the proton and electron in their independent states. Rather, it appears the electron and proton could potentially remain virtually intact and brought close together with the help of a little extra energy. While they remain fundamentally as electrons and protons, these particles could provide their own charge properties for the neutron, and do this in a sneaky way to make the physicists look foolish in thinking it is constantly uncharged.

Is this possible?

Well, certainly there is no reason to believe the neutron could not be charged. The problem for scientists is finding a sensitive enough instrument to measure the charge. All of our imperfect machines are designed to take an average result by sampling an environment over a given timeframe and displaying the average result on a screen. This timeframe can be very short, but in the case of measuring the charge of neutrons, it may not be short enough. Yet scientists rely on these imprecise instruments to tell them what the charge is, and if the instruments say "zero charge", scientists assume that this must be the case.

Think again. What if there is a chance that the electron and proton are separate charged entities spinning around each other to give the impression to an outside observer of a single uncharged particle? Consider this analogy: a boy and girl are holding hands and spinning around each other as quickly as possible. The boy's force keeps the girl standing, and her force keeps him standing. As long as neither releases the other's hands, the pair will continue to spin in place. But if either the boy or the girl weakens the force by suddenly letting go, the pairing will separate. Inside the neutron, a similar scenario is likely happening. Both the electron and the proton are engaged in an electromagnetic "dance" with each other, but if the dancing is disrupted, the neutron will be destroyed.

In that case, after seeing the neutron in this new way, we should probably eliminate the neutron from the hypothetical universe and focus on the last remaining energy: the anti-neutrino.

Could the source of gravitational fields come from this anti-neutrino? Highly unlikely. Like neutrinos, these are thought to have zero mass and travel at the speed of light, but others have claimed they can have a small mass value but cannot travel at the speed of light. If it is the former description, it is highly reminiscent of the properties of radiation, as if it could be electromagnetic in character. This is a bit like the gravitational field in the way it moves at the speed of light, except we are trying to find out here if it is electromagnetic in nature (i.e., created by radiation). Indeed, there can be no energy (or mass) capable of moving at the speed of light except radiation. So perhaps the neutrino and anti-neutrino are merely forms of radiation. If so, it is looking like we are returning to the idea that an electromagnetic field is controlling the gravitational field. However, if the neutrino does contain some mass and, therefore, cannot travel at the speed of light, it makes an even less likely candidate for controlling the gravitational field. The problem is that the neutrino (and, by implications, for the anti-neutrino) is especially hard to detect and study due to the energy packet being so dense and small—so small and so fast that it rarely interacts with other particles of matter. This is exactly the opposite of what you want when explaining the behaviour of a gravitational field. As we know, the gravitational field interacts with all matter. To give you an idea of just how rare the interaction between neutrinos and matter is, Katlyn Edwards said the following:

"In fact, a neutrino would have to pass through several thousand light years of solid lead before it would have a 50-50 chance of being absorbed."[8]

---

8   https://www.physicsforums.com/threads/why-neutrinos-rarely-interact-with-matter.434892/

If that is not enough, John F. Beacom and Nicole F. Bell from the University of Melbourne acknowledged the possibility that neutrinos could "decay into truly invisible particles"[9]. For all we know, these particles may be nothing more than photons merging with the rest of space. Gravitational fields, on the other hand, do not decay. They exist indefinitely.

So what remains? With neutrinos and anti-neutrinos not in the required numbers to account for the gravitational field or they appear indistinguishable from radiation, we are not left with many options. If there is any remaining energy in the universe to consider, it must be related to the strong and weak nuclear forces. These are the forces that somehow bind an electron to a neutron and keeps the protons together in the nucleus. However, they act over very short distances. Too short to account for the influence of the gravitational field over planetary and stellar distances.

With no other sources of gravitational fields to consider and all charges eliminated, including those residing in the neutron, it looks like we must conclude that the only thing capable of controlling the gravitational field has to be the electromagnetic field. The electromagnetic field is the only force of nature physicists know of that has the required range of influence on solid matter (which is infinite, just like the gravitational field), travels at the same speed (both the gravitational and the electromagnetic fields travel at the speed of light), and can influence solid matter in a remarkably similar way during a collision.

In that case, what is the gravitational field? Or maybe the real question we should be asking is, do we need a gravitational field to exist if the electromagnetic field can do all the work of the gravitational field in moving ordinary matter?

Now here is an interesting picture. It is suggesting that we are living in a purely electromagnetic universe. Whether this is true or not will greatly depend on whether the neutron is charged or not. Because if we do see the neutron as charged, there is a new electromagnetic explanation we can implement for the strong and weak nuclear forces, as well as gravity and universal gravitation, with radiation doing all the work of keeping solid matter together.

---

9   Beacom & Bell 2002.

Just to give an idea, we could say that the protons stay together in the nucleus of an atom not because of a mysterious strong nuclear force generated by the neutron, but rather because of the electron inside the neutron acting as an electromagnetic glue for the protons. As an analogy, imagine a circus clown who holds three bricks mid-air. All the clown has to do is press the bricks together with his hands and provide just enough lift to stop the bricks from falling down straight away. Once the bricks are momentarily suspended in the air, he takes his hands away, moves them back very quickly to grab one brick and put it in another position, and then quickly brings his hands together. The bricks are joined together and prevented from falling by the clown's hands. Once the bricks are lifted slightly to keep them in the air, he can start the whole process over again. Now imagine this happening at night with glowing bricks and the clown wearing black clothing throughout. The bricks will look like they are floating in the air and jostling with each other, vying for a position, and perhaps even trying to repel each other and move away, before something invisible somehow brings everything together. It would look as if a mysterious force is causing the bricks to stay attracted to each other.

Sure, we all know that it is the clown who is holding them together. However, if we didn't know this, we could say that the attractive force was gravitational. Or else we have to create a mysterious new force of nature to explain it. But even if the force is gravitational, how would we know for sure? With the advent of the Unified Field Theory, we learned that anything that is gravitational has to be electromagnetic in character. Therefore, in the atomic nuclei, the strong nuclear force holding the protons together is likely to be nothing more than the electromagnetic force that pushes the electrons and protons together. The electron is coming from the charged neutron. This is the force that makes the oppositely charged particles look like they are attracted to each other, rather than some mysterious strong nuclear force emanating from the neutron that physicists cannot explain.

As for the weak nuclear force, something is holding the electron to the neutron, but it can be overcome for the electron to emerge and turn the neutron into a proton. A simple electromagnetic solution is to imagine that the protons in the atom's nucleus are moving around. A proton can come within range of a neutron at the right moment to pick up an electron and have it bind to the proton, thereby turning the proton into a neutron. And the neutron that gave up the electron would turn into a proton.

As far as the Unified Field Theory is concerned, all forces of nature —gravitational force, electromagnetic force, the strong nuclear force, and the weak nuclear force—could be seen as manifestation of the one and only force of nature: the electromagnetic force. Everything might well have an electromagnetic explanation. Even the process of matter clumping together naturally in what physicists call the "gravitational effect" is probably electromagnetic energy using radiation as the prime mover.

Just to paint this picture, instead of accepting a mysterious gravitational field that can somehow pull matter together, we can envision the electromagnetic field (or radiation) pushing matter together. In the real universe, the charged particles we call electrons and protons are constantly doing electromagnetic dances—accelerating and spinning around atomic nuclei and within the chemical bonds holding atoms together. These constant changes in direction and/or speed result in the continuous absorption and emission of electromagnetic radiation. As this happens, the radiation emerging and arriving to influence solid matter can push against this matter ever so slightly in all directions. Hence, the object will vibrate. However, if two objects approach each other without touching, the absorption and emission of radiation on the surfaces facing each other must be reduced. This is because the mass of the two objects is acting as a radiation shield. Not so on the surfaces of the two bodies facing away into space. Therefore,. what we have is an imbalance in the "pushing" (and recoiling) force of radiation over the entire surface of each object. With nothing to shield the radiation on the outside of the objects for balance, the radiation with a higher energy density on the outside surfaces will push the objects closer together. The closer they are, the more effective the radiation shielding between them will be. With even less radiation between them to counteract the radiation force on the outside, the objects will appear to accelerate toward each other, leading to the so-called *gravitational effect*. The only thing is, there is no gravitational field. It is radiation that does all the work.

According to this new electromagnetic explanation of universal gravitation, radiation from space is the force pushing solid matter together. Where one mass shields another to reduce the amount and quality (or frequency) of the radiation arriving between them to influence each other and make them move, we should expect the two bodies to naturally move toward each other. Therefore, radiation is the force of gravity.

Now that we have an electromagnetic explanation for the gravitational field, the strong nuclear force, and the weak nuclear force, do we need any other force, energy, or particle to explain what is happening in the universe?

Indeed, all of this is encouraging the prospect of the universe being purely electromagnetic. It is suggesting that matter is constantly charged and comprised entirely of those two fundamental, charged atomic particles known as the *electron* and *proton*. As E.R. "Boblock" Le Clear wrote:

> "There are only two types of stable charged particles, the proton (sink) and the electron (source). All other stable matter is composed of protons and electrons including the neutron."[10]

Boblock further added that mass is defined as "being electromagnetic in character" and specified that "mass cannot exist without charge."

When it comes to explaining the mysterious forces of nature, we do not need to know about quarks or other exotic particles. Why would we need to when we can explain these forces from a purely electromagnetic standpoint using the two fundamental charged particles of nature—the electron and proton—and the radiation that is formed by them? The only reason physicists might continue to search for hypothetical particles is because they think there could be another way for these forces of nature to exist independently of each other. But let's face it. Are they really independent?

Furthermore, how likely is it that the physicists can find a stable and exotic particle to explain all the forces of nature in a non-electromagnetic way? Might there be an exotic uncharged particle, capable of surviving in a stable and independent state like electrons and protons do, that can produce the gravitational field and all the other forces of nature? Possibly. But you would think that after all this time, scientists would have already found this exotic particle acting in a very stable state, especially with regard to the gravitational field. To date, all

---

10   Boblock's email address as of June 1999 was boblock@montego.com. Miodrag Malovic now hosts his ideas on the Unified Field Theory at http://www.beotel.yu/~mmalovic/boblock/.

exotic particles discovered by scientists have very short lifespans of approximately 10 nanoseconds (0.00000001 seconds), making them very unlikely candidates. If a truly stable exotic particle does exist, we would have detected it almost as quickly as we found the proton, electron and neutron.[11] So far, the evidence is lacking. About the closest thing we have is the anti-neutrino, but even that exotic particle has trouble interacting with ordinary matter.

With nothing else to consider, we now face the unpalatable possibility (if you are one of those physicists who wants to stick to the old school of thought that gravity and light are not related) that the electromagnetic field is the gravitational field, as well as the strong nuclear force and the weak nuclear force. Indeed, why do we need an exotic force of nature or particle to explain what is happening in the universe? We only need to find the electromagnetic explanation for everything to fall into place.

This is a very subtle paradigm shift in our understanding. It involves jumping off the old gravitational and nuclear force bandwagon and boarding the supertrain of electromagnetism to see the kind of universe in which we could actually be living in.

The New Picture of How Radiation Moves Solid Matter

This brings us to an interesting new insight into the nature of radiation and how it interacts with solid matter. This insight has the power to pave the way for the development of a new electromagnetic technology —one that could explain much about UFOs and their unresolved electromagnetic observations.

When we discuss uncharged matter moving in the presence of radiation, what we are really talking about is radiation moving *charged* matter. Forget uncharged matter; this has nothing to do with the moving of matter according to Einstein's Unified Field Theory. As previously stated, radiation may seem to have mass by virtue of its ability to move what we originally thought was uncharged matter. Physicists assume this, and no one has ever questioned this observation. We simply assume that radiation is moving the mass side of ordinary matter. Now, we can no longer make this assumption. From now on,

---

11   Even the Higgs boson, or Higgs particle as it is known, used to explain why the weak nuclear force acts over a much shorter range than the electromagnetic field and why some fundamental particles have mass, is not able to last long enough to control the gravitational field with its estimated lifespan of 1×10^{-33} seconds.

we have to see the interaction of radiation with solid matter as moving charges.

Furthermore, when radiation moves solid matter, both negatively and positively charged particles move in the same direction, explaining why physicists previously assumed that radiation was moving the mass and not charges. Physicists understandably thought that, for oppositely charged particles to move in the same direction, the mass had to be what interacted with the radiation and the charges. As they had assumed, charges always have mass, so why not assume it is the mass that is doing the work? It makes perfect sense. However, in the electromagnetic theory, this is not the case. Radiation, when it is oscillating its energy, causes the negative and positive charges to move in the same direction.

If statements thus far are true, the force of radiation on solid matter is controlled by the following:

i)   the quantity of charges present on solid matter;
ii)   the charge density over a certain region;
iii)   the acceleration of the charges (also controlled by the frequency of the oscillating charge); and
iv)   the strength of the chemical bonds formed by the charges to create a solid crystalline structure[12].

In other words, we are effectively controlling the energy density of the radiation through the amount of charge present on solid matter, its density on the surface of solid matter, and how quickly we accelerate and decelerate the charge on the surface (which is just another term for frequency of the oscillating charge). Generally, a strong impact on (or the ability to grab hold and move) solid matter by radiation requires an increase in the charge (which is dependent on the number of electrons present in solid matter, or lack thereof when relying on the protons to provide the positive charge) and/or charge density. To emit radiation, the charge (i.e., the electrons) must oscillate on the surface at a certain frequency. Frequency is effectively acceleration/deceleration; so the higher the frequency, the faster the charge must move and hence the more it must accelerate and decelerate. Therefore, the greater the quantity of charge and the higher the acceleration of the charge, the higher the energy density of the emitted radiation. Moreover, since the radiation moves all charges in one direction, strong chemical bonds in

---

12   So as to allow the entire solid matter to move more effectively with the charges in the presence of radiation.

368

solid matter should allow the charges to be carried more effectively by the radiation without deforming the solid matter or losing atoms from the surface.

To put it simply, we should expect radiation to exert a stronger force on charged matter than so-called uncharged matter, as the former provides more charge against which the radiation can push. Similarly, the faster the charge oscillates (or higher the frequency), the greater the recoiling force is on all the charges present when radiation emerges from it.

With all that has been said, we must acknowledge the possibility that the radiation pressure, or force exerted on matter, should vary depending on the amount of charge present in solid matter, the charge density, and how the charge moves (or accelerates). In the scientific literature, this difference can be observed in the formulae used to calculate the acceleration and velocity of solid matter in the charged and uncharged (or lower amounts of charge) cases after the matter emits radiation.

Let us look at the formulas, starting with the uncharged case.

The Non-Relativistic Velocity of an Uncharged Object Emitting a Photon

If we focus on the emission of one photon from a particle, charged or otherwise, we know the particle will recoil in the opposite direction.

In the case of an electron moving in the orbits around the atomic nuclei (with the atom considered uncharged), experiments have revealed the magnitude of this momentum of the radiation exerted on the electron to be as follows:

$$p = \frac{hf}{c}$$

where f is frequency (in hertz); c is the speed of light; h is Planck's constant; and p is momentum, which can be converted into force (and eventually speed) as soon as the inherent momentum of the radiation pushes away the electron.

As the charge of an electron cannot be controlled (and appears to never vary) as well as the fact that an electron bound to an atom emits only one photon at a time, the only way to amplify the momentum of the radiation on solid "uncharged" matter is to increase the number of electrons emitting the radiation (a combination of multiple electrons at the same energy level within an orbit around the nuclei, and/or multiple atoms making up the solid matter), and/or to increase the frequency of the radiation by exciting those electrons in the innermost orbits of the atoms so they may emit a more energetic photon. As not all electrons in an atom will emit radiation simultaneously and the frequency may be limited, it is incumbent on us to increase the number of atoms for the radiation to be emitted roughly at the same time in order to push against the atoms.

To calculate the force of the photon on a single atom, imagine an *uncharged* object of mass m_o moving linearly through the vacuum of space with low initial velocity v_o. A moment later, the object emits a photon of momentum p from its surface. The object recoils in the opposite direction from the photon with a new velocity of $v_1 = v_o + dv$. Assuming the mass of the object remains the same as before, what is the velocity v_1 after emission of the photon?

According to the principle of conservation of linear momentum, which states that "when no resultant external force [such as gravity or air friction] acts on a system, the total momentum of the system remains constant in magnitude and direction"[13], the problem can be mathematically stated as:

$$m_o v_o + p = m_o v_1$$

or

$$m_o v_o + p = m_o (v_o + dv)$$

Rearranging terms gives us:

$$dv = \frac{p}{m_o}$$

Formally integrating the differential term dv,

$$\int dv = \frac{p}{m_o}$$

---

13   Zemansky et al 1982, p.147.

We get:

$$v_1 + C = \frac{p}{m_o}$$

To determine C, we must realise that just prior to the discrete emission of the photon (p = 0), the object had a velocity v_1 equal to its initial velocity v_o. Therefore, C = -v_o and so the non-relativistic recoil velocity for the object emitting just one photon is given by:

$$v_1 = v_o + \frac{p}{m_o}$$

If the uncharged object is the size of, say, a sodium atom, we can convert momentum p to show how the frequency f (or wavelength λ) of the emitted radiation from an electron orbiting around the atom's nucleus affects the speed of the particle. This is done like so,

$$v_1 = v_o + \frac{h/\lambda}{m_o} = v_o + \frac{hf}{m_o}$$

where h is Planck's constant, which has been determined from experiments conducted on electrons in orbit around an atom to be $6.62606957 \times 10^{-34}$ J s. (joules-second, or joules per hertz[14])

Generally, the higher the frequency (or shorter the wavelength), the greater the recoiling force of the radiation on the particle and the faster the particle moves.

EXAMPLE

Suppose a sodium atom of mass $m_o = 3.8 \times 10^{-26}$ kilograms, initially at rest ($v_o = 0$), emits a photon of wavelength $\lambda = 5.89 \times 10^{-7}$ meter. The recoil velocity of the atom is calculated as follows:

$$v_1 = \frac{6.6260755 \times 10^{-34} \text{J.s} / 5.89 \times 10^{-7} \text{m}}{3.8 \times 10^{-26} \text{kg}}$$

---

14   A joule (J) can be converted to kg.m²/s² where kg is in kilograms, m is in meters, and s is in seconds. Sometimes you may find in the scientific literature how Planck's constant can have the units of joules per second. However, this is a measurement of power, or the time derivative of energy.

or

$$v_1 = 0.029 \text{ meters per second}$$

Unfortunately, the radiation emitted by an electron inside an atom generates an electromagnetic force that is considered weak due to the fact that Planck's constant h is already a very small value. Therefore, if we have to rely on this formula and the radiation emitted by an electron in an atom to move uncharged matter, the mass of ordinary solid matter must be very lightweight (but still having enough atoms with electrons to emit sufficient numbers of photons), and the energy used to excite the electrons as needed to emit radiation must increase (e.g., raising the frequency).

Even so, a sufficiently lightweight spacecraft can be built to accelerate fast enough using photon propulsion from an uncharged surface to make a journey to the closest star after our Sun—Alpha Centauri—in roughly 200 years. The source for this energy to excite the electrons on a thin metal surface can come from the Sun to help push and recoil against the mass of a large number of atoms in a thin, lightweight metal sheet. With no additional charges added to the metal sheet that could help to add more radiation emissions, this would be just fine to create what is known as a *solar sail*.

The Solar Sail

The idea of light propulsion using sunlight to exert radiation pressure on large "sails" in space (known as *solar sailing*) has its humble beginnings in the 1920s. It began in 1924 when the Russian rocket pioneer Fridrikh Tsander suggested that interstellar travel could be achieved using mirrors.

The idea was not taken further until the 1970s when Dr. Louis Friedman suggested that solar sailing could be a relatively low-cost method of traveling to Halley's Comet. With not too many takers of the idea, he eventually left the Jet Propulsion Lab (JPL) in Pasadena and co-founded *The Planetary Society* with the late Dr. Carl Sagan and Bruce Murray. Since stepping down from his position as Executive Director of the Society after 30 years of distinguished service, he returned his attention to the solar sail concept in 2010, and his expertise in this field allowed him to lead the Cosmos 1 project.

Prior to making this first serious attempt at an electromagnetic spacecraft (outside of the U.S. military), NASA scientists published a paper in the mid-1980s showing the feasibility of a "solar sail" spacecraft called the *Yankee Clipper*. This idea was inspired by their observations of the tails of comets and how they are pushed away from the Sun by solar wind. The aim of the NASA spacecraft would be to use the continuous pressure in sunlight to "sail" across interplanetary space using a large, thin and solid metallic foil or "sail" attached to a spacecraft. However, for some reason, the project was never put into action.[15]

A more advanced version of a solar sail spacecraft was developed in the early 21st century when Cosmos 1 was designed on paper and constructed by the U.S. science-based media and entertainment company Cosmos Studios. The project was run by Ann Dryan, the widow of the late Dr. Carl Sagan, and with assistance from *The Planetary Society* (via Dr. Friedman).

After suffering several setbacks during the development and testing phases, Dryan and Friedman finally got to the stage of building the US$4 million 105-kilogram Cosmos 1 spacecraft. It was successfully launched on June 21, 2005 at the Babakin Space Centre in Russia on a modified Soviet missile with the help of a Russian nuclear submarine in the Barents Sea. The spacecraft was placed into orbit, and all systems were set to send it off into deep space using only the power of the Sun for propulsion.

Upon reaching orbit 825 kilometres above the Pacific Ocean, Cosmos 1 would have opened up its eight triangular panels comprised of 15 meters of very thin and lightweight Mylar. The light from the Sun would have continuously hit these panels, causing the spacecraft to accelerate. Slowly at first, but this would not matter as sunlight would exert a continuous force on the sails. After one hundred days of continuous acceleration, the spacecraft would have reached 16,000 miles per hour. The spacecraft would have gone into the history books as the fastest man-made object, overtaking the Voyager I and II to become the spacecraft that ventured furthest into interstellar space. However, inexplicably, after its successful launch[16] on June 21, 2005, no

---

15   Stemman 1991, p.115.
16   Despite the main engine of the Volna rocket first stage had shut down prematurely, the steering engines managed to put Cosmos 1 into orbit. Within a few hours after launch, portable tracking stations in the Kamchatka Peninsula and Majuro in the Marshall Islands received signals at the scheduled times of transmission from Cosmos 1. But suddenly soon after the signals were lost. Unable to locate Cosmos 1, the Russians have taken on the view that the spacecraft probably

signals were received from the satellite. The reason behind the mysterious loss in signal has yet to be determined.

If Cosmos 1 had not failed, it would have accelerated to the edge of the solar system within 2.4 years[17]. And after reaching a maximum speed of approximately one-tenth the speed of light (or 160,000 miles per hour) before the sunlight faded out too much, a journey to the nearest star to our sun, known as Proxima Centauri, could have been completed in a couple of centuries.

The Non-Relativistic Velocity of a Charged Object Emitting a Photon

However, unlike the uncharged (or more correctly, low charge) case in which the energy in the radiation disappears forever into space, and so more energy must be obtained from somewhere, a charged surface apparently does not require extra energy to be sourced from anywhere. Once the charge accelerates, it is somehow able to grab the energy from the radiation that the charge has emitted, use the energy to do something to prepare the charge in readiness to emit radiation again, and then goes ahead to emit the radiation only to repeat the process again and again while maintaining the emission of radiation. Furthermore, as the charge accelerates, it concentrates the radiation energy from the surroundings, thereby raising the energy density of the radiation. This has the effect of exerting an increasing recoiling force on the charge with each emission.

How is this possible?

More amazingly, mass is not considered essential for radiation to push against[18]. Sure, a heavy mass will accelerate more slowly at first compared to a lightweight mass. Rather, what is more important in this scenario is the charge and the acceleration of that charge. Frequency of the radiation is important in so much as to jolt the charge into motion. In other words, there has to be a minimum energy density set by the

---

never made it to orbit due to main engine failure.

17 Light from our Sun will diminish the further Cosmos I is from the Earth. Soon the solar winds of other stars will begin to interfere. If the spacecraft could wrap up its solar sails at maximum speed before rocks and ice particles destroy it and before other winds start to interfere with its speed and direction, it may take only a century or so to reach the nearest star after our Sun — Proxima Centauri — on the power of light alone.

18 The formula describing the motion of a charged object emitting radiation will be in a perfect vacuum. Otherwise, air and other radiation in space will restrict the initial motion of the charge. Also, the mass should be kept to a minimum to lower the energy requirements for initiating self-acceleration.

oscillating charge and frequency to create movement. But if there is any movement whatsoever, charge and acceleration are the only two factors to affect the recoiling force of the radiation on the charge itself according to the mathematics to describe this situation. Nothing else matters.

The formula for this special case assumes the charge is oscillating at the minimum frequency needed to move the charge. As we shall see, the solution to the formula does not quite reveal this oscillation of the charge. Instead it suggests a spontaneous acceleration of the charge can occur once the charge oscillates to emit radiation even though it is not clearly apparent in the formula, as we shall see.

In actual fact, according to the mathematical solution derived for this special case, the acceleration of a charged object emitting radiation in one direction (i.e., linearly) is said to be exponential. In other words, the acceleration will increase in a non-linear way and get dramatic as time passes.

Does this really occur in reality?

The Abraham-Lorentz Formula

Max Abraham (c. 1905)

The equation describing any electrically-charged object emitting radiation consistent with the conservation of energy is called the *Abraham-Lorentz formula*.[19] The formula was first published by Max Abraham (1875-1922) in 1905 in his book *Theorie der Elektrizität,* and since then, it has been the subject of intense debate among scientists, which we will discuss later.

In the words of Dr. David J. Griffiths of the Department of Physics at Reed College, in Portland, Oregon:

"According to the laws of classical electrodynamics, an accelerating charge radiates [electromagnetic energy]….The

---

19   Dr. Nunzio Tralli, former professor of physics at Long Island University, New York, USA, gives a full derivation of the Abraham-Lorentz formula in his 1963 book, *Classical Electromagnetic Theory*, pp.271–276. Derivation of the formula is also available in Jeans 1951, pp.559–592, and Panofsky and Phillips 1962, pp.377–400.

radiation evidently exerts a force back on the charge—a recoil force rather like that of a bullet on a gun...

The Abraham-Lorentz formula for [this] radiation reaction force...consistent with conservation of energy [is],

$$F = \frac{1}{4\pi\varepsilon_o} \cdot \frac{2}{3} \cdot \frac{Q^2}{c^3} \cdot \frac{d\mathbf{a}}{dt}$$

In this formula, Q is the electric charge, $\mathbf{a}$ is acceleration (in vector notation), $\mathbf{F}$ is the force (in vector notation), t is time, c is the speed of light, and ε_o is the permittivity constant of free space.

Dr. Griffiths also stated:

"...the Abraham-Lorentz formula has disturbing implications, which are not entirely understood 80 years after the law was first introduced."

What is so special about this formula? We only have to look at the accelerating solution derived from it for us to truly appreciate the controversy surrounding it.

If the charged object has mass m, then according to Newton's second law of motion:

$$m\frac{dv}{dt} = \frac{1}{6\pi\varepsilon_o} \cdot \frac{Q^2}{c^3} \cdot \frac{d^2v}{dt^2}$$

or

$$\frac{d^2v}{dt^2} - \frac{6\pi\varepsilon_o mc^3}{Q^2} \cdot \frac{dv}{dt} = 0$$

After solving this differential equation of the second order, we obtain:

$$v = a_o \frac{Q^2}{6\pi\varepsilon_o mc^3}\left(e^{t\frac{6\pi\varepsilon_o mc^3}{Q^2}} - 1\right)$$

If we assume the charge Q is constant[20], then let's simplify this solution by introducing the following constant:

$$k = \frac{Q^2}{6\pi\varepsilon_o mc^3}$$

Thus the solution becomes:

$$v = a_o k\left(e^{t/k} - 1\right)$$

This is the non-relativistic recoil velocity of a charged object of mass m, charge Q, and initial acceleration a_o. The acceleration of the charged object is, therefore:

$$a = \frac{dv}{dt} = a_o e^{t/k}$$

Interpreting the Runaway Solution

Here we have a truly remarkable equation (for both acceleration and velocity). It is remarkable in at least two respects. First, what the equation is telling us is that any charge emitting radiation will accelerate exponentially. And second, when the charge emits radiation, it does not lose the energy. Rather, it somehow has the ability to recycle the energy and re-use it again and again to emit radiation of ever increasingly higher energy density. As a result, the acceleration increases over time in a dramatic way.

But there is another aspect considered disturbing. The solution suggests that it would take only the slightest jolting effect to initiate movement for the charge to undergo its dramatic exponential acceleration. Even a tiny electron hitting a charged object the size of a space station would be enough for the object to accelerate. Slowly at first, but over time, you will notice it. So long as the radiation is being emitted on that side of the spacecraft that was hit by an electron to reinforce the movement, the spacecraft will perform its legendary exponential acceleration. If we did not know that the electron had

---

20   In reality, the charge has to initially oscillate to create the radiation in the first place.

exerted a force (perfectly understandable if no one saw it coming), we could be forgiven for thinking that the object had spontaneously accelerated without an external force being applied. It does not matter how large or small the object is, once the charge moves, it will maintain its acceleration without any further input of energy. So long as the energy in the charge is maintained (i.e., charge Q is constant, as appears to be the case for an electron, and is oscillating in a way that helps to emit radiation), the energy in the radiation is not only maintained but also exerts a greater radiation reaction force (i.e., the energy density of the radiation increases) on the charge over time as it moves faster and faster and so keeps recoiling the object with ever-greater force, and so pushing the object to dramatically increasing velocity and acceleration. This would start off slowly, but, as Richard T. Hammond, a physicist at the University of North Carolina, said:

"With no forces acting...leads to a self force that propels the particle forward in an extremely short time..."[21]

In other words, the acceleration would become so dramatic that it would be difficult for anyone to withstand the inertial forces if they were sitting inside this charged particle.

Of course, if there was a way to control these inertial forces[22], there would be absolutely no reason why anyone sitting inside the accelerating charged object could not reach nearly the speed of light in an "extremely short time". A journey to the nearest star to our sun—Alpha Centauri—in a charged object emitting radiation and following the pattern suggested by this mathematical solution would take a matter of days, hours or even less, depending on how much acceleration the person inside can withstand and how close we choose to get to the speed of light.

Now we have something interesting. Because here lies the solution to the UFO problem. There is a scientific way for UFO occupants to cover immense distances between the stars by building a technology based on this electromagnetic concept.

---

21   Hammond 2011, p.276

22   Fortunately, there is a way to control inertial forces. The Unified Field Theory with its interpretation that the electromagnetic field is the gravitational field means one could use a perfectly symmetrical metal box to reduce the electromagnetic field to zero and with it the gravitational field. Since the gravitational field is what controls the inertial forces, it is predicted the inertial forces should drop as well, thereby making it easier for pilots and passengers to cope with higher accelerating and decelerating forces, including changes in direction.

So where does this idea of no energy loss while moving faster and faster originate?

In the case of an electron, we know it has a very specific charge, measured by scientists as $1.60217657 \times 10^{-19}$ coulombs. This looks like a fairly precise number, which scientists claim never changes over time regardless of what the electron is doing. So, when the electron emits some energy to create radiation (somehow the charge must have varied to do this), there is presumably no reduction[23] in the charge of the electron (not even temporarily, as far as we can tell); it stays constant at all times.

In reality, any radiation emitted from a charged surface will reduce the charge. And to repeat the emission, the charge must be brought back up by some mechanism very quickly. Scientists can do this using an electrical generator to maintain the charge. However, in the case of the electron, we cannot detect any change in the charge of an electron when radiation is emitted. Furthermore, scientists are at a loss regarding the mechanism that the electron might employ for generating, or externally grabbing, energy in order to maintain the charge (let alone emit radiation).

The mass of the electron is not affected either, as far as scientists can tell. Not even the tiniest reduction in its mass—$9.10938291 \times 10^{-31}$ kilograms—has been observed to suggest that some mass is converted into energy to help maintain the charge and, with it, the radiation emission.

How can this be? Where is this energy for the radiation coming from?

If the electron's mass and charge are unaffected by the emission of the radiation, then the only logical conclusion one can reach based on this mathematical solution and without violating the principle of conservation of energy[24] is that the energy emitted by the electron as radiation to recoil the electron is not lost but is perfectly recycled by some unknown mechanism, preventing the charge from ever needing a "top-up" from an external energy source.

We are led to believe that radiation can be recycled.

Is this true?

---

23 The electron somehow maintains its own specific charge, but this could be a limitation of our instruments in not being sensitive and small enough to detect the tiniest of changes in the charge as the electron emits radiation.

24 Unless the derivation of the Abraham-Lorentz formula from the original equation of motion is incorrect, or the original equation of motion has not been structured correctly.

If we can rely on the Unified Field Theory for an explanation, we know there is a link between the electromagnetic field and the gravitational field. All it requires is a sufficiently high energy density of the radiation to produce a strong enough gravitational field to help bend the radiation back on itself allowing the electron to re-absorb the energy. How high the energy density should be is not entirely clear. An experiment will hopefully confirm this fact. But if it is possible, there is no reason electromagnetic energy could not be recycled using the right technology.

To be more accurate, this gravitational effect should be seen as a pushing force of radiation according to the Unified Field Theory. But that would imply a universal background radiation exists for the accelerating charge to concentrate this energy and use it to push things together, including a bending of the radiation back on itself as soon as it comes off the charge. Unfortunately, none of the equations shown above account for this extra radiation. The mathematical universe in which all these equations apply is a perfect vacuum free of all radiation except for the radiation coming off the charge. This makes it hard to imagine how the recycling effect would occur based on the Unified Field concept. About the only advantage in having a perfect vacuum is that it can explain how the charge can spontaneously accelerate from rest—all it takes is an infinitesimal initial force to set off the runaway solution. Beyond that, the perfect vacuum would also allow an infinite speed to be attained by the charge as it goes through its runaway exponential acceleration. However, in the real universe with radiation present everywhere, this is not possible. No charge can ever exceed the speed of light. And there is a minimum charge and frequency needed to initiate movement. So no such thing as a spontaneous acceleration in real life.

There is also another problem with the formulas shown thus far. The lack of universal background radiation means the mass and charge of an electron would not hold together. The background radiation is essential to push against other energy to help keep matter together and to supply an energy source to maintain the solid matter's internal structure, including the creation of mass and charge in electrons and protons. Yet, at the same time, the law of conservation states that this energy must be emitted into the environment to keep things in balance. This is why electrons and protons never get infinitely hot. Also, as the energy is emitted, it prevents things like other electrons from falling into the nucleus of the atom, and from becoming infinitely small points containing infinite energy density leading to a collapse of matter. There must always be a balancing act of energy going into matter and energy coming out in the real universe.

As the above equations are highly simplified and meant to represent a universe free of background radiation (i.e., a perfect vacuum) and the energy and mass of charges somehow stay intact under this impossibly cold conditions, physicists have to assume that an invisible and undetectable quantum vacuum field is likely present to keep things in balance.

As Peter W. Milonni writes in section 3.3 of *The Quantum Vacuum: An Introduction to Quantum Electrodynamics*:

"The fact that an accelerating charge loses energy by radiating implies, according to classical ideas, that an electron should spiral into the nucleus and that atoms should not be stable. The balancing of the effects of radiation reaction and the vacuum field..., however, suggest that the stability of atoms might be attributable to the influence on the atom of the vacuum field....We now know that the vacuum field is in fact formally necessary for the stability of atoms in quantum theory. As we saw..., radiation reaction will cause canonical commutators $[x,p_x]$ to decay to zero unless the fluctuating vacuum field is included, in which case commutators are consistently preserved."

The only thing is, under the Unified Field Theory, this quantum vacuum field is essentially radiation even though scientists cannot directly detect it.

Whatever the truth, if such a radical runaway solution can occur on a large scale, it would be the perfect solution for interstellar travel as there would be no further input of energy to maintain the acceleration and thus no need to carry fuel. Everything will be contained in the accelerating charge (acting as an energy grabber, recycler and storage) and electromagnetic field of the emitted radiation (acting as the engine and fuel for propelling the charge). Between charge, radiation and a bit of acceleration, this is all that is needed to accelerate charge to phenomenal speeds.

The Problem Scientists Have with this Solution

Despite the great promise of a new electromagnetic technology based on this runaway solution, scientists are not entirely sure it can occur in reality.

Part of the problem is explained in Dr. Griffiths book on electromagnetism. He says the Abraham-Lorentz formula is a poorly studied area of electromagnetism, suggesting that the few scientists who have looked at the formula believe that it is either a special case that cannot occur in reality (well, how can an electron constantly emitting radiation maintain its charge without losing energy? And even if it could, how is it that we have not seen the electron spontaneously shoot off into space in nature?). Or else, there has to be another problem with the formula that scientists cannot as yet figure out and put it into practice through a real-life experiment.

For example, in the 1987 edition of *McGraw-Hill Encyclopaedia of Science and Technology*, Volume 15, page 280, it is said that this self-accelerating solution derived from the Abraham-Lorentz formula indicates a "fundamental defect in the [classical electrodynamics] theory".

In *Mathematical Theory of Electricity and Magnetism*, British physicist and astronomer Sir James Hopwood Jeans (1877–1946), the author of the book, says:

"Many physicists now question whether any emission of radiation is produced by the acceleration of an electron, except under certain special conditions."[25]

Even though Carati and Galgani stated in their paper[26]:

"Everyone knows that a particle in uniform motion drags along with it an electromagnetic field",

---

25   Jeans 1951, p.577.
26   Some scientists question whether a uniformly-accelerating charge radiates electromagnetic energy. The argument is based on the Equivalence Principle (i.e., the equivalence between gravity and acceleration). In other words, an object can be both stationary and accelerating depending on who is doing the observing. If an observer sees the object accelerating and there is radiation being emitted, another observer sitting inside the object may claim no radiation is being emitted. How can this be? Or is the object always absorbing and emitting radiation from the universal background radiation all around the surface, but receives extra radiation energy on one side of the object when accelerating and it is this extra energy that is being emitted and seen by the observer?

the problem for a number of scientists seems to be more about whether a uniformly accelerated[27] charge will radiate even though we are discussing a non-uniform acceleration because of the exponential term.

For example, in 1919, the German physicists Max von Laue (1879–1960) thought it was abhorrent to allow the runaway solution to exist as he saw absolutely no evidence in nature to support it, so he claimed that a particle under constant acceleration does not radiate. He attempted to justify this through a careful application of mathematics to show his rational thinking about the matter. This was later supported by Wolfgang Ernst Pauli (1900-1958) in his 1921 book *Theory of Relativity*[28]. However, in 1949, D. L. Drukey adopted the opposite view by mathematically showing that a uniformly accelerating charge does radiate.

Then the arguments shifted to a concern whether an accelerating charge radiates under certain conditions. The example given is what happens to a charge that is accelerating in "free fall" by a gravitational field, or when a gravitational field is exerting a force on a charge that is resting on the ground. Does the charge radiate? Initially, it would seem logical to say that it should radiate because anything that is free falling must be accelerating, and anything being pushed to the ground by a gravitational field should emit radiation to counteract the gravitational force. However, this is not necessarily the case; it depends on the frame of reference of the observer. For example, when the gravitational field pushes on a charge, causing it to fall to the ground, any acceleration of the charge measured by an observer in his frame of reference will be seen to emit radiation. However, if the observer moves at the same rate of acceleration and direction as the charge, according to the principle of equivalence, it will be in the same frame of reference as the observer, and as such, would not appear to be accelerating. No apparent acceleration means that no magnetic field is generated by the charge and thus no radiation is emitted. Likewise, if the charge is resting on the ground, you would think that a force is being exerted on the charge to balance the gravitational field pulling on it and, therefore, it should be radiating. Not true. The overall net force on the charge is zero. To the observer standing on the ground looking at the charge, the charge is

---

27   This is just another way of saying "constant acceleration".
28   The original first edition written in German was translated into English and re-published in 1958.

not moving. That means that there is no magnetic field, and hence no radiation is emitted. However, the charge will be observed to radiate while at rest on the Earth's surface if the observer is free falling to the ground. This is because from his perspective, the charge will appear to be accelerating towards him. Gravitational fields are almost inconsequential to this analysis unless they help to create a measurable acceleration of the charge with respect to an observer. There must be acceleration present relative to the observer for that observer to see the charge radiate.

Thomas Fulton and Fritz Rohrlick from the Department of Physics at the Johns Hopkins University in Baltimore, Maryland, provided a fine explanation of this idea in their influential 1960 article, "Classical radiation from a uniformly accelerated charge" (published in the *Annals of Physics*):

> "A particle which is falling freely in a homogeneous gravitational field should appear to an observer who is falling with it, like a particle at rest in an inertial frame. When the particle is charged, the observer can establish the presence of a gravitational field by looking for radiation. If he observes radiation from the charge, he knows that he and the charge are falling in a gravitational field [or one is accelerating with respect to the other to allow the observer to detect a gravitational field]; if he observes no radiation, he knows that he and the particle are in a force free [i.e., no gravitational field] region of space [hence there is no apparent acceleration difference between the reference frames of the charge and the observer]."[29]

Basically, when a gravitational field is detected, this means that radiation is being emitted. As soon as the gravitational field disappears, no radiation is being emitted.

Some scientists see this as very similar to how inertia works. When you push on an object (to make it accelerate), it pushes back because of its inertial mass. In the case of a charge, by pushing on it, not only do you create a gravitational field by the accelerating mass, but also you make the charge radiate. As radiation is emitted, the charge recoils to create what physicists call the inertial effect that resists the force that

---

29   Fulton & Rohrlich 1960, p.499.

initially accelerated the charge, unless the charge can re-distribute energy in a manner that allows the radiating energy to reinforce the initial acceleration.

With all this in mind, some skeptical scientists think it is possible, depending on the observer's frame of reference, to "hide" this runaway solution if the charge can be made to look like it is not accelerating. Even so, any emission of radiation from an accelerating charge observed from a different frame of reference is still going to follow the runaway solution.

The problem remains.

Because of the thorny nature of this electromagnetic problem, a look at the scientific literature features several examples of scientists attempting to repress the runaway solution in the formula by adding some extra mathematical terms, or re-integrating over a different set of boundary conditions, in the hope the extraordinary mathematical solution would disappear.

On page 2 of a paper written by mathematicians Andrea Carati and Luigi Galgani at the University of Milan, titled "Recent Progress on the Abraham-Lorentz-Dirac Equation", the authors state:

> "...the scientific community essentially performed an action of psychological removal (or repression) [of the runaway equation], behaving as if that equation did not exist."[30]

This approach normally works best at low velocities of the accelerating charge. As Hammond confirmed:

> "To counter this [runaway solution], at least in the low velocity limit, one may integrate the [Lorentz-Abraham] equation using an integrating factor and apply a boundary condition to render it a second order differential equation. But in this case the solution is non-causal."[31]

In other words, you might be able to hide the runaway solution, but it seems not the ability of the charge to spontaneously self-accelerate, which turns out to be just as vexing and problematic as the runaway solution to scientists.

---

30   Carati & Galgani, p.2.
31   Hammond 2011, p.276.

Yet another group of scientists have questioned whether the original equation of motion used to derive the Abraham-Lorentz formula is correct. As Hammond said:

"What is the correct equation of motion [for a radiating charge]?"[32]

For some scientists, the mere mathematical existence of a spontaneous and exponential self-accelerating charge is enough to make them question whether the Abraham-Lorentz formula is correct. As Hammond said:

"For these reasons the LAD [Abraham-Lorentz-Dirac] equation is not considered to be the correct physical description, and for decades the correct solution has been sought."[33]

If that is not enough, other scientists, including Hammond himself, have attempted to mathematically show that the runaway solution probably exists, but only when a charge has reached a considerably high acceleration. This acceleration must be so high, in fact, that the runaway solution is technologically impractical to apply until another way to accelerate a charged object to the required relativistic level is developed. As Hammond stated:

"...in order for the self force effects to be observed, the acceleration must be so large that the relativistic equation must be used."[34]

Still, not everyone is in total agreement.

For example, Hammond has not accounted for the idea that, according to the unified field concept, radiation bends onto itself due to its own gravitational field at sufficient charge and frequency intensities. By charge, we mean the electrons, not the object carrying the charge, which in itself could be at rest. If this bending effect can occur at much lower energies and without the need to accelerate the charged object to a high enough level, it may be possible to observe the

32   Hammond 2011, p.276.
33   Hammond 2011, p.276.
34   Hammond 2011, p.280.

runaway solution at a much lower acceleration, and even perhaps from a rest position.

So, after all of that, it seems that there are only two things on which everyone can agree regarding this perplexing physics problem. As Dr. David J. Griffiths stated:

"The radiation reaction force is due to the force of the charge on itself."[35]

In other words, there is energy being recycled. That is quite literally the interpretation all scientists must make when looking at the runaway solution, no matter how unbelievable it may seem.

And the other common issue on every scientists' mind is whether the runaway solution occurs in reality, especially at low enough energies to make it a seriously practical consideration (e.g., to build a technology based on this very concept)?

As Dr. Griffiths sees it, the solution is probably not a defect or some other issue. Rather, there is a potential reality in the solution if there is a way to emit radiation in one direction from virtually *any* charged object. And because it seems that the electron doesn't lose energy as it emits radiation (in reality, we probably need an electrical generator to maintain charge), the energy in the radiation must somehow be recycled in order to produce the self-accelerating runaway phenomenon. How it does this is unclear from the mathematics, but there is a way under the Unified Field Theory to bend the radiation back on itself. All that is required is a method to initiate this runaway solution at a modest charge and frequency, even if the acceleration of the entire charged object is initially low or non-existent. What seems more important here is acceleration of enough electrons making up the charged object. Forget accelerating the entire charged object itself, at least in the initial stage. Start from a rest position and make these electrons accelerate on the surface of the object quick enough as part of the frequency of the oscillating charge and, technically speaking, we should be able to see this runaway solution take place so long as we know of a method to distribute a higher charge density to one side of the object compared to the opposite end.

This seems to be all we need to do.

---

35   Griffiths 1989, p.439.

For example, in Chapter 5 we discussed some UFO cases in which the witnesses have reported UFOs as being able to periodically become visible and invisible while hovering in the air (i.e., without acceleration). For any object to render itself invisible, this can only occur if the light from the object is being bent back on itself and the light from behind the object is bent around it to allow an observer to see what is behind the object, essentially causing the object to disappear. For this to happen, the Unified Field Theory simple requires an oscillating electromagnetic field of reasonable energy density to be emitted around the object at a charge and frequency that amplifies the gravitational field sufficiently to help bend the electromagnetic field back on itself. Acceleration of the UFO is irrelevant because high acceleration of charges needed to achieve this invisibility effect is strictly governed by the movement of electrons on a metal surface. To control the acceleration of these electrons is through frequency of the oscillating charge.

Does this mean there are UFO cases in which unknown entities flying these objects are trying to teach the witnesses a lesson in radiation recycling?

As we can see, the problem associated with the runaway solution is still far from resolved. Not even quantum theory has found a solution. As physicist Yaron Hadad said:

> "The problem of radiation reaction is still unsolved in the framework of quantum theory. This has striking consequences, as it means that a solution of the radiation reaction problem means a phase of revolutionary science that is going to be very revolutionary."[36]

This leaves us with the only way forward on this matter: to perform an experiment to test the runaway solution. With so much talk and rational justification using mathematics to convince those few scientists who have studied the problem to think that the runaway solution may or may not exist, it is extraordinary that no one has conducted an experiment to see what happens when a charged object emits radiation in one direction. As Hammond said:

36   http://www.yaronhadad.com/what-is-the-problem-of-radiation-reaction/

"In the past the entire question of radiation reaction was considered by some to be of only theoretical interest, since such effects could not be observed."[37]

By "observed" we mean something in nature that scientists can see and is undeniable evidence of the runaway phenomenon before they can be convinced that it might be worth the effort to re-confirm the observation through an experiment. Well, there are signs that this runaway solution is already appearing in front of their eyes. Only recently have some observations given scientists food for thought that perhaps the runaway solution might occur in reality, as this quote from Hammond indicates:

"...laser intensities of 10^{22} W cm^{-2} have been reached, and it has been shown how important radiation reaction is."[38]

And, as Professor James F. Woodward at California State University, Fullerton, said:

"In high energy elementary particle accelerators (like the ones at Fermilab or CERN) radiation reaction is an obvious fact of life. As the particles traveling at nearly the speed of light are bent into their circular paths by magnets, they are accelerated. And they radiate. The reaction force produced by the radiation slows the particles down, unless power is applied to replace the radiated energy and momentum."[39]

Nevertheless, the general belief among scientists today is that it would require a very high amount of energy to achieve the kind of acceleration rates and/or charge and frequency rates needed to observe the runaway effect in an accelerating charge. Add to this the perceived difficulty of how to distribute the charge in order to allow the radiation emission occur in one direction, and it seems the runaway solution has become an intractable problem for the scientists. Unless someone performs an experiment, no one will know for sure what is possible in the world of physics.

---

37   Hammond 2011, p.276.
38   Hammond 2011, p.276.
39   https://physics.fullerton.edu/~jimw/general/radreact/

Well, it might surprise some scientists involved in this area to know that someone else has already done an experiment, and noticed a runaway effect in action for a charged object starting from a rest position.

Before we explain an invention that would support the runaway solution, we have to ask how likely it is that UFOs could by applying the principle of radiation propulsion.

Are UFOs Large-Scale Radiating Electrons?

The details of a Russian UFO case were presented in Chapter 5 ("Observation 16"). Over one hundred reports by observers on the ground and in the air described a UFO with two bright flashing lights on its sides.

One witness was a Soviet Air Force pilot, Captain V. Birin. According to his description, the distance between the two flashing lights was between 100 and 200 meters. A less intense light between the two bright lights gave the appearance of a porthole-like window. He was struck by the manoeuvrability of the UFO, and the interesting pattern revealed by the flashing bright lights and the speed of the UFO. In his own words:

> "The trajectory depended on the flashing of the bright side lights: the more often they flashed, the faster the speed of the UFO and vice versa. While hovering, the object extinguished its lights almost completely." [40]

The above UFO case is interesting. For a UFO to reveal a large and bright light that flashed on-and-off at a frequency that controls the speed of the UFO, there is only one area of physics that can do this. We are dealing with radiation emissions. Get any object to emit radiation, and the frequency of the radiation will *always* control the speed of the object. That is a given fact.

But there is another intriguing observation that we should consider. In the Canadian UFO case described in Chapter 5, witnesses observed how the UFO slowly accelerated from rest, and then the acceleration became dramatic over time. This wasn't a linear form of acceleration (i.e., a straight horizontal line on a graph for acceleration, or a straight

---

40   Good 1991, pp.7–10.

angled line going up on a graph for velocity). It was something the witnesses considered uncharacteristic of any ordinary man-made flying object we know about.

Well, this accelerating behaviour can be explained by the exponential solution described above.

Does this mean UFOs are a kind of large-scale electron that emits radiation as a means of flying around in Earth's skies? And is the Abraham-Lorentz formula the clue we need to explain how UFOs move?

Frequency—the Missing Variable

Before we look closely at an invention that may well settle this UFO debate, there is another thing odd about the mathematical solution of the Abraham-Lorentz formula. It has to do with the frequency of the radiation emitted by the charge.

As we can see in the runaway solution as presented in the scientific literature, any adjustment to the frequency of the emitted radiation does not affect the acceleration (and hence the speed) of the charged object; only charge Q and time t are the variables affecting the acceleration. This contradicts the case described earlier for an uncharged case where a higher frequency of the emitted radiation does affect the object's speed.

Basically we know that the momentum of radiation in pushing matter always gets more significant at higher frequencies. That is why high-frequency types of radiation, such as gamma rays, are said to be more lethal in its ability to penetrate matter compared to, say, visible light. Similarly, visible light can push matter more effectively than radio waves.

Higher frequency simply means the electrons are accelerating faster, which in turn increases the energy density of the radiation they generate. Also, as the energy density rises, the gravitational field strength of the radiation increases, causing the electromagnetic energy to be compressed. Like squeezing a tennis ball to make it smaller, a stronger gravitational field in the radiation will compress the energy of the radiation to a smaller volume, thereby increasing the momentum of the radiation.

Therefore, in terms of the recoiling force that we see in charged objects emitting radiation, we should expect to see the force increase at higher frequencies of the radiation. As we do not see this energy density increase with frequency in the exponential runaway solution (which suggest some kind of over simplification has occurred in the derivation of the Abraham-Lorentz formula), the frequency component has to be considered and included in the solution. For a suggested correction to the Abraham-Lorentz formula, see Appendix G.[41]

Now let us look at a patented invention that will bring a sense of reality to the runaway solution.

The Invention of Thomas Townsend Brown

In the 1960s, American physicist Thomas Townsend Brown (1905–1985) patented an invention in the United Kingdom (British Patent Number 300,311) and the United States called the *Electrokinetic Apparatus* (U.S. Patent Numbers 2,949,550 and 3,187,206). The invention consists of two dissimilar electrodes separated at a fixed distance by an insulator.

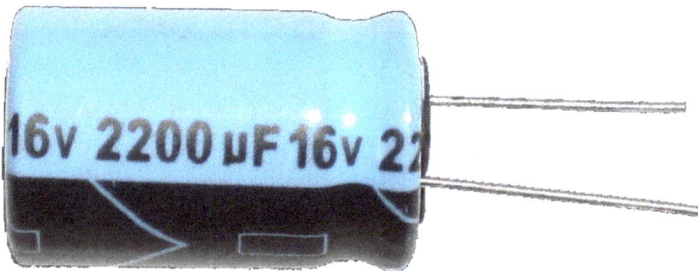

The more familiar capacitor we are used to seeing inside electronic devices

In electronics, we call this device a *capacitor*. A capacitor consists of two electrodes separated by an insulator. The principal use of a capacitor in our modern world is to temporarily store electrical energy. However, Brown discovered the capacitor can do more than just store electrical energy. It can also *move*. To make the capacitor move, he had to make sure the shape and size of the two electrodes are different from each other and apply an oscillating charge to them. As for achieving optimal movement through this interesting method, this was

---

41   The more complete solution shows the amplitude of the radiation (which is equivalent to half the diameter of a charged electrode) and frequency of the radiation are constants that are multiplied with the electric charge

something Brown experimented with over a long period of time mainly because the definitive scientific explanation for the movement eluded him.

As for the direction the capacitor moves, Brown made some interesting observations. For example, he stated to a group of keen listeners in 1977:

> "The basic Biefeld-Brown effect is quite simple. It is manifested as a departure from the Coulomb Law of electrostatic attraction, in that the opposite forces are not equal. The negative electrode appears to chase the positive electrode, so that there is a net force of the system (dipole) in the negative-to-positive direction. The Biefeld-Brown Effect states that in a highly-charged, two-electrode system, the positive electrode will 'lead' the negative electrode in the direction of the line between the two electrodes, or, the negative electrode will appear to be more attracted to the positive electrode than vice-versa. The negative electrode appears to 'chase' the positive electrode, so that there is a net force of the system (a dipole) in the negative to positive direction.
>
> This 'pure' force is a secondary effect and, therefore, somewhat difficult to isolate from ambient electrostatic forces which are much stronger by nature and tend to confuse any observations and their interpretation."[42]

However, we learn from his patents that it does not matter whether it is positive or negative for the electrode 'chasing' the other electrode. For example, in U.S. Patent 3,187,206, filed on May 9, 1958, Brown revealed a toroidal-shaped negative electrode and a small spherical-shaped positive electrode. He added arrows to indicate the direction of the thrust. For example, if the positive electrode is offset from the centre of the toroidal negative electrode toward the top, the movement of the capacitor is downwards:

---

42   Brown's statement originally published at
        http://www.thomastownsendbrown.com/stress/index.htm#docs as of July 1, 2013.

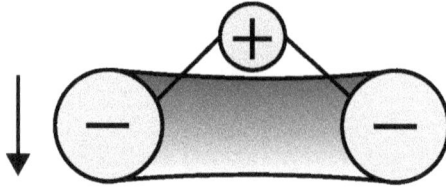

If the positive electrode is below the toroidal negative electrode, there is an upward thrust on the capacitor:

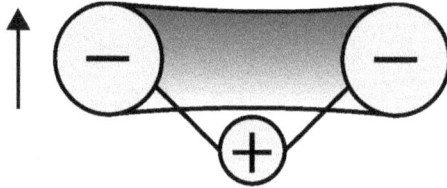

Should the positive electrode be positioned precisely in the middle, there is no movement of the capacitor:

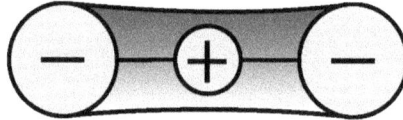

Thus, he have to say that, in this example, the positive electrode is the one "chasing" the negative electrode. In other words, the front end of the capacitor is effectively the larger negative electrode. What matters is the asymmetric nature of the electrode pair and the presence of an oscillating voltage applied to the electrodes for movement to take place. The only time the movement is diminished is when there is symmetry in the device through an appropriate positioning of the electrodes.

Indeed, the variety of different designs Brown came up with for the electrodes in a number of patents seemed to somewhat complicate things when he discovered how he could switch the electrode for leading the capacitor's movement. Thus it was possible for the negative electrode to be made to "chase" the positive electrode, and vice versa.

Confusing, isn't it?

So what is going on here?

The key must lie in the presence of emitting radiation, the size and shape of the electrodes that control charge density (which in turn affects energy density of the radiation), and how the two electrodes provide a radiation shielding effect on each other as the emitted radiation from both electrodes undergoes a certain amount of destructive interference and helps to reduce the recoiling force of the radiation on one end of the device compared to the opposite end. To put it in its most fundamental form, it seems the whole aim of Brown's device is to create an imbalance in the force of the emitted radiation on the device at opposite ends so as to create movement predominantly in one direction.

Looking at the above toroidal capacitor design, Brown stated that the positive electrode is the smaller of the two and is chasing the negative electrode. For this to happen, the recoiling force of the emitted radiation from the positive electrode has to be greater than the recoiling force on the negative electrode facing in the opposite direction because the surface area holding the positive charge is smaller compared to the same number of negative charges put onto the negatively charge electrode. It means the positive charge is at a higher charge density. That in turn is translated into a higher energy density radiation. And a higher energy density of the radiation is what gives radiation the ability to exert a greater recoiling force on this positive electrode.

But that is not the complete picture. Any radiation emitted from the surface of the electrodes facing each other is destructively interfered (as they have opposite charges) resulting in a reduction in the energy density of the radiation between the electrodes. So not only do you change the charge density through electrode area, you can also control the energy density of the emitted radiation through destructive interference of the radiation.

You can see this situation between the electrodes better by looking at the diagram on the previous page showing a variation on Brown's electrokinetic device and so revealing the importance of electrode area in controlling charge density and destructive interference in controlling the energy density of the radiation. As the diagram shows, unlike charges attract. Consequently, the closeness of the two electrodes helps to distribute the electrons on both electrodes whose surfaces are facing each other such that the electrons effectively balance the positive charge of the other electrode. A positively charged matter is never completely without electrons. The negatively charged electrode merely repels a certain number of electrons to create the positively charged electrode. But everything in between the electrodes facing directly at each other is more balanced. This ensures the radiation emitted between the electrodes is out-of-phase and of virtually identical energy density, thereby eliminating the radiation (the familiar Faraday's cage effect), or at least minimise any contribution the radiation could have had to the overall movement of the capacitor if it could come off the inner surfaces (especially around the edge). Therefore, the only contribution we need to worry about here is from the radiation being emitted to the environment on the outside surfaces of the electrodes. This is the critical point. Here, the emission of radiation on the outside causes the electrodes to recoil in the opposite direction. However, due to a change in the charge density on the outside surfaces (i.e., the positive electrode has a lower charge density on the outside surface compared to the negative electrode), there is an imbalance in the recoiling forces. Remember, charge density can be varied by changing the shape and/or area of one of the electrodes to be different from the other in a manner as to create a difference in the energy density of the emitted radiation from both ends of the electrodes. You are creating an imbalance in the recoiling forces, resulting in movement of the entire device.

This has to be the correct scientific explanation for how Brown's devices move.

However, almost as if the U.S. military did not want to see this explanation, two U.S. Army researchers who studied Brown's capacitor in 2002 (see Appendix F) claimed the larger negative electrode is the backend of the capacitor. More importantly, they felt that there was not enough movement when it was electrically charged to be of any use to society.

How do we explain this discrepancy?

On closer inspection of the capacitors built by the military researchers and the voltage used, we see a different set of conditions set up compared to the original work done by Brown. The main problem is that there is absolutely no oscillating voltage applied anywhere in the military experiment, yet we know Brown had used an oscillating voltage. He had to turn off and on the charge in order to create movement. Any contribution by the ionised air to exert a force on the charged plates must be negligible in comparison to the force of the radiation coming off the plates so long as there is asymmetry in the electrode pair.

The military researchers built a triangular or rectangular capacitor. The design is effectively a multiple of the basic original design by Brown as shown below:

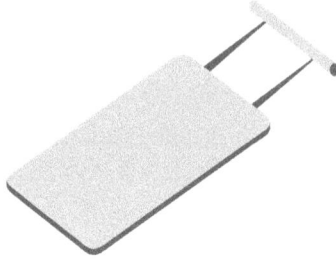

With this design, what the researchers have done is apply a high non-oscillating voltage to help strip off electrons from air molecules surrounding the positive charge electrode and, perhaps at the same time, emit some positive metal ions. All these positively charged metal ions and mostly ionised air molecules are electrostatically repelled away from the positive electrode. As these particles sweep over the larger negatively-charged electrode, the positive charges will be quickly neutralised and, in most cases, acquire a negative charge as extra electrons are added to these particles. As soon as the particles become negatively charged, they are repelled by the negative electrode. The Army researchers believed this repelling action and the subsequent recoiling force exerted by the general flow of ionised particles over the electrodes and onwards to form ionised exhaust fumes in one direction is what causes the capacitor to move in the opposite direction.

What is not noted in the experiment is what happens what the voltage is oscillating. It is odd that the military did not consider this in the experiment. Or perhaps we should not be surprised given what we know in Chapter 7?

In U.S. Patent No. 2,949,550, Brown seemed to acknowledge this "electrical wind" explanation by revealing a diagram of the path of positively-charged air molecules moving toward the negatively-charged electrode and later repelled when the molecules become negatively-charged. The diagram below shows the path above and below the above shown flat capacitor design viewed edge-on with the positive electrode positioned on the left:

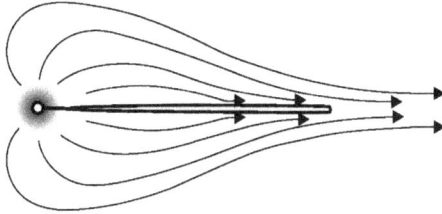

One should bear in mind that this is not meant to be an admission from Brown that his invention is based on the "electrical wind" theory. Far from it. Brown noticed something else when the voltage oscillates in certain capacitor designs. What he noticed was enough for him to realise that the movement was far more significant and he felt sure the "electrical wind" was not responsible for this. Another force was taking place.

To support his findings, Brown used a vacuum chamber at the French company SNCASO in 1955. Here it became clear to him how the ionised air could not be responsible because the movement was shown to be faster in a vacuum using a lower voltage. He even used a non-ionising oil bath and somehow the device would move. This could not be possible if the number of ionised particles had been reduced at a lower voltage. There could not have been enough ionised material to make the capacitor move so appreciably. Something else was happening. The question is, what was it?

Despite many years working on the problem, Brown never did succeed in finding the scientific explanation. As incredible as this may sound, without a proper theory to explain how his devices worked, Brown naturally came up with all sorts of innovative designs to help increase the speed of his devices. In one of those designs, as shown in the *Los Angeles Times* on April 3, 1952, his upside-down disks, when initially uncharged, were attached to strong wires supplying the power on a rotating pole where the negative electrode appeared at the top and the larger positive electrode at the bottom. Notice how this time Brown used a negative charge for the smaller electrode. The type of charge he used for the smaller electrode makes no difference. As published in the newspaper, the design of his capacitor was such that the movement was toward the larger positive electrode, so the negative electrode was chasing the positive electrode. As for the speed, he noticed that when the disks were given a high oscillating electric charge they could move at considerable speeds—far more than he had expected. There was no way in his mind that an 'electrical wind' could be responsible for such a significant movement.

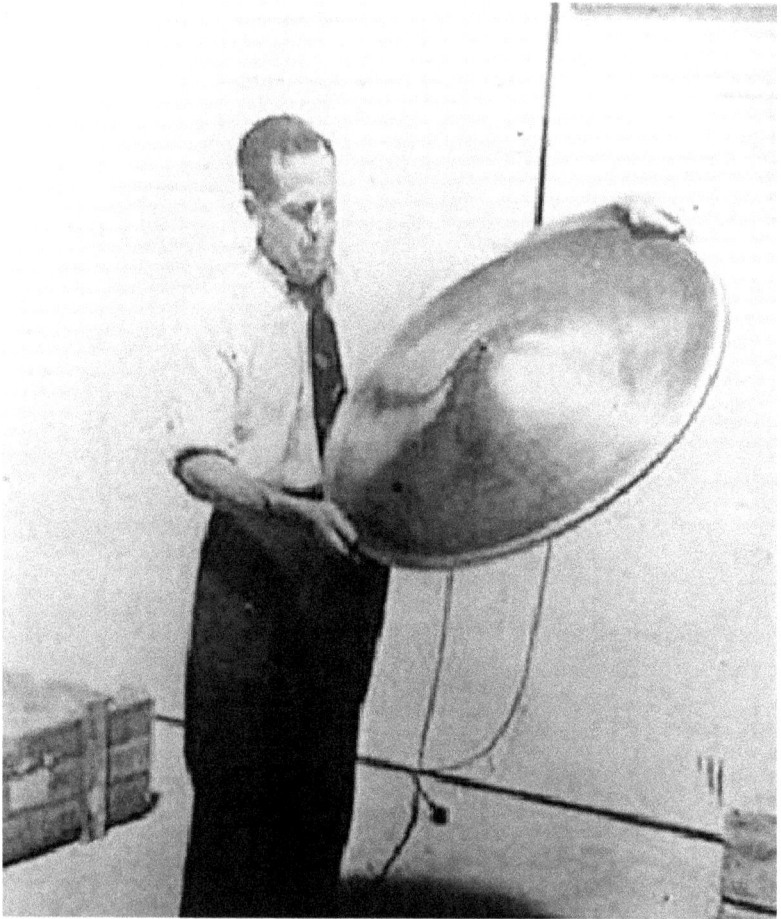

Brown holds up one of his moving capacitor disks developed in 1955 at the French company SNCASO (Photograph taken by Jacques Cornillon)

But there is one more observation Brown would make, and one that would put the final nail in the coffin of the 'electrical wind' theory. Brown claimed he saw something remarkable happen with his devices when a critical voltage and frequency is reached.

The Critical Discovery Made by Brown

As first noted when he was a teenager and later refined in the 1960s, Brown realized that when the voltage (or charge) on his capacitor reached a certain level (but no indications on the frequency he chose for the oscillating voltage), his device could be made to increase speed on its own for the same voltage. In other words, the capacitor would self-accelerate to higher speeds. To prevent his device from breaking the wires supplying the voltage and flying off, Brown had to reduce the voltage. And if he prevented the energy in the device from being lost to the surroundings, such as immersing the device in a vacuum or non-ionizing oil, he could use a lower voltage when he wanted his devices to self-accelerate.

Assuming this acceleration is non-linear (i.e., exponential), this self-accelerating behaviour in Brown's electrokinetic devices can be mathematically explained only by resorting to the classical theory behind the emission of an electromagnetic wave from the surface of a charged object.

In other words, the formula describing this interesting phenomenon is known as the *Abraham-Lorentz Formula*, and the self-accelerating behaviour of Brown's devices is indicating support for the runaway solution.

Simplifying the Invention

Electrode Size

Observing the invention more closely, we can see that the entire purpose of having two different electrode areas (or sizes) is to create an imbalance in the charge density and, with it, the energy density of the emitted radiation. In this way, the emitted radiation with the higher energy density will exert a greater recoiling force. For example, when the same amount of opposite charges are applied to the following electrode pair A and B (separated by an insulator), as both electrodes have equal area, the charges on the outside surfaces distribute evenly and are of *equal* density:

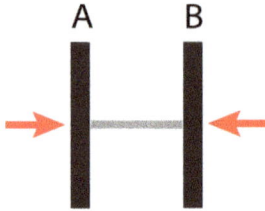

Therefore, the direction of movement (indicated by the red arrows) produced by the recoiling forces of the emitted radiation on the outer surfaces of each electrode is equal, which means the electrode pair will not move.

If you reduce the size, and hence the surface area, of one of the electrodes, the same amount of charge must distribute itself on the smaller surface area. This means the charge density on the outer surface of the smaller electrode A is higher than the charge density on electrode B.

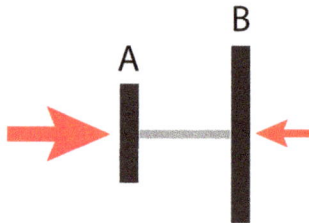

Therefore, movement is expected to occur while the electrodes have different charge densities. In the above diagram, the electrode pair will move from left to right.

Remember, the relative closeness of the electrodes helps to change the distribution of the charge on the inner surfaces of each electrode such that the charge density is roughly the same. And because the electrodes are of opposite charge, there is destructive interference of the emitted radiation between them. There is minimal, if any, contribution of the radiation emitted between the inner surfaces to the overall movement of the electrode pair. What is critical is the charge density on the outer surfaces of the electrodes.

Of course, you can again reduce the size of electrode A. This reduction will increase the charge density and, with it, exert a greater recoiling force on the smaller electrode by the emitted radiation. However, the trade-off in doing so is that the temperature of the electrode will increase. Unless the small electrode is composed of a tough and lightweight metal or alloy with an extremely high melting point, you would be better off using three or more small electrodes to carry the charge.

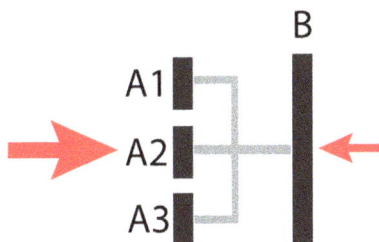

In the famous Belgium triangular UFO case in the 1990s, three bright lights were positioned at the apex of the UFO in a precise equilateral design. There was also a central and much larger glowing circular region directly below the body's centre of gravity. Its likely purpose was to provide lift. Whereas the lights in the apex positions controlled the tilt of the object. However, once the object gets into space, all four glowing lights could be computer-controlled and used to accelerate the object. There would be no need for additional glowing regions to emit radiation.

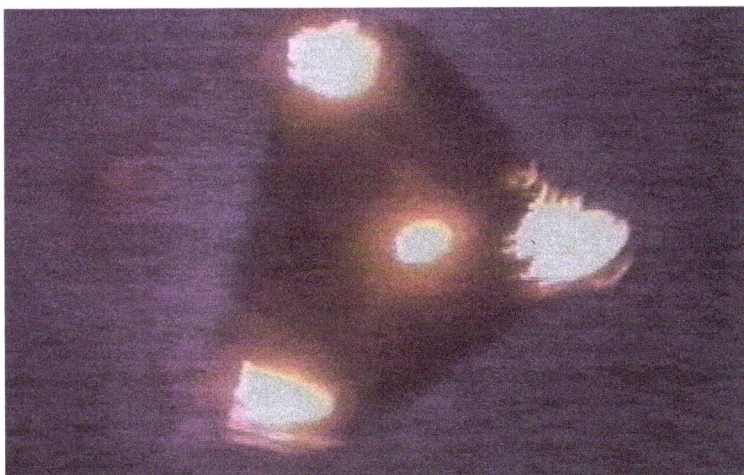

Photograph of a triangular UFO observed over Belgium in April 1990.

In another example, the turtle-shaped UFO from Brazil discussed in observation 13 in Chapter 5 with its multiple flashing lights around the rim or edge of the UFO could be a solution to rapid direction change or provide stability in the air when hovering. Computer-controlled injection of the charge over a small electrode area (i.e., the flashing lights) could emit radiation in a manner that ensures the object remains stationary in the air. Otherwise, the object's base can be charged to emit radiation to help provide the necessary and dramatic lift and acceleration.

Artist impression of the UFO observed by Almiro Martins de Freitas in 1970.

Electrode Curvature

The alternative method of increasing charge density is by inducing curvature into one of the electrodes. For example, if we bend electrode A in this design,

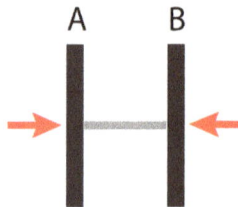

to look more like this,

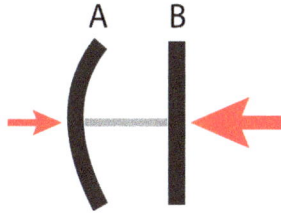

the smooth and even curve still distributes the charge evenly. All it does is increase the outer surface area of the curved electrode slightly. It just means the outer surface area of electrode B, being smaller, has a higher charge density. Therefore, the electrode-pair will likely move from right to left.

However, by creating a greater curvature over one smaller region, you will congregate the charge over this region and increases the charge density. The recoiling force of the emitted radiation on the curved region then increases and can potentially exceed the overall recoiling force on the opposite electrode.

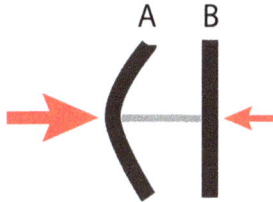

For example, in certain UFO cases, the technique for distributing charge or creating a higher charge density is more subtle and requires a sharp eye to discern the relevant details. These cases may also reveal a high level of engineering precision and refinement with minimal use of the materials and everything hidden inside that it may be possible to create a slightly curved bottom using a single spherical conducting object to press against a flexible "shape-memory" outer hull. As curvature is important in concentrating the charge to a higher density as needed to generate and emit higher energy density radiation over one end of the object, all it may require is a subtle variation in the UFO shape to make it asymmetric. If this is done with precision and accuracy and using a shape-memory alloy for its outer hull, there is no reason why the UFO could not move in a straight line. In which case, the entire surface of the UFO could be made to glow with no obvious signs of how the UFO moves unless you look very closely at the intensity of the glow beneath the UFO when it begins to lift off the ground and accelerate. When the UFO moves, the rear end may reveal a brighter and more intense glowing region to indicate more charge is present on the surface. Examples of these kinds of UFOs could be observed in Levelland, Texas, in 1957 and in the spherical glowing UFO from Japan seen by Masaaki Kudou in 1973.

Artist impression of the UFO observed near Levelland, USA, on the night of 2–3 November 1957.

However, if you want the flat electrode to have a greater recoiling force of the radiation than the curved electrode, do this:

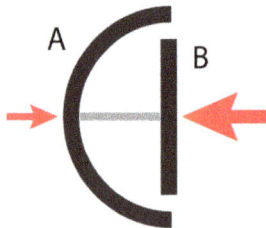

Actually, the above design is highly reminiscent of the insignia used by the makers of the UFO allegedly observed by police officer Lonnie Zamora on April 24, 1964 near Socorro, New Mexico:

SKETCH OF OBJECT
FROM MY POSITION.
AT APPROXIMATELY 103 FT.

UFO artist impression by Kari Rajanen.

407

Single Asymmetric Electrode

If everything that has been said so far is true, then we can simplify the designs to applying an oscillating charge to one electrode and shape the electrode at one end to ensure that there is a difference in charge density and, with it, an imbalance in the energy density and ultimately alter the recoiling force of the radiation emitted compared to that of the opposite end.

Imagine an electrode shaped like a sphere. The emission of radiation from an oscillating charge injected onto the surface of the sphere will not cause movement due to the recoiling force (the direction of which is marked by the red arrows).

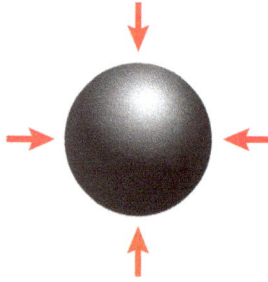

An increase in the metal surface's curvature will help increase the energy density of the emitted radiation. However, if the curvature results in an overall symmetric design, there is no movement in any specific direction.

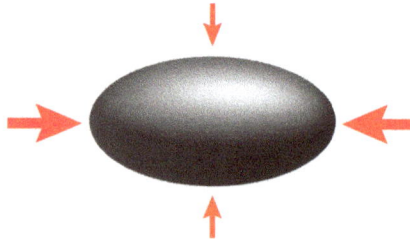

However, if we increase the curvature of the metal electrode over one end to create an asymmetric design, then the movement will predominantly be in one direction. Below is the simplest example (described as egg-shaped).

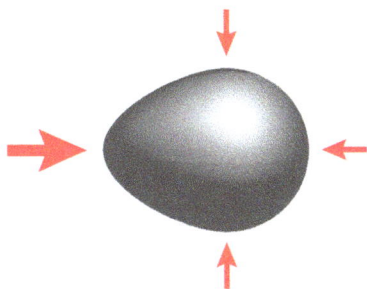

In this example, a greater recoiling force is exerted on the left side of the electrode due to the greater curvature of the metal surface. This greater force means the charge density will be higher over this region than anywhere else on the surface. Therefore, the higher energy density radiation being emitted from this region will push the object more strongly to the right.

MUFON Report Case 94668 mentioned details of a UFO observed by anonymous Australian witness from the town of Arncliffe. He reported watching a white, asymmetric egg-shaped (or tic-tac) object moving fast across the sky. He drew a picture of the UFO for investigators.

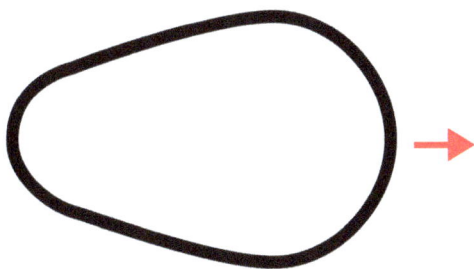

According to his testimony, he said he was in a vehicle waiting at an off-ramp red light at 1:00 p.m. On April 15, 2018. "Odd shape, trajectory and speed," the witness stated. "Thought it was a perhaps a balloon (like a party balloon) but there was only one. It wasn't the correct shape (more like a tic-tac or egg) and was not ascending (like a balloon with helium would); rather, it was moving horizontally across the sky, far quicker than a balloon would be moving."

The witness was surprised that an object like that would be in the sky so close to the airport. "Went outside of my field of view."

Using Multiple Curved Regions

You can also amplify the recoiling force on one side of the charged electrode by adding more curved regions. A minimum of three curved regions positioned in a near-perfect equilateral triangle over one end is necessary. More curved regions are possible and should be arranged symmetrically over one end and positioned at equal distances, but the amount of metal used will increase, which will increase the mass of the object.

No matter how many curved regions are created, for this radiation emission that propels the object to work properly, the position of this increased curved region(s) has to be extremely precise and positioned exactly over the centre of gravity of the electrode. If it is not, the object will spin uncontrollably. The goal is for the electrode to move in a straight line.

Suppose one is less confident in achieving this level of engineering precision in positioning the curved metal domes. Consider connecting each dome to an independently controlled computer system to deliver the right amount of charge, ensuring the object does not wobble and proceed into an uncontrollable spin.

In support of this "multiple curvature of the bottom metal plate" design, we can see from the UK UFO case near the town of Langenhoe in the 1970s (see observation 4 in Chapter 5 for further details) a flying object with multiple domes glowing intensely at the bottom end.

Artist impression of UFO observed by Mr Paul Green in the UK.
See observation 4 in Chapter 5 for more details.

Using External Fuselages to Create Asymmetry

Another approach is to use external fuselages with charged electrodes at the end and to position the fuselages around the central symmetrical body to achieve either an asymmetric configuration when accelerating or a cross or perfect equilateral triangle configuration when hovering in the air.

As an example, the UFO observed by Angus Brooks revealed a central circular flying object with four movable external fuselages. The fuselages likely distributed the charge. Should the fuselages be arranged in a cross configuration, there is no movement. However, as the witness had reported, moving three fuselages to the back end would result in acceleration due to the asymmetric distribution of the charge, with most of the charge pushed to the back end, even if the charge density is the same on all four fuselages.

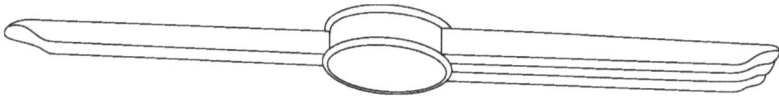

Acceleration to the left

In the Antonio Villas Boas UFO case in Brazil, the object's creators decided to use three fuselages at the back end that are fixed into position (for acceleration). An additional two fuselages resembling metal walking planks or winglets were attached to the side (possibly designed, in conjunction with the large glowing base of the object, to make it hover and prevent rotation and tilting of the object in the vertical plane). Another fuselage resembling a rudder was attached to the front end (possibly designed to help it rotate left or right in the horizontal plane, although the two outer backend fuselages under computer control could probably accomplish something similar, but perhaps without the finesse and refinement of the front fuselage). As for the large metal underside of the central body, this was likely to be where most of the lift of the object was achieved.

Artist impression of the UFO observed by Antonio Villas Boas.
UFO moved from left to right.

Again, such positioning of the fuselages and glowing effects seems reasonably consistent with the positioning of the movable fuselages of the Angus Brooks UFO case. Furthermore, technically it would probably meet the above-mentioned electromagnetic requirements for distributing electric charge at the right positions around and/or beneath the object in order to hover in a symmetry form of the charge distribution, to make turns in the air, or accelerate through a non-symmetrical form of the charge distribution.

A New Interpretation of Bob Lazar's Bizarre Claims

In Chapter 7, we learned of Lazar's rather complicated explanation of the way in which a UFO generates power from an impossible-to-manufacture heavy element material. Leaving that aside, Lazar did mention something intriguing. He reports that the UFO he observed at Area 51 was able to levitate and move using three movable metal spheres that were touching the external (and potentially flexible) hull of the object. He alleges that there were three *gravity amplifiers* and an *anti-gravity reactor* beneath the floor of one of these UFOs. He describes how all these items are arranged geometrically to achieve levitation. The gravity amplifiers are arranged in a perfect equilateral triangle, with the reactor positioned in the centre. Lazar claims these amplifiers can move to different positions on the underside of the craft to achieve forward motion.

If what Lazar has said is true, a more simplified explanation of the components he saw and their purposes is explained as follows:

GRAVITY AMPLIFIERS	These are likely to be regions where an emission of higher energy density radiation occurs. For example, a light beam is, according to Einstein's Unified Field Theory, a gravity amplified energy region since electromagnetic radiation creates its own gravitational field as the theory states. Therefore, radiation with a higher energy density has already effectively "amplified" the gravitational field in that region. In terms of a simple gravity amplifier device that can generate a higher energy density radiation, this can be achieved by creating extra curvature on the metal hull. Or, metal balls could touch a partially flexible shape-memory alloy hull to create this extra curvature.
REACTOR	This is likely to be the electrical generator/energy storage component for charging the "gravity amplifiers" (or metal domes, or any protrusion that directs extra radiation in an asymmetric design as needed for propulsion).
LEVITATION	Any talk by Lazar about emitting an anti-gravity wave to help achieve lift and levitation of the UFO is probably nothing more than extra radiation between the object and the Earth that the metal domes (or protrusions) emit to help the object recoil in the opposite direction. It is likely that the geometric arrangement to form a near perfect (if the distribution of charge is computer controlled) or perfect equilateral triangle by the "gravity amplifiers" (as he calls them) and positioned exactly in the middle of the centre of gravity of the UFO is there to prevent uncontrollable wobbling or spinning motion while achieving vertical lift or levitation.
ACCELERATION	As Lazar states, moving one or more gravity amplifiers to the side is how the UFO allegedly moves forward. Well, apparently the same thing could be done by moving the domes or protrusion to one side to ensure that a higher amount of charge and charge density emits a higher energy density radiation over one end of the object to make it move forward. Concurrently, any remaining domes or protrusions (or by using the entire underside of the object) can help to maintain lift when the UFO is in the Earth's atmosphere. Otherwise, all the protrusions can be employed to accelerate in the vacuum of space.

Can We Build this UFO?

At last, we understand it is the radiation that causes the movement in Brown's devices. No electrical wind is required to explain what happens (even if it plays a small role). And technically, we do not need to worry about Einstein's Unified Field Theory in terms of a gravitational field in the radiation or accept some kind of exotic "alien-like" gravity amplifiers from Bob Lazar. Only plain, simple radiation at work. It has all to do with the radiation and its *energy density*. Everything is controlled by the surface area and/or shape of the electrodes, the amount of charge generated, and any means to reduce radiation in between the electrodes for the imbalance in the recoiling force of the radiation on the outer surfaces to be more effective.

If all that has been written here is true, then we can start to see the engineering requirements needed to build one of these genuine glowing symmetrical-looking UFOs. More specifically, you need:

- a portable and lightweight high-voltage and variable high-frequency electrical generator device;
- a means of storing large amounts of electrical energy, especially when the object needs to take-off from the ground as well as provide a constant stream of reliable energy to the electrical generator during acceleration until the energy density of the radiation is high enough to be recycled and maintain the acceleration;
- a high voltage multiplier device to help enhance voltages/charges;
- an exterior metal shell to hold the oscillating high-frequency voltage;
- a mechanism to change the shape of the exterior metal shell in order for the charges to be distributed in an asymmetric form for acceleration and returned to its original shape for a more symmetric pattern when hovering in the air, or else some means of distributing the charge using movable or fixed fuselages or metal domes attached to the central symmetrical body is necessary;
- another metal box inside to protect the occupants and instruments from the radiation and high voltages outside (this time it must always maintain its perfectly symmetrical shape to

minimize radiation and the gravitational field entering the box[43]);

- a means by which the metal boxes can be separated, such as by using insulating materials; and
- to minimize the use of materials and to choose the most lightweight and strongest materials.

Essentially you are building an extremely lightweight re-useable weather balloon (to help minimise the amount of energy needed to move the object), except the materials used to construct it must be able to withstand high temperatures and have high strength so that it does not fall apart when traveling at high speeds. You need two metal "balloons" or shells. One must go inside the other. The outer shell carries the charge, and the inner one protects people and instruments inside from the full brunt of the electromagnetic fields outside and stops stray electric charges from entering the cabin(s). Also, the unified field concept of reducing the gravitational field inside by lowering the electromagnetic field helps control inertial forces on the body during acceleration.

To keep everything lightweight, the metal shells can be newspaper-thin. Charges only need to sit on the surface, so there is no need to have a thick hull. Of course, at this thin dimension, choosing the right high-strength alloy will be important when building a rigid and strong object.

In addition, there must be a gap or a means of separating the outer shell from the inner one. Thus, one must use flexible insulating sheets and less flexible plastic high-temperature and electrically resistant beams between the metal shells to provide the necessary structural support while ensuring the charge on the outside is prevented from touching the internal metal shell.

As mentioned before, the outer metal shell is designed to hold an oscillating electric charge and distribute the charge in an asymmetric pattern to create acceleration as the radiation is emitted and recoils against the charged metal shell. To achieve this pattern, the outer metal shell can be made with a shape-memory, or an inflexible, alloy. If using the shape-memory alloy method, a mechanism internally (say, a metal arm with a large metal sphere to inject the charge on the surface) can

---

43   This is how the inertial forces are controlled internally due to the fact that radiation drops to zero inside the metal box and with it, according to the Unified Field Theory, the gravitational field as well.

extend outwards to press against the shell over one end to help distort the shape slightly. This action has the effect of increasing the curvature of the metal surface and making the shape of the craft asymmetric. The charge will then congregate over this curvature, thereby increasing the energy density of the radiation emitted from this curved surface, which in turn exerts a greater recoiling force on this charged region. Later, the mechanism can be withdrawn to return the outer shell's original symmetrical shape, effectively stopping any further acceleration from taking place. The alternative method is, of course, to use a very hard outer shell alloy shaped permanently to be non-symmetrical. To control acceleration, the amount of charge and/or its frequency has to be adjusted internally (most likely computer-controlled), and the craft has to be turned to face the opposite direction to slow it down. Alternatively, one can use three or four movable or fixed protrusions around the central symmetrical body to help distribute the charge into the required asymmetric and symmetric pattern as needed for acceleration and hovering, respectively.

As for the mechanisms needed to collect, store and distribute charge, the energy in space can be collected using an antennae, stored into a battery, and some of the energy in the battery used to generate high voltages. When combined with a fast switching mechanical device to help create an oscillating charge/voltage, the energy can perform some kind of work (i.e., emit radiation for propulsion).

The use of an antenna to extract free energy from space is particularly useful. All it requires is a metal rod dangling out of the top of the spacecraft. As the spacecraft moves, the antenna can pick up energy in space as it cuts the planetary, solar and galactic magnetic fields. which can be tapped and injected into the battery using another set of coils to help keep energy storage in the superconducting ring topped up. The energy gathered by the antenna will naturally increase as the spacecraft gains speed.

Obviously, one should never be restricted to an antenna to collect energy. There are many other ingenious ways an electrical engineer could generate or extract energy to store it for later use. For example, piezoelectric crystals running from the top to bottom of the spacecraft and centrally located inside the main cabin via a column in the middle to provide structural integrity can be made to compress slightly with each emission of radiation (using a flexible alloy hull). As a result, the crystals can produce electricity in the order of millions of volts. To prevent further input of energy into the system, a mechanical handle on the central column can be turned to adjust the position of the crystals such that it can prevent the radiation emission from pressing against the crystals.

For example, we see a central movable pillar emerged from the central part of the UFO to touch the ground as if meant to discharge something in the French UFO case of the lavender farmer Maurice Masse in 1965 (see Chapter 5). This suggests that the pillar is probably part of the electrical generator and could potentially contain these crystals.

In addition, we can make use of the difference in temperature between the outer metal shell and inner metal shell to help generate more electrical energy in a process known as the *thermoelectric effect.*

In terms of energy storage, electrical engineers might build a large-scale superconducting ring (i.e., the battery). This ring can be cooled by the temperature of space, although advanced knowledge from experienced civilisations may find superconducting materials to work at or near room temperature. The aim of the ring is to store vast amounts of continuously circulating electrical current (with no electrical resistance to slow it down). To draw upon some of this energy and later inject some energy back into the ring, insulated metal coils are wrapped around part of the ring. A high-speed mechanical switch is used to turn this energy on and off. Thus when the energy injection phase is off, the energy extraction phase begins. and some of the energy from the battery can go through a voltage multiplier device. What emerges from the device is much higher oscillating voltages (or charge) ready to be injected onto the outer metal shell. When the switch is turned the other way, no energy is extracted. Instead, the energy obtained from the surroundings as the charged object accelerates and any remaining energy on the charged surface is introduced and returned, respectively, to help top up the battery storage.

Of course, we do not need to generate, store and supply the energy for accelerating the charged object constantly. Once the acceleration approaches relativistic speeds, the energy of space will already be attracted to or concentrated around the object. Once a critical energy density of the radiation is reached with the help of this extra energy from space, the emitted radiation can be recycled and made to continuously recoil against the object without any further input of electrical energy to maintain the charge. Alternatively, use a high-frequency oscillating charge to move the electrons at high speeds to achieve the same effect of concentrating the energy from space around the object.

Hence the object is essentially all of the following:

- a giant antenna for picking up electromagnetic energy from the surroundings as it accelerates;
- a giant battery to store this electromagnetic energy, not just temporarily on its surface (and hence perform as a giant capacitor), but for more longer-term use inside an internal storage device; and
- a giant electric light bulb as the outer metal shell glows brightly.[44]

What we have here is an electromagnetically "breathing" machine in which electricity is drawn in from the surroundings for temporary storage and expelled for propulsion at a very high rate of switching (or frequency). The entire purpose is to generate emitting radiation controlled in a manner that ensures the object moves in a straight line and follows the acceleration pattern described by the Abraham-Lorentz formula, which is exponential. That is basically what we are dealing with in those genuine UFO reports.

Can We Explain Common UFO Observations?

As the Abraham-Lorentz formula provides the angle we need to look at more closely and Brown's devices are the key to putting this formula into reality, how well can we explain common UFO observations from this electromagnetic perspective? In other words, can we find simple explanations for each of the most common observations discussed in Chapter 5?

The following table gives a summary of the observations in Chapter 5 and their explanations. For examples of UFO cases in support of the observations, see Chapter 5.

---

44   This glow is due to the electrons in the oscillating charge being accelerated and changing direction quickly resulting in emissions of radiation. So not only are you producing large-scale electromagnetic waves for propulsion, but smaller radiation emissions acting as noise over a relatively wide range of frequencies. Hence the UFO will not only be visible as it glows, but can be very hot to touch due to the emissions of infra-red radiation, and can also emit radio wave radiation that can interfere with reception on radio devices inside cars.

	UFO Observation	Likely Scientific Explanation
1	UFOs seen during the daylight hours are often described as silvery or metallic grey in color.	The colour may be a component of a metal or metal-alloy body designed to protect its occupants from extreme temperatures, retain an comfortable interior atmosphere, and to hold a high-frequency oscillating charge on the surface for emitting radiation.
2	UFOs have smooth surfaces.	An EM spacecraft carrying an oscillating charge on its surface must be designed to avoid sharp points and edges. Failure to do so may lead to a breakdown of the insulating properties of the air, causing a rapid loss of energy through electrical sparks with nearby objects. Also, a high oscillating charge can raise the temperature of sharp metal points, causing them to melt. Some extra curvature around the circumferential rim can help to ionise and deflect the air molecules as a means of reducing air pressure. Other than that, smooth outer surfaces will reduce drag as air flows over it.
3	UFOs seen at night at close range are usually described as being highly luminous.	An oscillating charge on the spacecraft's surface will heat up the metal and emit electromagnetic noise called *bremsstrahlung radiation*. This is because electrons vibrate intensely, or undergo rapid translational motion and suddenly collide with the atoms in the conducting material or with other electrons in a pinball-like fashion. It means the object will glow like an electric light bulb and emit radiation by way of noise in the radio, visible and infrared regions.
4	The position of the main glowing surface where the metal "heats up" indicates the direction of the UFO as it moves.	The position of the glow is essential to directing the emissions of higher energy density radiation compared to the opposite end, which in turn creates a stronger recoiling force of the radiation on the charged surface.
5	The glowing regions also appear on movable or fixed metallic fuselages.	Metal fuselages attached to the central symmetrical body of an EM spacecraft are there to distribute charge in a symmetric or asymmetric configuration around the object depending on whether it is hovering or accelerating, respectively. The makers of an EM spacecraft will likely employ a variety of innovative, efficient, and highly precise methods of distributing charge for emitting radiation for propulsion.

	UFO Observation	Likely Scientific Explanation
6	The glowing effect varies in luminosity and intensity depending on whether the object is flying over metals, water, or near other UFOs.	An EM spacecraft should maintain a safe distance or reduce its surface charge near metals, water, or other spacecraft, in order to avoid a breakdown in the electrical insulating properties of the air, potentially leading to an electrical spark between the oscillating charge on the surface of the spacecraft and an electrically conductive substance where electrons flow freely in metals and the weak electric dipole of water molecules can move around.
7	Glowing UFOs or UFOs with bright lights have been observed to blink on-and-off in a periodic manner.	The fluctuating luminosity of the glowing charged metal surface indicates an oscillating charge used to generate radiation. At low frequencies of the oscillation, time is available for the metal to cool down and reduce its glow. As soon as the charge is injected onto the metal surface, however, the metal heats up and the glow intensifies. When the process is repeated, the glowing effect can appear as a periodic on-and-off blinking.
8	Witnesses have reported seeing UFOs accelerate after the appearance of a flashing light or glowing effect on its surface.	A sudden acceleration that occurs immediately after an increase in the object's brightness indicates a half-cycle emission of high energy density radiation emerging from the oscillating charged surface, leading to a recoiling force that moves the object in the opposite direction by virtue of the radiation.
9	The rate (or frequency) of the UFO's blinking on-and-off effect controls its speed.	A blinking effect occurring at a higher frequency translates to an increase in the energy density (or momentum) of the emitted radiation due to greater acceleration of the electrons moving on and off the metal surface. Therefore, the recoiling force of the emitted radiation is greater leading to an increase in the speed of the EM spacecraft.

	UFO Observation	Likely Scientific Explanation
10	Flight paths of UFOs can take a zig-zag or sinusoidal pattern.	A low-frequency, high-voltage oscillating electric charge on the surface of the EM spacecraft generates radiation at a slower rate, meaning that there is time for gravity to pull down on an object, followed by an emission of the radiation to push it back up. This is not likely a comfortable mode of transport for occupants inside the EM spacecraft. Therefore, it is expected that the piloting of the object in this manner is done so in an attempt to teach observers on the ground about one aspect of the technology.
11	UFOs can exhibit bouncing drift and spiraling motions.	As an oscillating charged and glowing EM spacecraft is essentially a large-scale electron, whoever is flying the object may choose to mimic the behaviour of an electron while moving through converging and diverging magnetic fields, as one would find around a magnet (or the Earth, for that matter). Although this behaviour is not a normal mode of self-acceleration, the presence of these observations suggests that the pilot(s) of the spacecraft have a desire, again, to teach witnesses to record, learn, and hopefully understand the concept behind whatever intriguing electromagnetic technology is being applied.
12	UFOs cause nearby metallic objects to vibrate.	Whilst an EM spacecraft should technically avoid metal objects on the ground during normal operations, the use of a low-frequency high-voltage (or charge) on the spacecraft's surface can emit a large-scale radiation wave that can in turn cause nearby metallic objects to vibrate from side-to-side in time with the frequency of emission.
13	UFOs heat up nearby metallic objects.	High-frequency radiation emitted from the surface of a charged EM spacecraft would induce a strong current inside any nearby metallic object of such intensity as to cause them to rapidly heat up.

	UFO Observation	Likely Scientific Explanation
14	UFOs leave behind colored glows in the air.	The electrons coming off an oscillating, negatively charged surface of an EM spacecraft —especially where there is extra curvature, say, around the circumferential edge—can knock out electrons in the atoms composing the atmosphere. When the atoms are nitrogen and oxygen and they regain their electrons, they emit visible radiation reminiscent of an aurora, featuring greenish, bluish or yellowish colours. The smell of ozone may also be detected in the presence of the object.
15	UFOs can masquerade as cloud-like structures.	A rotating charged ring inside or outer charged hull section of the EM spacecraft can create a strong magnetic field that may be used to deflect surrounding ionised air. As a result, air pressure is reduced and, in the presence of moisture in the atmosphere, can lead to the formation of a cloud or fog. At first, there is a gap between the surface of the charged object and the torus region of ionised air and fog-like structure around the rim, but the fog or cloud-like structure can easily expand to obscure the entire object.
16	UFOs can disrupt man-made electrical circuits.	When ionised air surrounding an EM spacecraft reaches an electrical circuit, the electrified air can provide alternative routes for electricity to flow between the battery terminals of electrically powered devices that are not properly sealed (e.g., cars become immobilised, aircraft can lose power, etc.).
17	UFOs can bend light beams.	When an EM spacecraft emits high-energy density radiation into the surrounding are, the gravitational field of the radiation can be strong enough to bend the path taken by the radiation— as well as other forms of light in the vicinity, and within this high-energy density region.

	UFO Observation	Likely Scientific Explanation
18	UFOs can render themselves invisible in an intermittent or permanent way.	Emitting high-energy density radiation with the intention of generating a strong gravitational field can bend the radiation emitted by the EM spacecraft back on itself—and the light from behind the object can in turn bend around it to allow an observer to see what is behind the object. As a result, the object becomes invisible. As the radiation emerges from the surface in an intermittent manner, the object can render itself invisible, then visible again, periodically and in time with the frequency of the radiation being emitted.
19	UFOs exhibit unusual accelerating behavior.	This behavior is likely due to the exponential rate of acceleration when a charged object emits radiation.
20	UFOs are described as being highly symmetrical.	Due to the tremendous exponential acceleration of this self-accelerating charged EM spacecraft, it is imperative that occupants be protected from the inertial effects. This is achieved using a little-known unified field concept whereby a perfect Faraday cage (i.e., a symmetrical metal enclosure) reduces both the electromagnetic and gravitational fields within the interior of the craft. It is the gravitational field of radiation that creates the inertial forces to be exerted on the human body during acceleration and changes in direction.
21	UFOs are generally quiet, but at close range they produce sounds reminiscent of an electrical generator.	An EM spacecraft generating an oscillating charge requires a mechanical switch to turn the charge on and off very quickly. When the switch makes contact to allow charge to flow to the surface, a tiny clicking sound is emitted. At a high frequency, the slight clicking sound created by the switching effect can combine to sound like a buzzing sound. In addition, high-frequency radiation emissions in conjunction with a strong gravitational field can cause air molecules to vibrate and create a humming sound. At a distance, these sounds are usually undetectable.

	UFO Observation	Likely Scientific Explanation
22	UFOs can feature antennae protruding from their central bodies.	While accelerating through space, an EM spacecraft can tap onto the energy from its surroundings through the use of one or more metal antennas sticking out of the top of its central body. These antennae can help cut through planetary, interstellar, and galactic magnetic fields. Generally speaking, the faster the antenna moves through the fields and the stronger the magnetic fields, the higher the electrical charge generated at opposite ends of the antenna.
23	UFOs can appear to have rotating parts.	Any rotating charged metal components may help to produce a powerful magnetic field as a means of deflecting ionised matter in space, or in the atmosphere of a planet. The acceleration of the rotating charged parts can also concentrate a greater amount of electromagnetic energy from space, thereby increasing the energy density of the emitted radiation required for propulsion. The rotation can also form part of the electrical generator for creating an electrical charge.
24	Witnesses have experienced symptoms strongly suggesting radiation poisoning while in the presence of UFOs.	If the frequency of the emission of radiation is high enough (e.g., light beams), it could induce effects of radiation poisoning (with symptoms appearing soon after the encounter of nausea, diarrhoea, headaches, internal bleeding, loss of hair, etc.). However, the most likely source of this ionising radiation are the electrons streaming from the surface of the charged object and hitting, at high speeds, steel metal bodies such as cars and road-signs, resulting in the emission of ionising X-rays.
25	UFOs demonstrate remarkable maneuverability in flight.	A charged Em spacecraft will be constructed of the lightest materials, thereby requiring minimal energy to move the object and change its direction. Furthermore, the ability to control inertial forces inside the Faraday cage of the object's central body means that its occupants can withstand far more dramatic changes in direction and acceleration than occupants of an ordinary man-made aircraft, manufactured without the benefits of electromagnetic and gravitational shielding.

	UFO Observation	Likely Scientific Explanation
26	Materials to construct UFOs are extremely lightweight and tough.	When an EM spacecraft approaches the speed of light, the object's mass increases. The higher the mass, the more energy must be generated to push it even slightly closer to the maximum speed. To minimize this energy requirement at relativistic speeds, all materials and occupants must be as light as possible. This is not a critical consideration if the objective is to reach, for example, the star closest to our Sun. the very nearest star after our Sun. But for longer journeys, the reduction in the object's mass is imperative, especially if biological entities are to participate in the flight. Thus, from an engineering perspective, this means the outer metal skin that carries the charge should be paper thin and constructed of lightweight metals. Because of how thin the metal skin is, it should also be very rigid. Therefore, the right choice of an alloy is important, as it must be very hard or able to return to its original shape. The spacecraft must also withstand extreme temperatures and be corrosion-resistant. In fact, the entire charged object must be constructed in the manner of an extremely lightweight weather balloon—but with the strongest and most resilient materials available.
27	UFO occupants are described as physically thin and usually small individuals.	Again, this is a consequence of reducing mass in order to more easily and quickly approach the speed of light in order to minimize flight duration for those who participate in the journey. For very long journeys, extreme methods of reducing occupants' mass may be employed, such as removing body hair, eating a diet that minimizes excessive growth and muscle mass (similar to the dietary practices of horse jockeys), or for the ultimate extreme, a shortening or even removing the digestive tract in favor of providing essential nutrients through patches directly applied on the skin.

	UFO Observation	Likely Scientific Explanation
28	UFO occupants wear skin-tight and exceptionally smooth metallic suits.	The principle explaining the smooth outer surface of the EM spacecraft applies here to the occupants' flight garments. In other words, the design of both the garments and the spacecraft itself must take into consideration the possibility of electrical sparks caused by sharp points and edges. Furthermore, any buildup of charges on the surface of the body can quickly dissipate to the ground. Finally, the flight garment must be form-fitting in order to assist in pushing the occupant's blood to the region of the head, similar to man-made "G-suits". In this way, the occupant can withstand higher accelerating forces and avoid falling unconscious due to high inertial forces.

This table is not an exhaustive list of explanations for all UFO observations. For example, we have not mentioned how it is possible for ionised air around the circumferential edge of a charged metal hull to create a whirlwind effect. For instance, if a magnetic field is generated internally by the object in a way that is similar to Earth's, the point at which the electric and magnetic fields are perpendicular can move ionised air in a circular motion (especially around the edge). This can result in the creation of an intense whirlwind that can whip up dust on the ground and further aid in concealing the object near ground level.[45]

And after all of that, it is amazing what one can learn from an electric light bulb (with a little insight from Einstein's Unified Field Theory).

---

45   This is how Saturn's rings are formed and maintained.

CHAPTER 9

Conclusion

I know that most men, including those at ease with problems of the greatest complexity, can seldom accept even the simplest and most obvious truth, if it be such as would oblige them to admit the falsity of conclusions which they have delighted in explaining to colleagues, which they have proudly taught to others, and which they have woven thread by thread into the fabric of their lives.[1]

—Leo Tolstoy

BEFORE WE explain the results of this research, let us reiterate from Chapter 2 the quote by NASA engineer John Billingham. As he said:

"To be quite truthful, we do not pay significant attention to the UFO issue at all. We feel that the whole area is so debatable, uncertain, and unscientific that it is probably not going to help anybody very much. We don't do anything on UFOs, and we do not recommend that anybody else do anything on them."

---

1   Eisen 1999, p.41.

Why such a pessimistic view? There can only be two reasons:

1. The statistics on UFOs drawn from numerous studies in the past are seen as sufficient evidence for no UFOs mainly because it is assumed the remaining reports must have the same man-made or natural explanation, especially if given enough time to study each one; or
2. Scientists expect everything to be observed with their eyes and/or instruments before they will consider studying a phenomenon.

However, as we have seen in this book, there are indeed UFOs for which no man-made or natural explanation exist. When we keep an open-mind and look for the unusual observations, we discover the presence of portholes, unusual-looking UFO occupants, light bending, unusual flight behaviours, astonishing acceleration and changes in direction and so on. If one is going to advance science from the study of UFOs, one has to be prepared for the unexpected. We need to be looking for things that are unfamiliar and could potentially advance scientific knowledge.

Likewise, why should scientists expect everything to be observed by their own eyes and/or instruments before they can study anything? Can we not, at the very least, use our imagination to see the alleged observations being made by UFO witnesses and create a rational pattern to explain them without being so cynical or negative about the subject?

Are there UFOs?

As shown in this book, there were things in the skies of Earth that we did not initially understand. Various people from all walks of life have wondered what these mysterious flying objects were. For most people, the questions have settled on two explanations:

1. Could UFOs represent alien visitations?
2. Are UFOs nothing more than familiar man-made and natural phenomena?

For some open-minded scientists who were not aware of the electromagnetic technology lying at the heart of UFO reports, the question has been more along the lines of, Can we advance science from the study of UFOs, whether or not there is something alien? However, as we have seen from one NASA engineer, the general scientific view has been that there is nothing from the study of UFOs to advance science, which is pretty much in keeping with Dr Edward Condon's original conclusion of 1969. So to put it succinctly, there are essentially no UFOs, at least not ones that are unusual if given enough information and time to investigate.

One thing is undeniable after carrying out this research. Our universe is truly a mysterious place. Sure, this book may have unravelled one of the great long-held secrets of this universe, especially the ones with electromagnetic side-effects emanating from symmetrical and artificial-looking flying objects. These previously categorised UFOs are, at last, showing themselves to be IFOs, but of a kind that is new to science and with the chance for humankind to create a new technology based on the Abraham-Lorentz formula. Yet, despite this important discovery, we should not think for one moment that there are no more UFOs to be studied. Far from it. As William Shakespeare wrote in *Hamlet*:

"There are more things in Heaven and Earth, Horatio, than are dreamt of in your philosophy."

The universe is way too big and too mysterious to allow for certainty in our explanations for all UFOs. There must always remain UFOs more baffling and exotic than we can dare imagine. Without such mysteries, there would be nothing to question, nothing to study, and nothing to be curious about—and that would mean the end of science as we know it. The truth is, the universe will always have many more secrets to unveil and keep us astonished and wondering what else is possible.

Philosopher and physicist Albert Einstein understood the scientific importance of the mysterious when he said:

"The most beautiful experience we can have is the mysterious. It is the fundamental emotion, which stands at the cradle of true art and true science. Whoever does not know it and can no longer wonder, no longer marvel, is as good as dead, and his eyes are dimmed."[2]

We have to remember that we are nothing more than students in the great classroom of the universe—and our experience of it is our great teacher. Never can we be the teacher as we are not God. We are not the perfect all-knowing creatures. Rather, we are here to become like God or approach this ideal position through the work we do and to learn from our experiences. We are here to ask questions and to search for answers. No matter if a scientist is an expert in a particular field, he pr she is nothing more than a child learning from the teacher in the great classroom of the universe.

So, when we turn our attention to the skies, we should expect to see UFOs. Not just today, but for all eternity, no matter how well we can explain the current batch of electromagnetic UFOs observed by witnesses. Indeed, it would be incredibly foolish to think UFOs do not exist or will not exist as our scientific knowledge expands to explain more UFOs.

The term "unidentified" in the acronym UFOs means just that; by their very nature, UFOs are a mystery to solve. They have to exist like shining beacons in the night sky (especially those that glow at night), beckoning for our attention and daring us to learn something new. Those who argue that there are no UFOs or assume UFOs have a familiar explanation are not true scientists. They will only say this because if they cannot see natural or man-made explanations for all UFOs, they naturally assume there is "insufficient information" or they must all be hoaxes. But this is not the scientific approach. The only true scientific approach is that we must remain curious children if we are to fully experience and understand everything about the universe. There is nothing more we can do. Until we have all done the work of understanding UFOs, we cannot afford to make assumptions, especially if genuinely curious scientists suddenly discover something no one else has seen before.

Still believe in the "insufficient information" view for those electromagnetic UFOs of a symmetrical design, more affectionately known as flying saucers? Think again. We have presented detailed examples in Chapter 4 of unknown flying objects. If, for any reason

---

2   Rassam 1993, p.33; originally in *The World as I see It — An Essay by Einstein*. This quote and the full essay can be read at http://www.aip.org/history/einstein/essay.htm as of February 2010.

this is not enough, Chapter 5 compiles the most common UFO observations, revealing far more detail. An open-minded and curious scientist will be able to see an interesting pattern emerge from these observations straightaway. We now know we are dealing with something artificial.

Chapters 6 and 7 reviewed ball lightning and secret military experiments as possible explanations. Careful analysis revealed serious shortfalls in relying entirely on these two possible explanations for all UFOs.

Then, in Chapter 8, we reveal an area of physics that can support and explain common UFO observations for electromagnetic flying objects in a way that has never been done before. We are now at the stage of confirming the concept by building a prototype to see the concept in action and soon we can build our own electromagnetic flying machine.

It really is that simple.

Indeed, we really can find something new to science from the study of UFOs, which is in direct contradiction to the 1969 conclusion reached by Dr. Edward Condon. We can finally observe a new electromagnetic technology emerging from the UFO reports. Amazing for something that was originally described a hoax, hallucination, and all the rest. How can that be? After all this time, it seems that someone must have figured it out and kept quiet, as the common observations slowly but surely revealed this technology before our very eyes. This is a truly astounding discovery, yet Dr Condon had no idea what was hidden in the UFO reports, especially those describing symmetrical flying objects with electromagnetic side effects. Or are we all unimaginative and lazy people who have no clue what we have hidden in those UFO reports?

And how long has it taken to discover this? If Dr Condon and others had not been so pessimistic about the prospects of finding something new in UFO reports, this mystery could have been solved in the 1950s, and by now we might all be flying around in electromagnetic vehicles.

Extraordinary.

Now, at last, after going through a number of UFO cases, highlighting the most common UFO observations, and carefully examining the laws of electromagnetism, we have identified an area of physics that would permit UFOs and their occupants to reach our planet. According to the laws of electromagnetism, there is a poorly understood concept of exponential acceleration by a charged object emitting radiation. Exponential means dramatic acceleration. Combined with the fact that there is a way to control inertial forces and recycle electromagnetic energy using Einstein's Unified Field Theory, it is amazing that scientists have overlooked this obvious area of science when explaining UFO observations. For a long time, it seems people have not realised interstellar travel is possible. Remember, exponential acceleration will provide the necessary speeds to cover the immense distances between the stars. It can actually be done. Yes, people can participate in the flight. There is nothing in the laws of electromagnetism to tell us it is impossible. In fact, it has been like this for millennia if reports of UFOs from ancient times are anything to go by.

Even more shocking is the revelation that this new technology is something that we can build today—not thousands of years from now. It is not so alien that our minds cannot comprehend how it works. Far from the nonsense we have been told from some USAF personnel, we can build a UFO right now. This probably explains why witnesses have observed certain kinds of UFO behaviour, suggesting that whoever was flying them was trying to teach humans a lesson in UFO technology. For example, we have seen the Russian UFO case where multiple witnesses observed a UFO demonstrating the relationship between the rate of flashing of the bright lights and its speed. Why do this if the technology is too complicated and beyond human comprehension? Why do UFOs go to the trouble of showing humans how they fly in the air when they perform spiralling motions, bouncing and drifting motions, or show human abductees (e.g., Antonio Villas Boas) an escorted look at the internal structure of a UFO? Are the UFO occupants thinking we can figure it out? If they think we can figure it out, surely the technology is not so alien that we cannot figure it out, too.

It is within our scientific and engineering capabilities to do the work and build one of these electromagnetic vehicles. This is in direct contravention to the views of some USAF personnel who have recently emerged to convince UFO researchers and investigators that the reason for the continuing UFO secrecy is because the USAF has not been able to reverse-engineer the crashed disc in its possession. What a load of "you know what!". Now we can see the claims for what they are: artful deception. In fact, the technology is so simple that it could have been built by a 14-year-old kid with some physics knowledge in the 1950s, or perhaps as early as the beginning of the 20th century when the electromagnetic concept of the radiation reaction force was first published.

Indeed, a young practical-minded scientist like Thomas Townsend Brown, if we had known about the scientific concept behind UFOs, could have built in his teenage years just one prototype, without all the fanfare of creating a multitude of different designs as he did, and convinced the world of what he had. At the same time, he could have even advanced scientific knowledge in areas such as gravity and universal gravitation, and how radiation moves solid matter. Then we wouldn't be where we are today.

This seems to be the only question that remains: Can we implement the idea in reality? Remarkably, while scientists are debating the reality of the mathematical concept, an American inventor has already found the crucial observation needed to prove the concept: a runaway acceleration effect. And no, initiating this effect does not require phenomenal acceleration of a mass, as one scientist claimed in 2011. We see no dramatic initial acceleration when witnesses report seeing certain UFOs render themselves invisible (which is necessary to recycle radiation in order to create the runaway acceleration effect) and then move off at phenomenal speeds from a rest position—nor do we see it in the invention. All the inventor needed was a sufficient charge and frequency to help jolt a stationary charged object into motion, and afterwards it undergoes the "runaway" exponential acceleration result. The acceleration part just merely enhances what is already there: energy density of the emitted radiation is controlled by an accelerating charge; therefore, it will naturally increase the density as the charge moves faster.

If another experiment determines this to be true (now that we know what the concept is—astonishingly, the inventor had not worked it out despite being on the right track to consider Einstein's Unified Field Theory), what would it mean regarding the alien nature of those electromagnetic UFOs with odd-looking humanoid occupants? Can we now say that Earth has been visited by aliens?

Leaving aside the possibility of an alien connection, this technology is not exactly a new development. It is almost embarrassing to hear that a patented device by an American physicist has been supporting the concept for so long—since the 1950s to be precise. Whatever we were doing back then not to notice, we now have no excuse. Clearly the inventor observed an interesting accelerating behaviour, not explained by him or any other scientist. The concept behind this important observation had eluded him, as he was more a practical physicist who relied on experiments to observe certain things. As for other scientists, they tend to be described as theoretical physicists who rely on mathematics for an answer. However, the problem with the latter is how skeptical they are of anything that looks too amazing to be true. Even if the mathematics tell them the idea of exponential acceleration could be possible, every effort is made to deny the mathematical solution. Yet, these theoretical physicists choose not to do the one thing that would bring a sense of closure to the matter and highlight the potential reality of the mathematics: an experiment.

Clearly the different points of view in the scientific world are stark and laid bare for all to see from this UFO study. No experimental testing from the theoretical physicists' perspective to confirm the concept. No going back to advanced theoretical concepts in electromagnetism by the inventor to explain the experimental results. Well, not anymore. The theory and experiments have been brought together in this book and we can now see how it works. Better still, we can explain the common UFO observations with such alarming clarity that it is almost scary given how realistic and easily achievable those observations are for science. In fact, we have to say that the solution to the UFO problem is finally at hand. All we need to do today is recreate the invention, subject it to the necessary high frequency and voltage needed to initiate movement, and watch how the device accelerates. Then, we can answer the all-important question: Does it accelerate exponentially? Apparently it should according to the solution of the Abraham-Lorentz formula. All the UFO evidence is pointing to the fact that it should. The observations support the concept. Well, how else can UFO occupants reach our planet in their ridiculously small flying objects? No massive spacecraft with cryogenic or hibernating pods has ever been reported by witnesses through any alien abduction case (see Appendix J for a detailed alien abduction case in which the witness was escorted around the inside of the craft and never reported anything to suggest the occupants are preserved in some technological manner to survive a long time in space). Somehow the occupants must have some way of reaching us, and staying alive through the entire journey. We can also certainly discount meditation as the secret to slowing down metabolism and breathing as the solution to interstellar travel. It is unlikely anyone, not even the aliens, can hold their breath long enough to reach the stars if the UFOs could not move at exponential speeds.

Now the final piece of the electromagnetic jigsaw puzzle has fallen into place. The concept that explains the "runaway" observation is known, and is available in any university textbook on advanced electromagnetism. No more scientific debates about whether the mathematical solution is really possible. There is enough evidence to tell us that it already exists—and the time has come to build a prototype and see the object in action.

Do We Have to Observe the Universe Directly All the Time?

Knowing what we do now, we have to wonder why the scientific community has not solved the UFO problem. It is not as if UFOs are a new phenomenon. They have been around for a long time. Even if the information was scant in Biblical and medieval times, after 1947, detailed UFO cases on record were ready to be analyzed by the right open-minded scientist. This naturally has us wondering what has stopped the scientists from conducting a proper investigation of UFOs? Why has it taken so long to identify the technology indicated by the reports?

Part of the problem is that, because some scientists are unable to "accept even the simplest and most obvious truth," as Leo Tolstoy put it, it seems that UFOs go against "reason" and logic. Observations, such as invisibility and light-bending, seem so incredible that scientists prefer to ignore the phenomena, no matter how many times they are reported.

Funny how scientists think this way. This is exactly how other scientists think about the mathematical solution of the Abraham-Lorentz formula. If something looks too incredible, they will ignore it, or find clever and sophisticated mathematical ways to "sweep it under the rug" so to speak.

However, the other problem for scientists is that many of them expect something to be directly observable (with their own eyes or instruments) before they can properly study it. As American physicist Dr John Archibald Wheeler said:

437

"No phenomenon is a phenomenon until it is an observed phenomenon."[3]

Even then, despite the derivation of the radiation reaction force that all scientists can "see" and acknowledge exists, there is a continued disbelief that the equation could occur in reality. Without performing an observable experiment, anything that seems too fantastic using the laws of electromagnetism and mathematics has to be ignored or interpreted as unlikely to occur in reality simply because scientists cannot understand why it should exist. So just imagine how scientists will view the UFO phenomenon.

But is this truly the right way to be doing science?

Or should we perhaps do the same as electromagnetic energy? You know that thing called radiation. While scientists can't directly see and haven't the foggiest idea precisely what this energy is in reality, perhaps we should follow the previous arguments and assume it does not exist. But, of course, as we all know, experiments have been done on this energy to show it exists and is ubiquitous throughout the universe. Should we not apply the same effort of doing an experiment to test this "runaway" mathematical solution, which is the proper scientific approach?

The same is true for UFOs. Why should everything be observed directly with our eyes and instruments? Just because UFOs evade our direct attention does not mean we should stop studying them. Nor should we ignore them because the observations look too incredible. What about the witnesses? Especially those who have seen and experienced electromagnetic side effects. We are not talking about ball lightning because too many witnesses reported seeing portholes, alien occupants, and other "artificial" features. As for those witnesses who have suffered radiation poisoning, are we to infer from their descriptions that they are all making it up? Surely not all of them are faking it, especially if some witnesses have died from radiation exposure. It does not make sense for some people to sacrifice their own health just to make a point to society that aliens might exist. In Chapter 4, we learned about one chap who died from a lethal dose of radiation from a UFO because of his unfortunate decision to shoot at one of the aliens. It seems rather pointless to try and create a UFO hoax by dying prematurely. Clearly to die so soon after the UFO encounter tells us

---

3   Wheeler 1980, p.25.

that the witness did indeed see something, and it was not a hallucination. As for the number of these sorts of cases, there are enough of them for scientists to give their due consideration to the available indirect evidence.

What this phenomenon should be teaching us is the value of using people as the instruments of observation, and for us to compile the common observations and to use our imaginations to identify new patterns from the observations. Science is not just about using our eyes, as doing so would lead us to end up like those scientists who ignore their own mathematics, regardless of whether the mathematics and the conservation of energy principle are applied correctly. If the results seem too amazing to be true, scientists who rely only on their eyes will choose to ignore the results or see them as a flaw in the theory of electromagnetism. As they say, "I don't see this amazing result occur in reality, so it must be untrue!". Then, they fail to do the one thing that could resolve the issue once and for all: perform an experiment.

Just find out. Surely that is the essence of a true science: to be astonished by the unexpected and to be curious enough to go out there and find out.

In the case of UFOs, it is notoriously difficult to see them through the eyes of a scientist. Thus, before you can perform an experiment, you have to use your imagination to see the common UFO observations and piece them together into a rational and coherent scientific explanation using physics.

A true science always requires us to open not just our eyes but also our imagination to explain the mysterious. No scientific inquiry into any mysterious phenomenon should begin without imagination. Indeed, we need imagination to begin developing theories of how something might work. Also, no scientific inquiry should end prematurely simply because something cannot be detected or measured directly by the eyes of a scientist or his instruments. We must allow others to do the observing and for us to compile the common observations and apply our imagination to see what we have and whether a consistent and rational pattern exists in the observations. By listening carefully to what other people have to say, by asking the appropriate questions, and by using the mind as the laboratory to imagine rather than observe nature at work, we may have the opportunity to uncover a hidden pattern in people's statements. And who knows where this might lead us? To the stars? Given what we know now in this book, it certainly is possible.

American physicist, mathematician and philosopher Charles Sanders Peirce (1839–1914) commented on the importance of imagination in science when he said:

> "When a man desires ardently to know the truth, his first effort will be to imagine what the truth can be…imagination unbridled is sure to carry him off the track. Yet nevertheless, it remains true that there is, after all, nothing but imagination that can ever supply him an inkling of the truth. He can stare stupidly at phenomena; but in the absence of any imagination they will not connect themselves together in any rational way."[4]

If anyone in the scientific community could have appreciated the importance of the imagination and open-mindedness to UFO reports, it should have been the SETI scientists. Unbelievably, these people have scoffed at the idea of finding anything unusual or alien in the UFO reports; quite remarkable, considering these are the people who are searching for alien radio signals on the slim chance that they exist. If this is not true, why haven't there been at least one SETI scientist looking at the UFO reports? Because if he or she did make an effort, we would know what is in the UFO reports by now. Clearly no one has done the work. We must again ask, how successful have the SETI scientists been in finding the evidence through their preferred and often heavily promoted traditional method of using radio telescopes? The evidence isn't overwhelming, to say the least. Or should we apply Dr Wheeler's view of alien radio signals in the sense that if we cannot see the signals arriving on Earth, then it isn't an observed phenomenon? In which case, why carry out scientific work on something that we cannot see or detect? But if not and if SETI is meant to be a legitimate area of scientific study, then why are these scientists not investigating UFO reports?

Why Has It Taken This Long to Solve the Mystery?

Sure, some SETI scientists will claim that they can't be expected to investigate and study every single reported UFO observation due to a lack of time, money, and other resources to do this work. Also, some scientists believe that because the American inventor did not succeed in

---

4   Hall 1988, p.153; Maney & Hall 1961, p.169.

selling his invention, it must have failed, so there is no reason to study it.

These arguments are merely excuses. For a start, the research for this book was done with virtually no money or support—just a curiosity to find out why people have been reporting UFOs and a desire to understand those cases with electromagnetic side effects. It does not mean no effort was made to seek support from universities to do the study. Unfortunately, the very nature of the subject is so controversial and amazing that no university has the imagination to see what is possible from the UFO reports. On top of that, too many people were not expecting UFO-related discoveries that could advance science. Preconceived ideas of hoaxes, hallucinations, and natural or man-made explanations have pretty much scuttled any chances of making a discovery in this field of study. Combine this with the government's aim to force universities to quickly find practical solutions in order to sell them in an economy, and it is no wonder that, if people cannot see the end game soon enough, no funding will be provided. So just imagine what we have been missing out on over the decades from not doing the work.

But does one need the funding to study UFOs? Some people will say that lots of money and other resources are needed to do a thorough scientific job on anything. But as we have seen, sometimes even the U.S. government paying a university to study UFOs, as had occurred in 1969 under the direction of Dr. Condon, does not necessarily produce a better result. Quite the contrary, depending on the personal views of the scientists involved, paying more may only reinforce what people want to believe and expect to see in the study, but not necessarily get to the truth of the matter. Instead, one needs to find someone who is more open-minded and curious about UFOs and will work with much fewer resources other than the time to look at the reports and a brain to imagine and apply rational thinking to the problem. The result? You can often achieve a lot more and see things that no other scientists have seen before, without all the personal bias getting in the way.

It is all a question of time. The resources are already available in terms of one's brain to imagine and apply rational thinking. The rest is a personal decision to spend a little of one's free time reading about the subject. The time we are talking about here is nothing more than reading a chapter in a respectable UFO book or reading an original UFO report, as opposed to reading another article in a scientific journal. Detailed information on hundreds of UFOs has already been gathered from witnesses and is available from government, military, and civilian archives. We call them UFO reports—which means there is no real need for scientists to travel to another country to talk to all the witnesses. And it doesn't have to be done during work hours; scientists can study UFOs in the privacy of their own homes, if they so choose. All that is needed is an open mind, a willingness to be astonished by the amazing and seemingly inexplicable common UFO observations, and the curiosity to go ahead and study these observations until a scientific explanation is found.

As for the idea that the American invention was unsuccessful, this is not true. In fact, the U.S. government was determined to prevent the inventor's work from being successful. It is not because the inventor did not see a runaway effect in his invention. He clearly did. The reason seems to be that—if people were to find out what this runaway effect is, could see it occur in reality, and could make the connection to the inventor's devices and the Abraham-Lorentz formula in electromagnetism—electromagnetic UFOs will have a solution. Then, society would have been very different from what it is today had this invention been built and put to use. For the USAF aimed at keeping the UFO truth under wraps, this is not the ideal outcome. It seems better for the USAF to do everything it can to prevent the inventor from succeeding in his work.

Now, with the work already complete, we can identify the electromagnetic idea and technology emerging from the UFO reports and can see the connection to the invention and the mathematical "runaway" solution. Everything makes sense. It is rational and simple. Now imagine what a little support could achieve this time around. Forget the time spent understanding whether UFOs could have a new technology and a means of advancing science. We are long past that stage. We are at the final stage of this research to build and test the invention.

But, if time and money truly are too limited to take this final step, then people must ask what it means to provide support to the creative and curious people in our community. Must this support always take the form of money? With the way society is structured today and tying people's survival to money, it seems we have no choice. People need money. Scientists are no different, for they too have to go where the money is. No doubt the USAF fully understands this, which is why scientists often end up in jobs paid for by the U.S. government in areas considered to be more "pragmatic" (such as in military programs designed to build nuclear weapons, nerve agents, and new contagions to annihilate other human beings). Otherwise, those SETI scientists who want to search for aliens will have to stick to radio searching or join the NASA team to find evidence of alien bacteria on one of the moons or planets of our solar system through the promise of continued funding —and if you don't like doing any of these things, well, no funding for you! You can suffer and fight for your survival or find another job paying a pittance (and still struggle to make ends meet). That's life. And all the while UFOs are made to look unscientific by the CIA and the USAF by releasing numerous fake alien photographs and films, and by the claims of one alleged Area 51 employee, as we have seen in Chapter 7. How convenient. Is this because the U.S. military and government do know the truth but are afraid to tell the world, lest it create a better future for everyone? If so, perhaps another quote, this time from Emile Zola, might be in order for those who believe they can keep the whole secret under wraps forever:

"If you shut up truth and bury it under the ground, it will but grow, and gather to itself such explosive power that the day it bursts through, it will blow up everything in its way."[5]

As for supporting curious people in their quest to uncover new insights about the universe without influence from others, maybe it would be better if there was a new world order with a different way of rewarding people for their contributions to society rather than through money. In fact, the real issue is why do we tie people's survival to a reward system? Do we not know that, when we properly support people with what they need to survive, curious people will naturally reach certain goals when they see what needs to be done for the benefit of everyone? Curious people only need access to the things they need, and if there are other resources available to borrow or use to achieve the goals faster, that is a bonus. Otherwise, you can be sure that curious people will always achieve their goals for the benefit of society. The only tricky bit about this approach is that it does require members of society to show faith in these people in achieving great things. In a world where 95 per cent of the population is religious, it is amazing

---

5   Le Poer Trench 1974, p.129.

how little faith is shown to those who are curious and believe they can achieve certain things.

As for those who are not religious, we should take heed of the wise words from the Beatles' songwriter and singer Paul McCartney:

Let it be, let it be, let it be, let it be.
There will be an answer, let it be.
Let it be, let it be, let it be, let it be.
Whisper words of wisdom, let it be.

We just have to remember, curious people will always get to the truth (and no they don't have to die to get there) and achieve their goals. That is how it works, and it is in their nature to do so. Once they see what needs to be done, they will make it happen for the benefit of everyone.

Still too expensive to support these people? Then do not use money. Find another way to ensure their basic needs are met and let their curiosity take them to wherever it may lead them in the end. They will find a solution. Just let it be.

Even if society still values money like it is the only thing that matters, we should still provide the funding needed to do the work. We should not let our prejudices about a controversial subject decide the fate of an individual's or group's work when they can see a solution at the end of the tunnel. Likewise, we can never quash curious spirits. Even without the support, a handful of open-minded and curious scientists will still surely study privately and find out the truth. That is something authorities can never stop. Yet, at the same time, if we do not encourage others to be curious and appropriately reward them, imagine the loss in benefits for others from the work when we do not know what is possible. So much is lost when people lack the confidence to try something different and others are just simply locked into a certain way of thinking that can never be broad enough to see what else is possible in reality.

If you ever need to see evidence of what is possible by curious people free to pursue their goals, then take a look at this book. This is one example of what can be achieved without support. Imagine what could have been done, and how much sooner it would have all happened, and what more could have been achieved, if the support was there in the first place.

No wonder it has taken this long to solve the UFO mystery.

All this should tell us something. This isn't a question of not having the time or resources to independently study UFOs. Scientists can choose to study them if they so wish. Good science should not be dictated by money simply because society has created a system that links money to our survival, and certain people think another area is more sensible, makes more money, or helps to hide the truth. How do you know what is sensible? How do they know that where the money is going will help people and the planet in the long-term? How do they know they are not inadvertently helping others to prevent a big secret from coming out so that they can benefit from the current economic system in their own selfish way while selling aspects of the secret through new supposedly man-made technologies? For all we know, in supporting those who seek to uphold the current system, we could be creating more world problems rather than solving existing ones. To avoid this and to instead balance the situation, we must give more freedom to the curious people to find solutions without anyone else's personal bias, secret agendas, or monetary influences to dictate where the work should go. Because when we do, we will quickly realise things like how UFOs are looking perfectly sensible and realistic thanks to the radiation reaction force and the availability of the one important invention.

Good science should not be controlled by money and those who have it. It never has been, and never will be. Money is only useful in speeding things up when doing the work and getting to a solution in terms of the equipment and other tools that can come in handy for curious scientists. But if all you need is imagination and the information is already at hand ready to be analysed and understood, there is absolutely no need for money. It is a personal choice to study something and find out the truth. If we are truly curious and astonished by the mysterious, we should not need money to ask questions and find the answers. We simply need to be open-minded and use more of our imaginations and other tools at our disposal to see where UFO observations can take us, and do the work ourselves. And for that, we must be curious.

That is the essence of a true scientist: curiosity.

Fortunately, this book has well and truly started the process and should give other people the impetus to find out whether what has been discovered in this book is true.

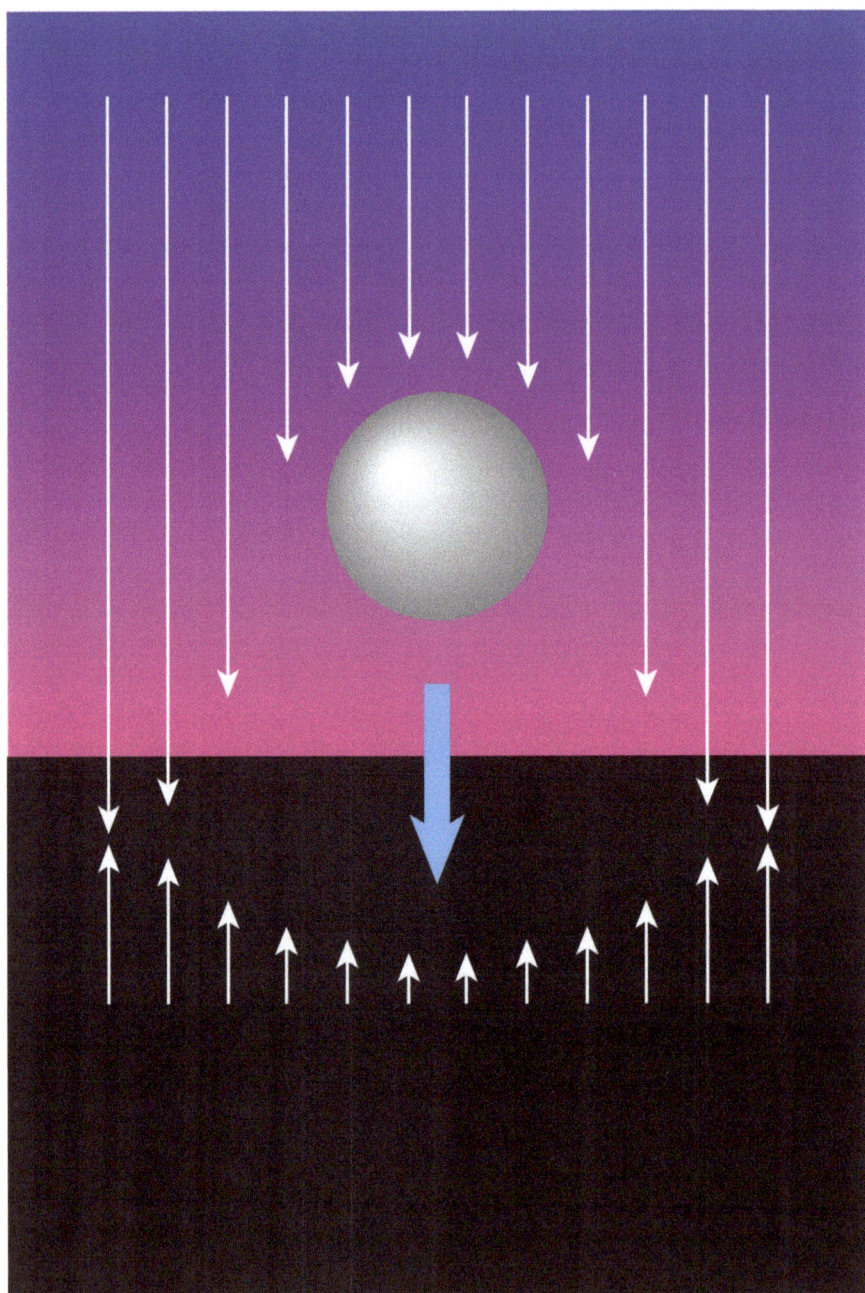

Gravity is explained by the Unified Field Theory as an imbalance in the pushing force of radiation due to radiation shielding by both objects on each other.

Can UFOs Advance Science?

As for Dr Condon's remark that nothing in science can be advanced from the study of UFOs, think again. There are areas of electromagnetism and how radiation interacts with solid matter that require closer attention. Furthermore, Einstein's Unified Field Theory tells us of a new electromagnetic universe, ready and waiting to change how we view everything, as well as a way to recycle radiation, which is useful for radiation propulsion according to the Abraham-Lorentz formula. Just look at electromagnetic UFOs for a clue as to how we can amplify the radiation force through electric charge and to use charge and frequency to raise the energy density of radiation to the point of being able to recycle radiation. Then we can appreciate what will be expected to be a phenomenal acceleration that we have not seen before, as UFOs have been displaying to witnesses for a long time. And now, at last, we have an invention that can not only solve the UFO mystery and support the radiation reaction force, but also has the potential to take us to the stars and, at the same time, advance science in greater ways we can only dream of.

Need some inspiration on how UFOs can advance science? Consider this: We know that Brown's invention consists of two asymmetrical electrodes separated by an insulator. Now imagine these electrodes are two planetary bodies and there is nothing between them. What happens next? According to Sir Isaac Newton, the two bodies have mass, so technically speaking, they should accelerate toward each other. Why? Presumably because there is a mysterious gravitational field generated by each body that influences the other and makes the bodies move toward each other as if they are attracted. Now, let's go back to the electrodes once again. Remove the insulator that separates them. What happens now? The plates will still move toward each other. Why? Because the plates have created a radiation-shielding effect between themselves. In other words, the density of the radiation energy is reduced in the region between the plates due to destructive interference of the reflected out-of-phase radiation. Then, the radiation from the environment (which is at a higher energy density) outside the plates pushes the two plates together. This is the intriguing but well-known *Casimir effect*.

Okay. So imagine if we turned the plates into two planets. We still have the same radiation-shielding effect taking place to lower the energy density between the two bodies. Surely, the same attraction-like effect will occur as the two bodies naturally come closer together to create the so-called gravitational force of attraction. However, it isn't an attraction force. Rather, we have a pushing force by radiation in bringing matter together.

How interesting? Does this mean gravity is a purely electromagnetic phenomenon? To put it another way: Is radiation doing all the work of the gravitational field by pushing matter together rather than pulling matter together?

Speaking of gravity and electromagnetism, Albert Einstein, the great theoretical physicist of the 20th century, had no trouble seeing this link. He was quite happy to show the connection based on his imagination and observations of the behaviour of radiation on so-called uncharged matter. What he saw was enough to convince him that he had to develop his final scientific masterpiece: the Unified Field Theory, which was completed in 1929. So, all that is required now is for one brave soul to put forward a reasonable interpretation of the theory in order to test the validity of Einstein's most ambitious work. Clearly it has to be the idea that an oscillating electromagnetic field, or radiation, generates a gravitational field. In which case, the question scientists should be asking next in order to advance science is whether the electromagnetic field is the gravitational field.

Not too sure about this? Well, let us try to reach a simpler goal. Is there any link between the gravitational field and the electromagnetic field.

Need a clue?

According to Dr. Jessup and a careful examination of UFO reports and Einstein's Unified Field Theory, the link appears to be based on the idea that an oscillating electromagnetic field (or radiation) generates a gravitational field. In other words, when we see UFOs become invisible to the naked eye, this observation suggests that the energy density of the (oscillating) electromagnetic field may be magnified to such an extent that the gravitational field is strong enough to bend the electromagnetic field back onto itself, and light from behind UFOs is bent around the objects. What we see is not the UFOs, but rather what is behind the objects. To put it more simply, UFOs can become invisible, and this can be achieved in a periodic manner depending on the frequency of the radiation that these objects seem to emit when in operation, exactly as some witnesses have seen of UFOs.

So instead of using high amounts of mass and/or to accelerate mass to relativistic speeds, how about accelerating the charges through an application of frequency on the oscillating charge, and using enough charges to see what the effect might be on the gravitational field allegedly generated by the electromagnetic field.

If scientists are looking for an area in which to advance science, why not amplify the oscillating electromagnetic field and see what happens to other radiation in this region of high-energy density? If an object placed in this region becomes invisible, scientists will prove there is a connection between the two fields. But if scientists need definitive proof that the gravitational field is the electromagnetic field, consider going the opposite direction. Instead of amplifying the oscillating electromagnetic field, use a Faraday cage to eliminate the electromagnetic field inside. Now ask yourself, what happens to the gravitational field inside the cage? Would it not get eliminated as well? Because, if this is possible, imagine the applications of this concept? The chance to control inertial forces on the human body accelerating inside a Faraday cage would open up the opportunity for more dramatic acceleration to be applied using the radiation reaction force. In terms of the implications to UFOs, we would have the scientific explanation not only for how UFO occupants can reach our planet, but also for why UFOs are designed as symmetrical metallic boxes and can change directions and accelerate so effortlessly.

Now can we say that UFOs can advance science? Of course we can.

And in terms of the new electromagnetic technology, the recycling of radiation at a high enough energy density is likely to provide insight into how a charged object emitting radiation accelerates exponentially. It could be all that is needed to finally prove the reality of the runaway solution.

Why stop there? What about the question of whether matter is truly uncharged? The UFOs are telling us a high charge is needed on the surface to amplify the recoiling radiation force. Even without the extra charge, scientists now have an opportunity to test whether radiation moves the existing charges in a so-called uncharged matter. Previously, it was thought that radiation only moves the mass of uncharged matter. Under Einstein's Unified Field Theory, however, we must change this picture. There is no such thing as a truly uncharged object at all times. Everything is charged. The things we see with our eyes are the result of radiation spuriously being emitted from the charges (namely, the electrons) constituting so-called uncharged matter. Surely we cannot have uncharged matter if this radiation is entering our eyes. Everything we see is charged. So, when radiation collides with this matter, or when this matter emits radiation, it is probably closer to the truth to say that radiation is moving the charge making up the matter—not the mass. If we want to amplify the radiation force on matter, we must increase the charge applied to the matter.

That is half the secret to UFO technology. The other half is what happens when the energy density of the radiation is high enough to cause it to recycle itself. Combine that knowledge with a method of distributing the charge of a higher density to one end of the object, and you will have the complete and unrestricted knowledge behind UFO technology.

Not even the USAF can stop this knowledge from reaching your eyes through this book.

But why stop there? If the universe is indicating it might be electromagnetic in nature, can we not find a way to unify physics under the umbrella of electromagnetism? Use radiation as the prime mover for all things and all the mysterious forces of nature—gravitational, strong nuclear and weak nuclear—should have an electromagnetic explanation.

As you can see, we have the potential to not only introduce a new technology and thus explain UFO observations, but we can also advance science in many areas of physics.

Amazing.

UFOs *can* advance science.

Even if scientists are too flabbergasted to be sure about these scientific ideas, don't worry about them. At the very least, they should enjoy the technology for what it is. Just build the invention, and see what happens.

Are we ready to take the leap and see where this new and extraordinary technology will take us?

Appendices

APPENDIX A

Common IFOs

Aurora Borealis

456

Sun dog

457

Comet Lulin

Lenticular clouds

459

Birds

Light flare

461

Fireworks

462

Weather balloon

Kite

464

Drone

465

Parachutes

466

APPENDIX B

Breaking the Time Barrier

Albert Einstein's Special Theory of Relativity

DO YOU want to know how realistic it is to break the time barrier for interstellar travel without all the impossible-to-achieve, science-fiction stuff of creating perfect vacuum tunnels in space called *wormholes* to allow for infinite speeds and near instantaneous journey times to be attained? World-renowned German rocket engineer Dr. Wernher von Braun (1912–1977) knows the answer. In an interview on BBC radio in the late 1970s, he said:

"Will men ever be able to go on interstellar travel? Well, I have learned to use the word 'impossible' with utmost caution.

Theoretically, if you travel at the speed of light—which of course you can never reach—time would stand still completely. If you could travel at say 99 per cent of the speed of light, then only a week or so elapses in the spaceship while a year passes for the outside world...so you could travel to a star 1000 light years away, and return to the Earth only a few years

Albert Einstein (1879–1955), Princeton, New Jersey, in 1947.

older. The trouble is that when you return, several thousand years have elapsed on Earth. So you are still a young person, but you meet your own great, great, great…grandchildren—and for them you are almost a prehistoric animal."[1]

For a clue of how this is achieved, it is well worth your time to check out the rather interesting work of famous theoretical physicist Dr Albert Einstein.

As Einstein discovered at the turn of the 20th century, if a spacecraft reaches a constant speed of v metres per second, each passing second (t_o) onboard the spacecraft would appear to an outside observer to dilate to an amount t_1 given by the following equation:

$$t_1 = \frac{t_o}{\sqrt{1 - \dfrac{v^2}{c^2}}}$$

To give an example, if an observer on Earth could look inside a spacecraft travelling at 90 per cent of the speed of light, 60 minutes on Earth would be dilated to 137.65 minutes inside the spacecraft. For a pilot or passenger onboard the spacecraft, time goes by at the usual rate. However, the occupant will notice how time has miraculously sped up for everyone else sitting at home or at the destination as if everything along the direction of motion is moving into the future at a rapid pace.

In terms of distances, both the occupant onboard the spacecraft looking at his destination (or where he came from) and the observer on Earth measuring the length of the spacecraft and everything else attached to it along the direction of motion will notice how the distance will be reduced or contracted by an amount given by the equation:

$$L_1 = L_o \sqrt{1 - \frac{v^2}{c^2}}$$

where L_1 is the distance travelled by the occupant in the spacecraft, and L_o is the distance measured by a person sitting at home or the destination.

---

1   George 1978, p.277.

For example, suppose a spacecraft travelled to Proxima Centauri—a star located about 4.25 light-years from Earth—at 99 per cent the speed of light. The new distance measured in this moving reference frame is 0.60 light-years. Travelling at this speed, it would take a few months to reach the destination. Forget the 4.25 years from the standpoint of an observer sitting on Earth if the spacecraft could reach the speed of light. Only the occupant of the spacecraft would benefit from short journey times.

Thus, the following statement reveals just how perfectly feasible and within the laws of physics it can be when you travel close to the speed of light:

"If a spacecraft approached Earth at [very close to] the speed of light from a distant galaxy, the inhabitants aboard that ship could see every event that occurred on this planet from the moment it began to form from cosmic dust to the present. They would have an instant overview, in a matter of hours, minutes or seconds, like a motion picture seen in fast forward until they slowed to 'Earth Time'."[2]

But why is it not possible to travel at the speed of light? The answer lies with the mass of the spacecraft and everything else inside of it. If the mass at rest is m_o, Einstein realised the mass increases with speed to m_1 as measured by the outside observer by the formula:

$$m_1 = \frac{m_o}{\sqrt{1 - \frac{v^2}{c^2}}}$$

From the perspective of the occupant, any extra mass built up is not felt so long as the spacecraft is designed as a perfectly symmetrical metal box to prevent whatever extra mass by way of electromagnetic energy formed outside from penetrating the box and flowing through the body. Otherwise every time the spacecraft accelerated, the occupant would feel the extra mass outside as a form of extra inertial forces applied on his body.

All these equations have been experimentally verified. Numerous experiments conducted with atomic clocks flown in high-speed aircraft

2   http://www.nancyredstar.com/newslinks/quantumsight.htm.

as well as measuring the mass of atomic particles such as electrons inside a particle accelerator, have proven time and time again the impeccable accuracy of the equations in the real world. What this means is that any observer, no matter who they are or where they are located, looking inside, say, a spacecraft travelling at nearly the speed of light will observe all the above relativistic effects, as scientists call them.

As intriguing as all this may seem, some readers might wonder why the outside observer sees these things occur. Indeed, what does this mean for the occupant sitting in the moving spacecraft when we watch him? Will he be crushed by the extra weight, squeezed to death by the length contraction, and appear to suffer a slow speech impediment when he speaks? Don't worry. Everything is fine for the occupant. If you could go there to sit right next to him, you would see that. When we talk about time dilation, length contraction, and mass increase, we are referring to how the mass-energy of space surrounding the spacecraft is compressed and the way this affects the light (or radiation) emitted by the spacecraft to tell an outside observer what appears to be going on, and likewise when light is received by the spacecraft from the outside. We need light to observe everything, and from this energy make measurements. The movement of the spacecraft at its dramatic speed simply distorts the light to such an extent that it creates these strange effects.

You see, the mass-energy of space (made mostly of electromagnetic radiation, or light in its most general sense) is compressed at the front of the moving spaceship, not unlike the way in which a plough moves dirt or snow in front of it. At the same time, this extra mass-energy collides to a higher degree with the atoms that comprise the moving object. This results in the atoms vibrating more vigorously and emitting more mass-energy to the surroundings by way of radiation. This in turn helps to gravitationally attract more radiation. According to Einstein's Unified Field Theory, the electromagnetic field generates a gravitational field. This means that any increase in the electromagnetic energy in a certain portion of space will amplify (or increase) the gravitational field of that portion. It is this increase in the strength of the gravitational field that gives the impression to the outside observer of a more massive object.

For the occupant, the increase in the gravitational field of the spacecraft will draw in more of the invisible mass-energy of space to congregate just in front and eventually flow to the back of the spacecraft. If the spacecraft is not designed as a near-perfect metal cage to protect the occupant, he will experience greater inertial forces during acceleration.

As a result of the displacement of this energy from the rest of space, there is another region of space further ahead where the energy has been displaced to congregate on the spacecraft. The energy density in this region is lower than the rest of space. However, the Universe does not allow this lower energy density to exist. This means the Universe will balance the situation as best it can by pulling (gravitationally) on the compressed mass-energy and spacecraft in front[3] or by pushing (electromagnetically) on the compressed mass-energy and spacecraft from behind, forcing all this mass-energy to quickly come in and fill this low energy density region. From a gravitational perspective, the result is comparable to a stretched rubber band, where the spacecraft is pulled by the mass-energy of the Universe to fill the low-density region. But because the spacecraft is still moving at high speed and constantly attracting mass-energy, this effect continues.

This means that the spacecraft will travel faster than the speed of light as the Universe pulls on it. Of course, the occupant of the spacecraft will not be aware of this, because the view in front would suggest that there has been some kind of length contraction along the direction of motion. In other words, everything in front will appear closer than if the spacecraft was not moving at all. However, this is actually because the compressed mass-energy of space directly in front of the spacecraft acts like a powerful magnifying lens, bringing things closer to the spacecraft. If you, as the occupant, did not realise this situation and accepted whatever an outside observer claimed was the speed of your spacecraft, you would find you reach your destination much quicker than the outside observer's calculations suggest. For you, this is presumably because of length contraction along the direction of travel. In reality, the distance to the destination has not changed. Instead, light in front of you has been magnified while your spacecraft has been made to travel faster than the speed of light, thereby covering the distance measured by the outside observer much quicker than you expected.

---

3   According to the Unified Field Theory, this pulling effect described as the gravitational force of the mass-energy of space is really an electromagnetic pushing force from the radiation coming from behind the spacecraft.

Remember, the true distance to any destination in space is always the same whether you are at rest or moving at high speeds. Never does the spacecraft and everything inside it get squashed along the direction of motion[4], nor do the stars and planets along the direction of motion become magically jolted from their position and brought closer together by the moving spaceship. Everything in the Universe and inside the spacecraft is located in the same place. It is just that the displacement and compression of the mass-energy towards the front end of the spacecraft causes the rest of the Universe to gravitationally pull on the spacecraft from in front (or electromagnetically pushed from behind) to a much higher speed to balance the lower energy density of space created just in front of the compressed mass-energy. The speed we are talking about here is well beyond the speed of light. The person inside the spacecraft and the outside observer will not know this. From the frame of reference perspective of each observer, either the spacecraft is travelling near the speed of light (for the outside observer) or the distance to the destination and where the spacecraft came from will appear shorter (for the occupant sitting inside the spacecraft). Furthermore, the wavelength of the light emerging from the spacecraft will be shortened by this compressed energy along the direction of motion to give the impression that the length of the spacecraft (as seen side-on) as well as the magnifying effect of the mass-energy that the occupant has to use to measure distances along the direction of motion is compressed. This gives the effect that things are closer to the spacecraft. Not true. What the occupant sees is an illusion caused by the compressed mass-energy along the direction of motion to distort space-time, and so causing the light to bend and get compressed.

As for time, the dilation effect seen by an outside observer is due to the radiation emerging from the compressed mass-energy in front or behind the spaceship. As it emerges, the wavelength of the radiation is suddenly much longer (or redshifted) than when it was initially emitted from the antenna of the spaceship. Part of this is due to energy loss in the light as it moves through the compressed mass-energy, and this is

---

4   Although you will feel greater inertial forces on your body and spaceship during acceleration, there is a way to reduce the inertial forces of anything sitting inside the spaceship when accelerating, as well as gravity itself. All you need is a symmetrical metallic box to lower the electromagnetic fields penetrating the box. And as Einstein's Unified Field Theory tells us, the gravitational field generated by the electromagnetic field, and the thing that causes us to feel inertial forces, is also reduced. It is interesting that such a concept is probably being employed in UFOs with their unmistakable symmetrical metal designs for the central body

translated as a stretching of the light (or a lowering of the frequency). The other part is simply because as the radiation moves away from the spacecraft, the density of the compressed mass-energy decreases, causing the wavelength to naturally increase. By the time the radiation emerges from this compressed mass-energy and moves through the normal density of mass-energy for the rest of space, the radiation will have a much longer wavelength than it did when it was emitted by the spacecraft. Consequently, the outside observer, on receiving this radiation, must wait longer than usual for all the information in the light to arrive to see what is happening inside the spaceship. If he does not, he will be watching you inside the spaceship moving as if you were in slow motion.

From the standpoint of the occupant, the reverse is true. If you were sitting inside the spaceship looking back at home, or even at the destination planet for that matter, the higher mass-energy density of space created by the spaceship just in front of and behind it (that is, along the direction of motion) will force the wavelength of light arriving to the spacecraft to shorten. There is some energy loss in the radiation travelling through this compressed mass-energy, but not sufficient to bring it back to the normal frequency sent by the outside observer. Therefore, if you, while sitting inside the spaceship, were to pick this up with a receiver and translate it into a moving image on a screen to see what was happening, it would be like watching a movie being played at high speed.

As you can see, the key factor in breaking the time barrier is in finding a technology to help you accelerate to a speed approaching the speed of light. To make it easier on the technology to achieve this, it is important to remove every ounce of unnecessary mass from the spacecraft and everything inside (including the occupant)[5], and use a

---

5   This mass minimisation technique can be relaxed for short journeys between neighbouring stars and/or if the propulsion technology is sufficiently refined to achieve the required high speed for the occupants. For example, the UFO observed by Antonio Villas Boas (which later disappeared to the south) showed an unusually complex design with six fixed protrusions attached to the central body and several different compartments or rooms for housing the occupants inside, all of which adds quite a bit of mass to the object. Together with numerous alien occupants and the unusual decision to transport a female alien to be inseminated by the male abductee during the sexual encounter (why not extract the sperm by other means and transport them back to where the occupants have come from?), this would have had a much more significant impact on the overall mass when accelerating to nearly the speed of light. Unless the UFO design is highly advanced (why so many protrusions?), it is most likely the UFO travelled a short distance (perhaps a neighbouring star to our sun). However, given that the UFO observed by Angus Brooks utilized fewer protrusions that moved around the central body when hovering and accelerating and was simpler in design than the Antonio Villas Boas case, one could argue that the Angus Brook's UFO was probably a more refined and possibly more advanced design. But

method to prevent the mass-energy of space from penetrating the spacecraft, where it can create extra inertial forces on the occupant's body during acceleration.

For example, to reduce the inertial forces, Einstein's Unified Field Theory tells us we can use a Faraday cage. As the theory states, the gravitational field is generated by the electromagnetic field, specifically an oscillating electromagnetic field (or radiation). Now, if the spacecraft is a perfectly symmetrical metal box, it reduces the electromagnetic field inside to zero and, with it, the gravitational field as well. And since inertial forces are created by the gravitational field of the radiation acting like extra mass to jolt the individual atoms in your body to create the sensation of inertia, all this supposed extra mass of the radiation miraculously disappears. There are no more inertial forces, and suddenly you can withstand a much greater acceleration. It would all seem like you had pressed a switch and inertial forces had suddenly been nullified.

Considering UFOs seen at close range invariably have a central symmetrical metal body, one cannot help but wonder whether this is an attempt by the occupants or whoever manufactured the objects to reduce the inertial forces inside.

In conclusion, and assuming the mass is minimised and there is enough energy available to move the mass, any UFO occupant can make the journey to Earth from a distant world beyond our solar system if it so wishes. There is nothing in the laws of physics that would forbid anyone from making the journey. As the formulas show, the natural laws of time dilation, increasing mass with speed, and length contraction along the direction of motion are all part of the Special Theory of Relativity as devised and published in a scientific journal[6] by Einstein in 1905.

The late Professor Stephen Hawking (1942–2018) has thrown his weight behind the theory (in fact, scientists now describe it as a law,

---

then again the length of the protrusions seems to negate some of the mass-saving benefit of this design. However, once we get to those almost ridiculously small UFOs with no protrusions (or appearing as nothing more than metallic domes on the underside or bright lights at specific positions), the description of single cabins, bucket seats, control panels and nothing else, and that the occupants driving these objects are described as extremely thin and short, we could be dealing with extreme mass-minimisation techniques being employed for the simple aim of travelling much longer distances and needing every refinement to the UFO technology and the occupants to ensure that minimal journey times are achieved.

6   The scientific article published by Einstein on the Special Theory of Relativity is titled "Zur Elektrodynamik bewegter Korper" (translated as "On the Electrodynamics of Moving Bodies"). The article may be found in *Annalen der Physik*. 4th Series, 1905, Volume 17, pp.891-921.

showing how confident they are). He said to British media on 2 May 2010 that it is possible for humans to travel into the future using any spaceship capable of travelling near the speed of light, allowing the crew to reach distant stars or galaxies within a human lifetime. As Hawking said:

> "[At] 98 per cent of the speed of light…each day on the ship would be a year on Earth. At such speeds a trip to the edge of the galaxy would take just 80 years for those onboard.
>
> Time travel was once considered scientific heresy. I used to avoid talking about it for fear of being labelled a crank."[7]

After making this statement, Hawking remained comfortable with the idea.

---

7   "Time travel a possibility: Hawking": *7 News/Yahoo!* 3 May 2010.
http://au.news.yahoo.com/queensland/a/-/technology/7153027/time-travel-possible-hawking.

APPENDIX C

Breaking the Energy Barrier

Choosing the Right Kind of Energy

AT PRESENT, most, if not all, scientists dispute the notion that another civilization in the Milky Way can visit Earth and be sighted as a random, mysterious UFO. The argument is that a spacecraft would require unbelievable amounts of energy to move it to sufficiently high speeds in order to achieve short journey times. This is especially true for a number of UFO reports allegedly revealing alien occupants.

As Dr. Ronald Drayton Brown (1927–2008), former professor of chemistry at Monash University, Melbourne, Australia, commented:

> "All my training as a scientist tells me the spacecraft theory is extremely unlikely...It is possible that life forms exist elsewhere in the universe, but I do not believe other creatures

would be able to shift a solid object such as a spacecraft at such enormous speed. An incredible amount of energy would be required to propel such a craft, and science already knows that the universe contains only a limited amount."[1]

To better appreciate the considerable difficulties scientists are faced with on this issue, below is a table showing the estimated maximum speed of several flying machines from our civilisation. The column marked "Time (Years)" indicates how long it would take for humans to travel to the nearest star outside our solar system—Proxima Centauri, lying 4.25 light-years or approximately 41,000,000,000,000 kilometres away—using the machine.

Vehicle	Speed (km/h)	Time (Years)	% Speed of Light
Jumbo	930	5 million	8.61×10^{-5}
Concorde	2,250	2 million	2.08×10^{-4}
Space Shuttle	28,000	170,000	0.0026
Apollo 11	40,250	119,000	0.0037
Voyager II	100,000	10,000	0.0093
Cosmos 1	160000	200	0.02
Daedalus	108 million	45	10

As the table reveals, the fastest spacecraft to date would take over 10,000 years to reach the nearest star after our sun. Even if we rely on nuclear energy to propel a hypothetical yet technologically feasible machine proposed by scientists in their study of Project Daedalus, it would take at least 45 years to cover the distance. Clearly it is not the kind of journey you would want to spend your whole life waiting for, or even to participate in if you wanted to remain sane and not look like skin and bones at the end of it, unless we could somehow find a way to expend more energy to move faster and make the journey times shorter.

---

1   Blundell & Boar 1984, p.43.

Dr. Enrico Fermi, taken in the late 1940s.

Yet there is no doubt from the UFO reports that witnesses claim to see small symmetrical flying objects with occupants inside. How is this possible?

Maybe it does not matter. There is the possibility that aliens could be already close to Earth and living in space or on some unexplored planet or moon in our solar system. All that is needed is a spacecraft of modest speeds and adequate time to make the immense journeys between the stars possible.

To explain what we mean by this, scientists involved in the theoretical design of a nuclear-powered spaceship known as Project Daedalus in the 1970s claimed that such an interstellar machine could be built by anyone with at least the same technological capabilities as our own. Perhaps we won't build it, but if just one civilisation in our galaxy was to make the decision to build such a machine and let it travel at 10 percent of the speed of light, a trip to Alpha Centauri would take less than 45 years. Even if we allowed the travellers centuries to get to the destination and centuries more to establish themselves on a lifeless new world, build another Daedalus spaceship and repeat the journey, there would be more than enough time for the travellers to cover every corner of the Milky Way galaxy and colonise all the solid rocky planets, given the age of the universe. This is precisely what the mathematical analysis of Sir Frederick Hoyle, a former professor of astronomy at Cambridge University, and Professor Chandra Wickramasinghe, of University College, Cardiff, Wales, has revealed. Therefore, the problem scientists have at the moment is why we haven't seen these aliens by now. Or, as

Italian-American physicist Dr. Enrico Fermi (1901-1954) put it: "If they are out there, where are they?"

Perhaps the real reason is that alien civilisations are visiting Earth but are playing the hide-and-seek game with us as part of some sociological principle of non-interference in the hope of preserving our culture while watching us as part of their own scientific studies? Sure, all this might be pure conjecture as there is no direct proof of aliens visiting our planet, unless we want to look at UFO reports. And from the reports we do see an active effort by the UFO occupants to keep well away from us unless they need to examine us and other living things. But just because UFO occupants do not want to be contacted does not mean they do not exist, or the technology to reach us is not possible. In all likelihood, we just haven't figured out a way to physically travel these distances quickly enough. Well, the UFOs have to travel very fast as they are described as small objects containing what we presume to be reasonably healthy-looking occupants carrying very little by way of instruments, food and other materials. Not even cutting back on a little bit of food to look as thin as these aliens do would be enough to survive the journey, unless they could somehow reach much higher speeds approaching the speed of light. So the problem must be finding the right technology to achieve these speeds, and the right type of energy to propel it. Until we can figure this one out, we cannot assume interstellar travel is an impossibility. How would we know? Already we have seen in Appendix B how it is feasible to break the time barrier of interstellar travel. All we need to do now is find a way to break the energy barrier of interstellar travel.

Can we achieve this?

Breaking the Energy Barrier

Ever since the first aircraft was developed at the beginning of the 20th century, we all have been indoctrinated with the idea that solid fuel, by way of chemical or nuclear energy, is the only means by which we can propel objects to high speeds, according to scientists.

In the chemical propulsion technology still used by the spacecraft of the 20th and early 21st century, as shown in this diagram, (liquefied) hydrogen (A) and oxygen (B) are carefully combined in limited amounts per unit of time and ignited safely inside an engine chamber (C). The explosion generated by the spontaneous chemical reaction between hydrogen and oxygen forces a rapid expansion of energy and hot gases by way of the wasted by-products (e.g., water) of the reaction, which are then allowed to escape in a controlled fashion through a small opening (D), causing the object to recoil in the opposite direction. As multiple explosions are performed inside the engine, the object is able to accelerate.

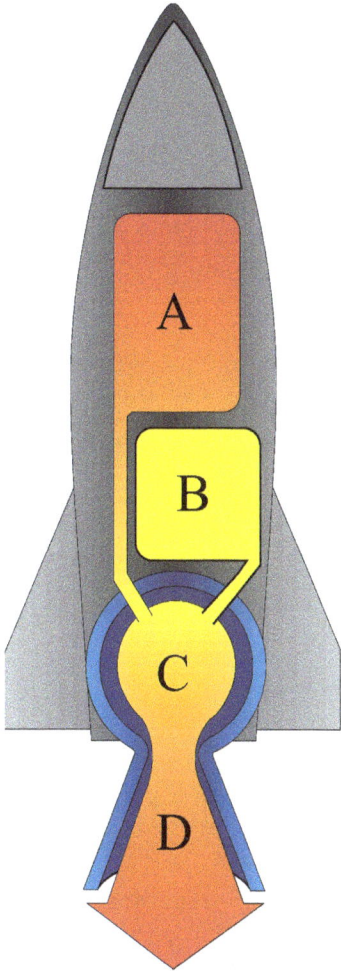

We see the same principle applied in the nuclear-powered spacecraft designed by a team of scientists for the British Interplanetary Society, known as Project Daedalus. The only difference is that a lot more energy is released at the back end of a spacecraft during the explosive fusion or fission reaction that takes place inside the engine because the nuclei of certain atoms in the nuclear fuel are being reorganised, and in the process emit a lot more radiation to help push the nuclear waste products out the back end.

Despite the tremendous energy released by this nuclear-powered machine when in operation, the spacecraft, weighing over 68,000 tons (much of it by way of the nuclear fuel), would still take at least 15 years to accelerate to a tad over 10 percent of the speed of light. Then the fuel would be exhausted, by which time the speed attained would see the spacecraft take about 45 years to travel to the nearest star system outside our own—Alpha Centauri.

We can already see where the problem lies. The solid nature and quantity of the fuel, as well as the size of the spacecraft needed to carry this fuel and the payload, adds up to an extraordinarily significant amount of mass (the greater the mass, the greater the energy that must be expended to accelerate the spacecraft to near the speed of light). Welcome to the reality of

U.S. Space Shuttle about to take off using its solid fuel for propulsion.

Artist impression of Cosmos 1.

interstellar travel. If we are going to break the energy barrier, it makes perfect sense to do something about this mass.

Clearly we need to reduce the mass to its absolute minimum. Sound familiar? It may remind you of the crashed disks mentioned in Frank Scully's book, as well as the world's most well-documented UFO crash near Roswell. Witnesses have said that the materials were extremely lightweight, yet strong. This is consistent with the laws of physics, as anything that can travel near the speed of light must have its mass reduced. Furthermore, there are combinations of elements available to ensure high strength and hardness for any thin and lightweight materials used.

Dr Paul C. W. Davies

The importance of reducing mass to increase speed for the least amount of energy is well understood by scientists, as Dr. Paul C. W. Davies said in his book titled *God and the New Physics*:

"A light particle is more easily moved by a given force than a heavy one. If a particle becomes exceedingly light it will be accelerated by any stray forces, and so will tend to travel very fast. In the limiting case that the mass dwindles away to nothing, the particle will always travel at the fastest possible speed, which is the speed of light."[2]

This quote tells us how critical it is to reduce the mass when approaching the speed of light. According to Einstein's Special Theory of Relativity, as a spacecraft approaches the speed of light, its mass and the mass of everything in it will increase dramatically. This won't be noticeable, however, until the moment the spacecraft begins to

---

2   Davies 1990, p.149.

accelerate, at which point a much stronger gravitational field will pass through the spacecraft and push its occupants backwards, creating what scientists call the *inertial force*. The only way to control this inertial force is to either reduce the mass of the spacecraft and everything else inside to a bare minimum, or find a way to control the gravitational field penetrating the spacecraft.

In Appendix B, we explained one way to reduce inertial forces. This leaves us with ways to reduce the mass. In the case of the UFO occupants, it is interesting that witnesses have never said they've seen an overweight creature, but always thin and mostly small ones. The low-mass concept again gets carried through to the construction of UFOs, if crashed-disc cases in the late 1940s are anything to go by. Combine this with the apparent high manoeuvrability of the UFOs in flight to support another way of controlling the inertial forces, as observed by witnesses, and there is every possibility that UFO reports have always revealed this "low-mass and control of the inertial forces" secret but have not garnered the attention of scientists and made them wonder why that is so.

So how can we reduce the mass and still travel to the stars?

Well, you can certainly reduce the payload of a spacecraft to just a couple of "bucket seats" and a metal cage to hold the air inside for its occupants. Well, this is precisely how the crashed discs appear to be constructed. No cryogenic or hibernating lifepods inside the discs to preserve occupants for very long timeframes traveling space (as we often see in Hollywood science fiction movies). It is clear we are dealing with phenomenal acceleration to quickly reach nearly the speed of light and then the journey to a distant star is measured in days or less for the occupants. But to do so, requires all sophisticated techniques in minimising mass, for both occupants, the craft and anything they may carry. Thus, sophisticated technical equipment for analysing foreign materials could be left at home, while the occupants carry a few glass vials to hold and preserve samples from an alien planet. Short journey times at close to the speed of light would ensure that the samples always remain fresh.

As for the spacecraft, you could use a strong, lightweight, newspaper-thin alloy foil made of the right materials to help enclose an atmosphere inside.

What about the occupants?

Given the proper diet, genetic engineering and a bit of surgery, the mass of the occupants participating in the flight can also be dramatically reduced. Want to be hairless, short and very slender? Easy. Science can achieve it. We already see a glimpse of this possibility in horse-racing jockeys, who have shown that diet alone is usually sufficient to inhibit growth, if the process is begun at a young-enough age.

Seriously, there is no limit to how far science can achieve extreme low mass in the occupants. You could, for instance, potentially reduce the length of the intestinal tract (or remove it altogether) in return for an increase in the quality of food or the use of patches on the skin to transmit nutrients directly to the bloodstream. This is certainly not beyond the realms of science.

You can even remove the reproductive organs, and the occupants can still live a normal (or probably a prolonged) life. It just simply means the occupants may not be able to reproduce normally and would have to rely on cloning techniques, or else the occupants will have to reproduce the normal way before having radical surgery to remove the reproductive organs.

Need low-tech solutions?

How about keeping fingernails trim and so on? Or using a massive laxative to blow everything out before starting the journey to the stars? No special scientific skills required there.

Yet reducing the mass of the spacecraft and its occupants is not quite enough. We still have the biggest hurdle to contend with: the fuel. If we are forced to rely on chemical or nuclear reactions to generate vast amounts of energy, the necessary high quantity of fuel to be carried will undoubtedly bog down the spacecraft. And this fuel has mass.

Another problem to consider is how a great deal of energy generated by the chemical and nuclear reactions inside man-made propulsion technologies is lost into space and can never be retrieved again for reuse.

Tackling the Fuel Problem

To solve this fuel-mass problem, let us look at what the chemical and nuclear propulsion technologies have in common:

1. There is an emission of not just hot (radioactive or non-radioactive) gases (the mass component), but also pure electromagnetic energy by way of heat and light (the massless component).
2. The object utilising chemical or nuclear propulsion technology recoils in the opposite direction after emitting the hot gases *and* electromagnetic energy.

Now that is interesting. We actually have a massless form of energy called radiation capable of pushing a solid object to a certain speed. Is radiation the solution?

Let us entertain ourselves on the following thought: suppose we eliminated the hot gases and retained only the heat and light. Heat and light are forms of electromagnetic energy that, as the laws of electromagnetism tell us, have no mass (even when generated on the spacecraft). Yet, as Einstein and his colleagues have found out from experiments such as with the photoelectric phenomenon, the electromagnetic energy can exert a force on ordinary matter (irrespective of whether it is electrically charged or not) as if it has mass. Whether or not radiation is a form of mass, the fact that it does not register as mass until it actually hits solid matter can have enormous benefits to interstellar travel. In particular, there is absolutely no need to contribute anything to the overall mass of the object just to generate this kind of energy for emitting light as a form of massless fuel for propulsion.

Fair enough. But how do we generate electromagnetic energy? And where does this energy come from? Fortunately, in the world of electromagnetism and the universe we live in, there are plenty of ways to generate the energy for creating a voltage (or charge) as needed to generate radiation for propulsion. In the most basic scenario, the energy can be obtained from the radiation-filled energy of the universe lying between the stars.

For example, we could use several large piezoelectric crystals arranged in a cylindrical fashion within a central pillar of the spacecraft (which, incidentally, would aid in providing structural support as well). Each time radiation is emitted at the bottom end, the electrons of the atoms composing the crystals would quickly grab the electromagnetic energy from their surroundings (including space) and store it in the chemical bonds, ready for use. Then, as soon as the radiation has been emitted, the recoiling force would compress the hull of the spacecraft and the crystals themselves in the process, resulting in energy being generated that can be tapped and used to re-generate more radiation for propulsion. If the radiation increases in frequency or more charges are generated to produce the radiation, the recoiling force gets stronger and with it, more energy is generated from the crystals through this positive feedback system. The kind of energy that can be generated in the crystals from a simple compression will be in the order of millions of volts (which is more than enough to electrically charge Thomas Townsend Brown's electrokinetic device to achieve reasonable acceleration).

Alternatively (or to supplement the previous power-generation system), we could attach an antenna to the spacecraft and let it pass through the natural planetary, solar and galactic magnetic field lines to generate voltages at opposite ends of the antenna. Generally, the faster the spacecraft moves, the more energy is generated by the antenna as it passes through the magnetic fields.

As you can see, the energy source would come from the surrounding environment of space. There is plenty in space we could utilise if we so choose. But if not, a small, portable and reliable nuclear-powered electrical generator could be employed with all the energy contained inside the nuclear fuel. Certainly you won't need the amount of nuclear fuel seen in Project Daedalus. That is a bit of an overkill. You only need enough energy to generate a few million volts to help generate radiation for propulsion.

If you think the energy from the surroundings is not enough, how about sending the energy electromagnetically in the direction of the star you wish to travel to, so that the spacecraft will be able to tap into it along the way?

Philip Norem, the Canadian director of engineering for the Peninsular Research and Development Corporation, proposed a similar system, whereby a 1,000-ton spaceship, known as the "laser galleon", could be accelerated to speeds of about 100,000 kilometres per second using an array of highly intensified laser beams on Earth to push the ship along. The "wind" pressure associated with the laser beams would be picked up by a huge parachute-shaped sail connected to the front of the ship by a 30-kilometre cable. It is estimated that a journey to the nearest extrasolar star—Proxima Centauri—would take 60 years to complete using such a propulsion system.[3]

Or better still, if the Cosmos 1 spacecraft, being much more lightweight, had not failed on its maiden flight in space using the power of sunlight (and would have been the fastest spacecraft humankind has made to date), similar microwave beams could have aided in its propulsion by having the energy directed at the metal sails. Now just imagine how much quicker the spacecraft could have reached Alpha Centauri. Certainly not 200 years as was estimated by the makers. It would have been a lot less.

Or why not use radio waves?

Nikola Tesla (c. 1896)

In 1899, Dr. Nikola Tesla (1856–1943) succeeded in developing a means of transmitting electricity without the use of a conductor. In his experiment, Nikola claimed to have lit up hundreds of lamps over a distance of 40 kilometres using a large 61-meter dome-like antenna for transmitting electromagnetic energy for conversion into electricity. Yet surprisingly, owing to a lack of funds and limited interest by business persons at the time[4], the idea has remained largely forgotten by the general public and scientists alike.

Wireless power transmission was proven feasible, however, by the electrical engineer Sir Raymond P. Phillips of Texas who, in 1988, patented[5] a simple electronic circuit for converting the energy in electromagnetic waves in the radio spectrum directly into electrical current. Phillips' circuit behaves like a solar cell, the only difference being that the circuit taps into radio waves rather than the visible region of the electromagnetic spectrum to generate direct current

The radio frequencies the circuit can utilise start from the high-frequency (HF) band right down to the extremely low-frequency (ELF) band, and may even pick up and convert the "seismic vibrations of the

---

3   Stemman 1991, p.46.
4   Probably for good reasons too as it would be difficult for businesspersons to charge consumers for the cost of using radio waves from a local radio station as a source of electricity.
5   Publication No. 4685047A / U.S. Patent No. 4,685,047.

Earth's magnetic field", as the patent claims. The circuit consists of diodes to modify the radio frequency signals to direct current, and a plurality of "electronic cells" (which act like a battery), consisting of capacitors, diodes and coils to store the electromagnetic energy received and eventually to increase the output of the current when powering electrical circuits, such as radios and battery chargers. The current can also be increased by using a larger-size antenna.[6]

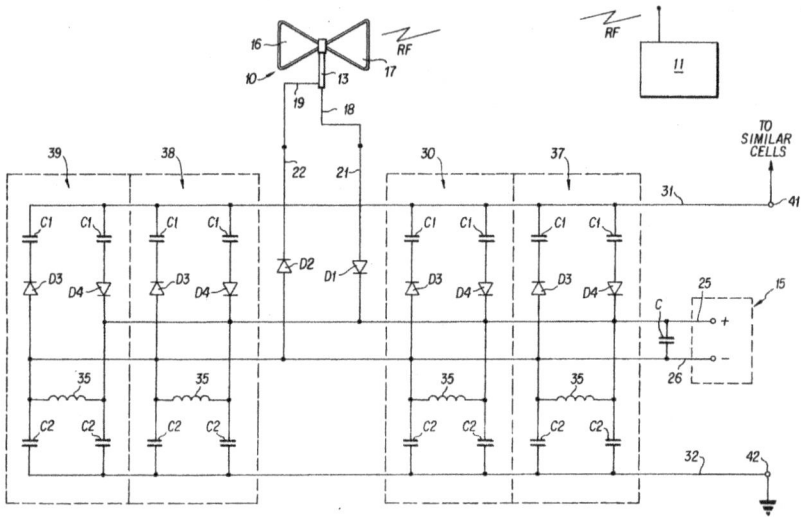

Sir Raymond Phillip's RF-to-DC converter circuit.

Whatever the power source, it seems altogether likely that Tesla may have been correct when he said in 1891:

"[In] many generations past ours, machinery will be driven by power obtainable at any point in the universe....It is a mere question of time when men [and women] will succeed in attaching their machinery to the very wheelwork of Nature."[7]

Not quite enough energy?

6   World Patent Number WO 88/00769.
7   Childress 1990, p.145.

Well, there is a further, as yet unconsidered, benefit to electromagnetic radiation: the possibility that we could recycle electromagnetic energy for propulsion.

According to the runaway solution of the Abraham-Lorentz formula, the radiation emitted for recoiling a charged object and the electromagnetic field generated by its acceleration seem to create a recycling effect in which the radiation somehow is able to "exert a force on itself" as Dr. Griffiths puts it, thereby causing the object to recoil again and again without a further input of energy. Whether this is true or not, this interpretation of the solution suggests that there is something else about the electromagnetic field that scientists have not yet realised.

As scientists have theorised regarding collapsed stars known as black holes, a powerful gravitational field can bend light (or electromagnetic radiation) back on itself. The star then becomes invisible, as the light behind the star can bend around it, allowing an observer to see what is behind the star.

The same is true of UFOs. As we have seen previously, people have reported seeing UFOs render themselves invisible and visible in a periodic way. Whether this is the way UFO occupants want to demonstrate some aspect of their technology and to reveal the scientific concept to us, and knowing there is a strong possibility that genuine UFOs are electromagnetic vehicles emitting radiation for propulsion based on the evidence in this book, it makes sense to consider a similar "gravitational field in the electromagnetic field" concept taking place.

As Einstein's Unified Field Theory is telling us, if you increase the electric charge, raise the frequency of the oscillating electric charge that produces the radiation, or increase the acceleration of the electric charge, something interesting happens: as the energy density of the radiation increases with the higher charge density and frequency, the strength of the gravitational field increases as well. Once the gravitational field reaches a critical level, the radiation can bend back on itself.

If this is true, why would we need energy from the surroundings to propel a spacecraft when electromagnetic energy can be recycled by its own gravitational field? Just let the radiation do all the work of recoiling a spacecraft and, potentially, recycling itself, and the recoiling forces would be amplified with increasing speed without a further input of energy.

In summary, there is no reason whatsoever why any object cannot be accelerated to nearly the speed of light. Just use the massless energy of radiation to move solid objects, and then accelerate in accordance with the solution of the Abraham-Lorentz formula. Any energy required for the system should be readily obtained from the surrounding environment. For example, a metal antenna passing through the planetary, solar and galactic magnetic fields at reasonable speeds can induce and separate an electric charge at opposite ends of the antenna. The energy can then be tapped for the purposes of radiation propulsion. Need more energy? Consider recycling the energy. Such a technique should effectively remove the impediment of traditional solid-based fuels in accelerating a solid object to speeds approaching that of light.

APPENDIX D

UFO *Statistics* *1933-1969*

Swedish Investigations

1933–1934

	No. of Reports
Explained	96
Unexplained	15
TOTAL	111

1946

	%
Explained	80
Unexplained	20
TOTAL	100

USA Investigations

Project Sign

	%
Explained	
Astronomical phenomena	32
Human-made objects	32
SUBTOTAL	64
Unexplained	20
Insufficient information	16
SUBTOTAL	36
TOTAL	100

Project Grudge

	%
Explained	
Astronomical Explanations	32
Balloons	12
Hoaxes, Aircraft, etc.	33
SUBTOTAL	77
Unexplained	23
TOTAL	100

Condon Study

	%
Explained	70
Unexplained	30
TOTAL	100

494

The Robertson Panel

	%
Explained	90
Unexplained	10
TOTAL	100

Project Blue Book

	No. of Reports
Explained	11917
Insufficient information	701
TOTAL	12618

Project Blue Book 1947–1969

Year	Total Sightings	Unidentified
1947	122	12
1948	156	7
1949	186	22
1950	210	27
1951	169	22
1952	1501	303
1953	509	42
1954	487	46
1955	545	24
1956	670	14
1957	1006	14
1958	627	10
1959	390	12
1960	557	14
1961	591	13
1962	474	15
1963	399	14
1964	562	19
1965	887	16
1966	1112	32
1967	937	19
1968	375	3
1969	146	1
TOTAL	12618	701

APPENDIX E

Thomas Townsend Brown's British and U.S. Patents

British patent 300,311, shown on the following page, is believed to be the first one granted to the inventor. Subsequent patents for the same device (such as 2,949,550 and 3,187,206) were made in the United States and show different designs for his invention, including one where two devices were attached to wires supplying a high voltage and strung to a central pole, which eventually revealed an unusual accelerating behaviour, forcing Brown to lower the voltage to prevent his devices from flying away. The question is: what is the rate of acceleration of his devices? Is it exponential or linear?

After Brown received funding from millionaire A. H. Bahnson, Jr., Bahnson patented one of Brown's devices under U.S. patent number 3,227,109.

PATENT SPECIFICATION

Application Date: Aug. 15, 1927. No. 21,452/27

Complete Accepted: Nov. 15, 1928.

COMPLETE SPECIFICATION.

A Method of and Apparatus or Machine for Producing Force or Motion.

I, Thomas Townsend Brown, of 15, Eighth Street, in the City of Zanesville, State of Ohio, United States of America, a citizen of the United States of America, do hereby declare the nature of this invention and in what manner the same is to be performed to be particularly described and ascertained in and by the following statement:

This invention relates to a method of controlling gravitation and for deriving power therefrom, and to a method of producing linear force or motion. The method is fundamentally electrical.

The invention also relates to machines or apparatus requiring electrical energy that control or influence the gravitational field or the energy of gravitation; also to machines or apparatus requiring electrical energy that exhibit a linear force or motion which is believed to be independent of all frames of reference save that which is at rest relative to the universe taken as a whole and said linear force or motion is furthermore believed to have no equal and opposite reaction that can be observed by any method commonly known and accepted by physical science to date.

The invention further relates to machines or apparatus that depend for their force action or motive power on the gravitational field or energy of gravitation that is being controlled or influenced as above stated; also, to machines or apparatus that depend for their force action or motive power on the linear force or motion exhibited by such machines or apparatus previously mentioned.

The invention further relates to machines and apparatus that derive usable energy or power from the gravitational field or from the energy of gravitation by suitable arrangement, using such machines and apparatus as first above stated as principal agents.

To show the universal adaptability of my novel method, said method is capable of practical performance and use in connection with motors for automobiles, space cars, ships, railway locomotion, prime movers for power installations, aeronautics. Still another field is the use of the method and means enabling the same to function as a gravitator weight changer. Specific embodiments of the invention will be duly disclosed through the medium of the present Specification.

Referring to the accompanying drawings forming part of this Specification :

Figure 1 is an elevation, with accompanying descriptive data, broadly illustrating the characteristic or essential elements associated with any machine or apparatus in the use of which the gravitational field or the energy of gravitation is utilized and controlled, or in the use of which linear force or motion may be produced.

Figure 2 is a similar view of negative and positive electrodes with an interposed insulating member, constituting an embodiment of the invention.

Figure 3 is a similar view of a cellular gravitator composed of a plurality of cell units connected in series, capable of use in carrying the invention into practice.

Figure 4 is an elevation of positive and negative electrodes diagrammatically depicted to indicate their relation and use when conveniently placed and disposed within a vacuum tube.

Figures 5 and 5₁ are longitudinal sectional views showing my gravitator units embodied in vacuum tube form wherein heating to incandescence is permitted as by electrical resistance or induction at the negative electrode; and also permitting, where desired, the conducting of excessive heat away from the anode or positive electrode by means of air or water cooling devices.

Figure 6 is an elevation of an embodiment of my invention in a rotary or wheel type of motor utilizing the cellular gravitators illustrated in Figure 3.

Figure 7 is a view similar to figure 6 of another wheel form or rotary type of motor involving the use of the gravitator units illustrated in Figure 5, or Figure 5₁.

Figure 8 is a perspective view partly in section of the cellular gravitator of Figure 3 illustrating the details thereof.

Figures 9, 10 and 10a are detail views of the cellular gravitator.

Figure 11 is a view similar to Figure 8

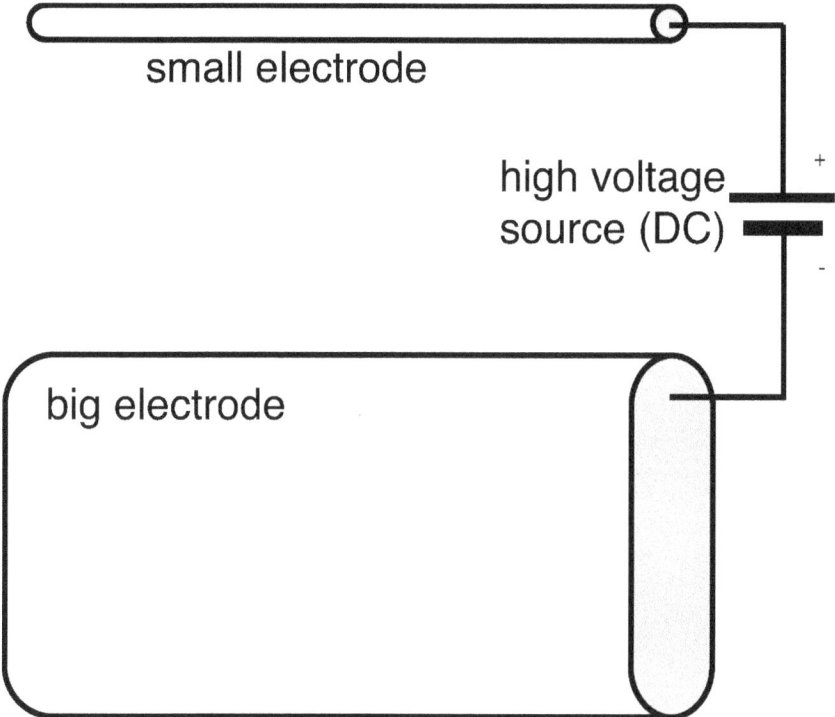

small electrode

high voltage source (DC)

+

-

big electrode

1

2,949,550

ELECTROKINETIC APPARATUS

Thomas Townsend Brown, Umatilla, Fla., assignor to Whitehall-Rand, Inc., Washington, D.C., a corporation of Delaware

Filed July 3, 1957, Ser. No. 669,830

12 Claims. (Cl. 310—5)

My invention relates to electrokinetic apparatus, and more particularly to a method and apparatus for utilizing electrical potentials for the production of forces for the purpose of causing relative motion between a structure and the surrounding medium.

This invention was disclosed and described in my application Serial No. 293,465, filed June 13, 1952, which application has become abandoned. However, reference may be made to this application for the purpose of completing the disclosure set forth below.

The invention utilizes a heretofore unknown electrokinetic phenomenon which I have discovered; namely, that when a pair of electrodes of appropriate form are held in a certain fixed spaced relation to each other and immersed in a dielectric medium and then oppositely charged to an appropriate degree, a force is produced tending to move the pair of electrodes through the medium. The invention is concerned primarily with certain apparatus for utilizing such phenomenon in various manners to be described.

Priorly, intervening electrokinetic apparatus has been employed to convert electrical energy to mechanical energy and then to convert the mechanical energy to the required force. Except for the insignificantly small forces of electrostatic attraction and repulsion, electrical energy has not been used for the direct production of force and motion.

Since any conversion of energy from one form to another is accompanied by losses due to friction, radiation or conduction of heat, hysteresis, and the like, as well as serious reductions in the availability of the energy by increases in the entropy of the system, it is apparent that great increases in efficiency may be achieved through the use of the direct production of electrical energy and force and motion made possible by my invention. Likewise, the elimination of the machinery for the intermediate conversions results in great savings in first costs, maintenance, weight and space, the latter two being of great importance in self-propelled vehicles including mobile vehicles such as aircraft and space craft.

It is therefore an object of my invention to provide an apparatus for converting the energy of an electrical potential directly into a mechanical force suitable for causing relative motion between a structure and the surrounding medium.

It is another object of this invention to provide a novel apparatus for converting an electrical potential directly to usable kinetic energy.

It is another object of this invention to provide a novel apparatus for converting electrostatic energy directly into kinetic energy.

It is another object of this invention to provide a vehicle motivated by electrostatic energy without the use of moving parts.

It is still another object of this invention to provide a self-propelled vehicle without moving parts.

It is a feature of my invention to provide an apparatus

2

for producing relative motion between a structure and the surrounding medium which apparatus includes a pair of electrodes of appropriate form held in fixed spaced relation to each other and immersed in a dielectric medium and oppositely charged.

It is another feature of my invention to provide apparatus which includes a body defining one electrode, another separate electrode supported in fixed spaced relation by said body, and a source of high electrical potential connected between the body and the separate electrode.

It is also a feature of my invention to provide apparatus having a body which is hollow and a source of potential contained within the body.

It is another feature of my invention to provide apparatus having a body and an electrode connected to the body, which combination comprises a vehicle.

It is also a feature of my invention to provide apparatus which comprises a plurality of assemblies, each including a body and an electrode secured in side-by-side spaced relation to each other.

It is another feature of my invention to provide vehicular apparatus which includes a pair of electrically conductive body portions joined by an insulating portion, whereby said electrically conductive portions constitute the electrodes.

Other objects and advantages of my invention will be apparent from a consideration of the following specification, read in connection with the accompanying drawings, wherein:

Figure 1 is a side elevational view illustrating diagrammatically a simple form of apparatus embodying and functioning in accordance with the principles of my invention;

Figure 2 is a plan view of the apparatus shown in Figure 1;

Figure 3 is a perspective view illustrating the manner in which a plurality of devices of the character illustrated in Figure 1 may be interconnected for joint operation;

Figure 4 is a diagrammatic view similar to Figure 1 illustrating a modified form of the invention providing a means for reversing the direction of the propulsive force produced;

Figure 5 is a perspective view illustrating diagrammatically a self-propelled device utilizing the principles of this invention;

Figure 6 is a perspective view of one illustrative embodiment of this invention showing a pair of electrokinetic propulsion devices suspended from a rotatable arm which arm is supported at its midpoint;

Figure 7 is a side elevational view of a mobile vehicle with parts broken away to show the interior construction;

Figure 8 is a side elevational view illustrating diagrammatically the arrangement of parts used in an alternative form of mobile vehicle.

Referring to the drawings, I have illustrated in Figure 1 a simple form of apparatus which is readily adaptable for use in demonstrating the principles of my invention, and which is utilized in this application as a simplified representation to facilitate an understanding of the principles involved. The apparatus illustrated in Figure 1 constitutes one electrode which is preferably in the form of a body member 20, said member preferably comprising a relatively thin flat plate. A second electrode 21 in the form of a wire or other suitable form of electrical conductor is held as by means of insulated supports 22 in fixed spaced relation to the body 20, the wire 21 being disposed in the plane of the body 20 and preferably substantially parallel with a leading edge 23 of the body 20. A source 24 of high voltage electrical potential is provided and connected as shown at 25 and 26 to the two electrodes 20 and 21, respectively.

FIG. 1

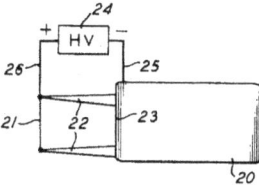

FIG. 2

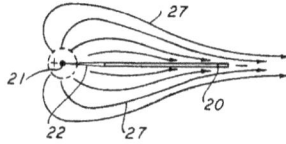

FIG. 3

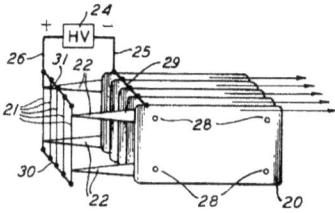

FIG. 4

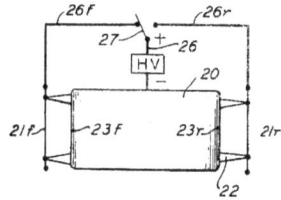

FIG. 5

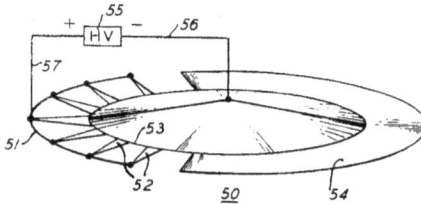

INVENTOR
THOMAS TOWNSEND BROWN
BY
Watson, Cole, Grindle & Watson

ATTORNEYS

501

FIG. 6

FIG. 7

FIG. 8

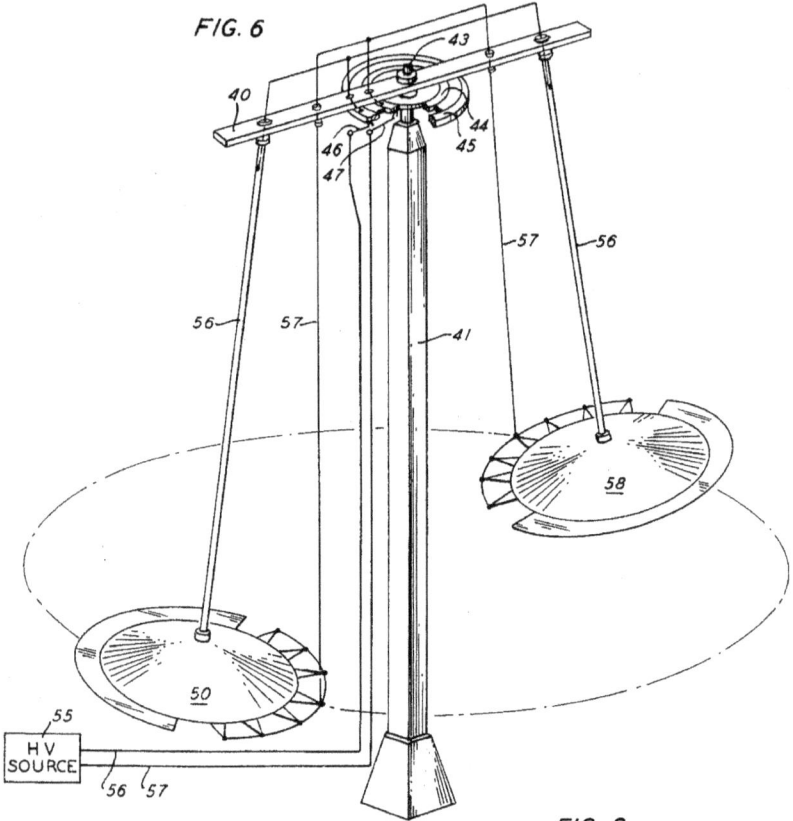

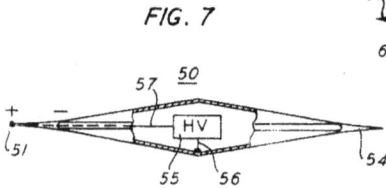

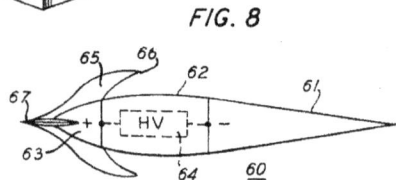

INVENTOR
THOMAS TOWNSEND BROWN
BY
Watson, Cole, Grindle & Watson
ATTORNEYS

502

1

3,187,206
ELECTROKINETIC APPARATUS
Thomas Townsend Brown, Walkertown, N.C., assignor,
by mesne assignments, to Electrokinetics, Inc., a corporation of Pennsylvania
Filed May 9, 1958, Ser. No. 734,342
23 Claims. (Cl. 310—5)

This invention relates to an electrical device for producing thrust by the direct operation of electrical fields.

I have discovered that a shaped electrical field may be
employed to propel a device relative to its surroundings in
a manner which is both novel and useful. Mechanical
forces are created which move the device continuously in
one direction while the masses making up the environment
move in the opposite direction.

When the device is operated in a dielectric fluid medium, such as air, the forces of reaction appear to be
present in that medium as well as on all solid material
bodies making up the physical environment.

In a vacuum, the reaction forces appear on the solid
environmental bodies, such as the walls of the vacuum
chamber. The propelling force however is not reduced
to zero when all environmental bodies are removed beyond the apparent effective range of the electrical field.

By attaching a pair of electrodes to opposite ends of a
dielectric member and connecting a source of high electrostatic potential to these electrodes, a force is produced
in the direction of one electrode provided that electrode
is of such configuration to cause the lines-of-force to converge steeply upon the other electrode. The force, therefore, is in a direction from the region of high flux density
toward the region of low flux density, generally in the direction through the axis of the electrodes. The thrust
produced by such a device is present if the electrostatic
field gradient between the two electrodes is non-linear.
This non-linearity of gradient may result from a difference in the configuration of the electrodes, from the electrical potential and/or polarity of adjacent bodies, from
the shape of the dielectric member, from a gradient in the
density, electric conductivity, electric permittivity and
manetic permeability of the dielectric member or a combination of these factors.

A basic device for producing force by means of electrodes attached to a dielectric member is disclosed in my
Patent 1,974,483. In one embodiment disclosed in my
patent, an electrostatic motor comprises devices having a
number of radially directed fins extended from one end
of the dielectric body and a point electrode on the opposite end of the dielectric body. When this device is supported in fluid medium, such as air, and a high electrostatic potential is applied between the two electrodes, a
thrust is produced in the direction of the end to which
the fins are attached.

Other electrostatic devices for producing thrust are disclosed and described in detail in my British Patent 300,-
311, issued August 15, 1927.

Recent investigations in electrostatic propulsion have
led to the discovery of improved devices for producing
thrust by the use of electrical vectorial forces.

Accordingly, it is the primary object of this invention
to provide an improved electrical device for producing
thrust.

It is another object of this invention to provide a device
for producing modulated thrust in response to varying
electrical signals, which device produces a greater effect
than the prior type devices mentioned above.

It is another object of this invention to provide a device
which shapes or concentrates electrostatic flux to produce
an improved thrust.

Broadly, the invention relates to shaping an electrical
field to produce a force upon the device that shapes the
field. The electrical field is shaped by the use of an elec-

2

trode of special configuration whereby the electric lines-
of-force are made to converge at a distance from the
electrode. One illustrative embodiment of this invention
which satisfies the above requirement is an arcuate surface or, alternatively, a system of wires, tubes or plates
embedded in a dielectric surface and forming a directive
array. One such highly-charged electrode acting within
and upon an ambient of different electrical potential will
move in response to the forces created by the shaping of
the electrostatic field. If a smaller electrode is added at
or near the focus of the field-shaping electrode and me-
chanically attached to that electrode, both electrodes as a
system will move in a direction of the larger or field-shaping electrode. As is mentioned above, the field-shaping
electrode alone, when charged with respect to its electric
ambient, will move or possess a force in the direction of
its apex. If another electrode carrying a different charge
is added at or near the focal point of the field-shaping
electrode, then the field becomes more concentrated, i.e.
shaped t   greater degree and the resulting thrust is
greater than that which exists when the field-shaping electrode alone is employed.

Briefly in accordance with aspects of this invention,
an electrode is connected on each end of a dielectric member and one of the electrodes defines a large area flat or
preferably arcuate surface which is curved in such a direction to produce, usually in co-operation with the other
electrode, a shaped electrostatic field.

Advantageously, if the arcuate electrode is in the form
of a parabola or hyperbola, the length of the dielectric
member may be such that the other electrode is located in
the region of the focus of the parabola or hyperbola, as
the case may be. If the arcuate electrode is hemispherical,
the other electrode is located near the center of the hemisphere.

In accordance with other aspects of this invention the
dielectric member supporting the two electrodes may have
electrical conductivity and/or dielectric constant which
varies progressively between its ends so that the dielectric
member contributes to the non-linearity of the field gradient and causes a greater thrust to be developed.

In accordance with still other aspects of this invention,
an annular electrode member is secured to an electrode
mounted in the region of the axis of the annular electrode.
If the second electrode is located at the center of the annular electrode and the two electrodes are energized,
such force is not detected. However, if the second or
innermost electrode is displaced from the center of the
annular electrode in the region of the axis of the annular
electrode and the electrodes are energized, then thrust
will be produced by the two electrodes. The annular
electrode may either be a flat ring, a toroid, or a section
of a cylinder.

In accordance with still other aspects of this invention,
tapered dielectric members having electrodes secured to
opposite edges thereof may be employed to produce a
thrust in response to the application of potentials to
these electrodes. The thrust produced by these tapered
dielectric members may be further augmented by em-
bedding massive particles, such as lead oxide, in the
wedges, which particles are usually more concentrated
near the points of the wedges.

Accordingly, it is a feature of this invention to provide
an electrical device for producing thrust which includes
a dielectric member and electrodes supported at each end
of the dielectric member, one of which electrodes is
located in the region of the focal point of the arc of the
arcuate electrode.

It is another feature of this invention to provide a
device for producing thrust having a dielectric member
and a pair of electrodes secured to opposite ends of the
dielectric rod or member, one of which electrodes de-

FIG. 1

FIG. 2

FIG. 3

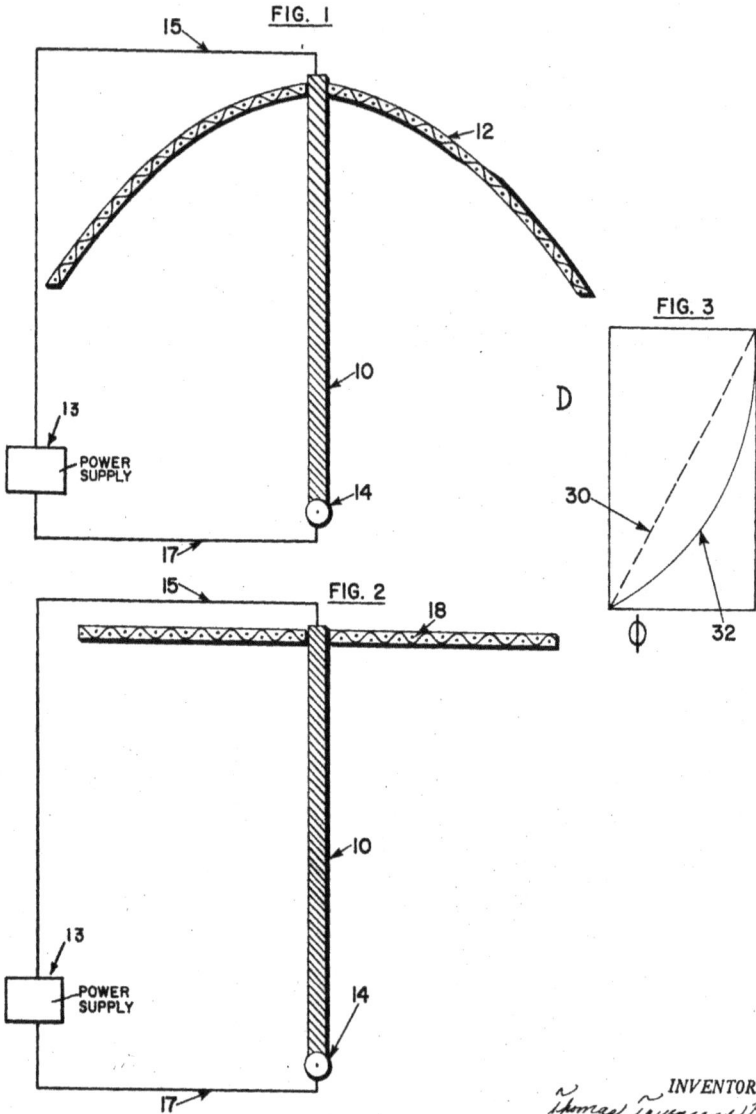

INVENTOR.

Thomas Townsend Brown

BY

Watson, Cole, Grindle & Watson
Attorneys

504

FIG. 4

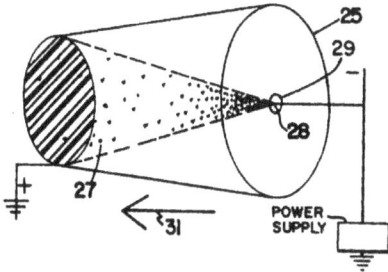

FIG.5

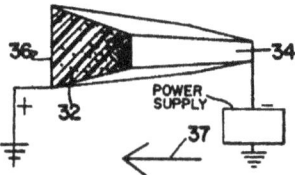

FIG. 6

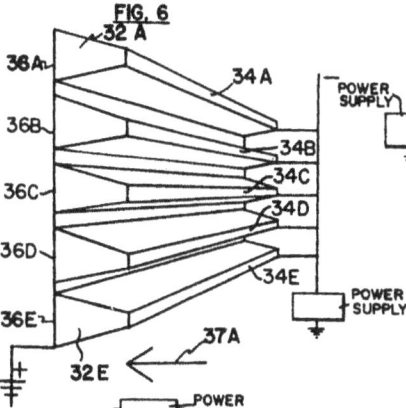

FIG. 7

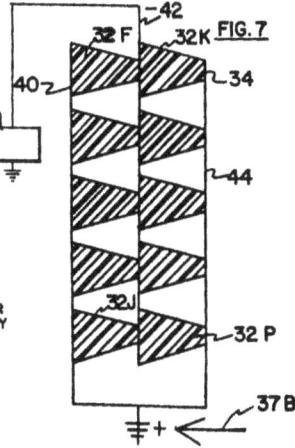

FIG. 8B

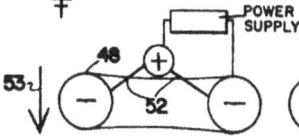

FIG. 8A

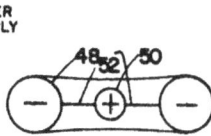

FIG. 8C

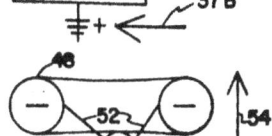

INVENTOR
Thomas Townsend Brown

BY
Watson, Cole, Grindle & Watson
ATTORNEYS

505

1

3,227,901
ELECTRICAL THRUST PRODUCING DEVICE
Agnew H. Bahnson, Jr., 1001 S. Marshall St.,
Winston-Salem, N.C.
Filed Sept. 15,1961, Ser. No. 138,493
5 Claims. (Cl. 310—5)

This invention relates to a device for directly producing thrust in response to the application of electrical energy and is a continuation-in-part of my co-pending applications Serial No. 745,652 filed June 30, 1958, now abandoned and Serial No. 34,522 filed June 7, 1960, now abandoned.

In my aforesaid parent application, there is disclosed and claimed an improved arrangement for producing thrust wherein the thrust producing device is comprised of a pair of main electrodes supported in spaced relationship with each other by means of a dielectric member and a difference in potential is applied to the electrodes to effect a thrust on the device in the direction of the longitudinal axis of the assembly of support and electrode members.

The present invention is directed to a further organization of elements by which one is enabled to obtain a thrust laterally of said longitudinal axis in any one of a plurality of selected lateral directions. In accordance with one embodiment of the invention which will be hereinafter described in detail, an auxiliary cylindrical electrode is secured to the periphery of the dielectric supporting member for the main electrodes and a plurality of other auxiliary electrodes are spaced in different radial positions with respect to the auxiliary cylindrical electrode. These electrodes are selectively energized to produce a lateral thrust in a desired direction. Advantageously, the auxiliary electrodes having different radial positions may be constituted by conductive arcuate portions of a plurality of rings, the conductive portions of each ring being separated by arcuate portions of dielectric material, and the rings being arranged concentrically with respect to the auxiliary cylindrical electrode and encircling the support and also being carried by the support.

More specifically, the embodiment of the invention which will be described is comprised of three rings, each such ring being provided with a pair of oppositely disposed arcuate conductive portions and the respective pairs of conductive portions of each ring having their centers mutually displaced by 60°. Each electrode is selectively energized from a source of direct current potential relative to the inner cylindrical electrode so as to develop a lateral thrust in any selected one of six different directions having their directional thrust axes spaced 60° apart. Such an embodiment is illustrated in the accompanying drawings, in which:

FIG. 1 is a view in vertical central section of the thrust producing device; and

FIG. 2 is a view showing the detailed arrangement of the three auxiliary lateral thrust producing electrode rings together with the circuit means for their selective energization.

With reference now to FIG. 1, a thrust producing device is depicted which comprises a dielectric member 10 which is in the form of a hollow cylinder and may advantageously comprise a high dielectric material. Mounted on the lower end of the dielectric member 10 is an arcuate electrode 11 made of electrically conductive material. Mounted further about the body of the dielectric member 10 is an arcuate electrode member 12 which surrounds the member 10. The arcuate electrode 12 which has an expanded surface relative to that of electrode 11 may be comprised of a series of fine electrically

2

conductive wires 13 embedded in a suitable dielectric material 14.

A source of high direct current voltage indicated schematically at 15 and which can be supported generally within the dielectric member 10, if desired, has its positive (+) terminal connected to the arcuate electrode member 12 and its negative (−) terminal connected to electrode 11. The resulting forces which develop as a result of the application of this high direct current potential to electrodes 11 and 12 cause a thrust to be developed on the dielectric member 10 longitudinally thereof in the direction of the arrow 16.

In accordance with the present invention, thrust on the member 10 may also be developed in a selected lateral direction, i.e. in a direction normal to the longitudinal axis of the dielectric member 10. This lateral thrust is brought about by arranging auxiliary electrodes laterally of the dielectric member 10 and in different radial positions. In one practical embodiment of the invention, as shown in detail in FIG. 2, the auxiliary electrodes are arranged in the form of rings which surround and are concentric with the dielectric member 10 and are carried by the latter by means of suitable supporting structure not shown. There are three of these rings denoted by the numerals 17, 18 and 19. Each ring is composed of two 90° arcuate portions of electrically conductive material and two 90° arcuate portions of dielectric material, and the dielectric portions alternating with the electrically conductive portions so that the portions of electrically conductive material are thereby oppositely disposed with respect to the center of the ring. The conductive portions of ring 17 are denoted by 20a and 20b and the dielectric portions by 21a and 21b. The conductive portions of ring 18 are denoted by 22a and 22b and the dielectric portions by 23a and 23b. The conductive portions of ring 19 are denoted by 24a and 24b and the dielectric portions by 25a and 25b.

Located within these rings 17–19 is an auxiliary cylindrical electrode 26 which is concentric with the axis of the dielectric supporting member 10 and which is conveniently mounted on the periphery of the latter.

In order to develop six different directions of lateral thrust upon the dielectric member 10 from the three rings 17–19 depicted in FIG. 2, and the cylindrical electrode 26 it will be noted that the center of the conductive portion 20a, 20b of inner ring 17 is displaced by an angle of 60° from the center of the conductive portions 22a, 22b, of the intermediate ring 18, and that the center of the conductive portions 24a, 24b of the outer ring 19 is displaced by an angle of 60° from the center of the conductive portion 22a, 22b of intermediate ring 18. To develop the thrust in a direction laterally of the longitudinal axis of the dielectric member 10, arrangements are made to apply an electrical potential in an alterantive and selective manner between electrode 26 and any one of the conductive portions 20a, 20b, 22a, 22b, 24a and 24b. The potential source 15 used for energizing the main electrodes 11 and 12 which produce the thrust longitudinally of the dielectric member 10 can be used for this purpose. The circuit for energizing the electrodes can include conventional rotary switch means 27. Thus one arcuate electrode is always charged positively with respect to the negatively charged cylindrical electrode 26 and the direction of the lateral thrust from the center will be towards the particular electrode connected to the positive terminal of the direct current voltage source 15.

Thus, for example, if a lateral thrust in the direction of the arrow 28 in FIG. 2 is desired, electrode 20a will be energized, in which event, electrode 20a will be connected through switch 27 to the positive terminal of the

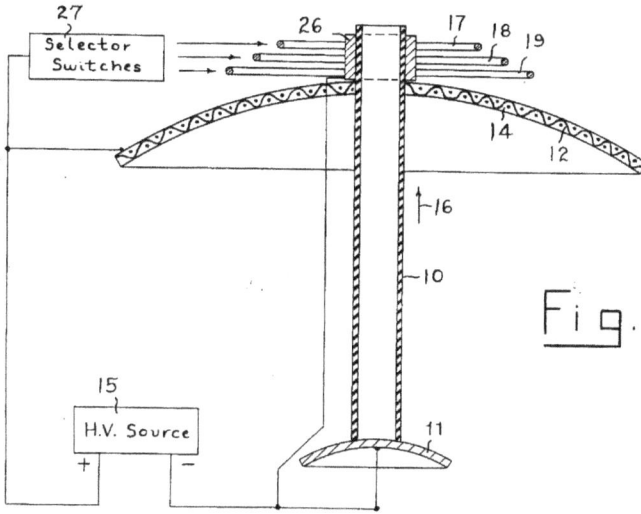

Fig. 1

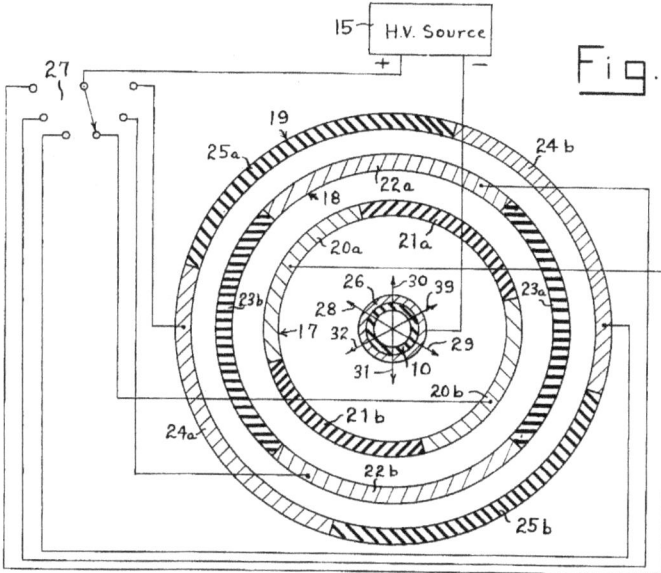

Fig. 2

This design was based on an improved version of Brown's patented device after the physicist received funding from Mr Agnew Hunter Bahnson, Jr.

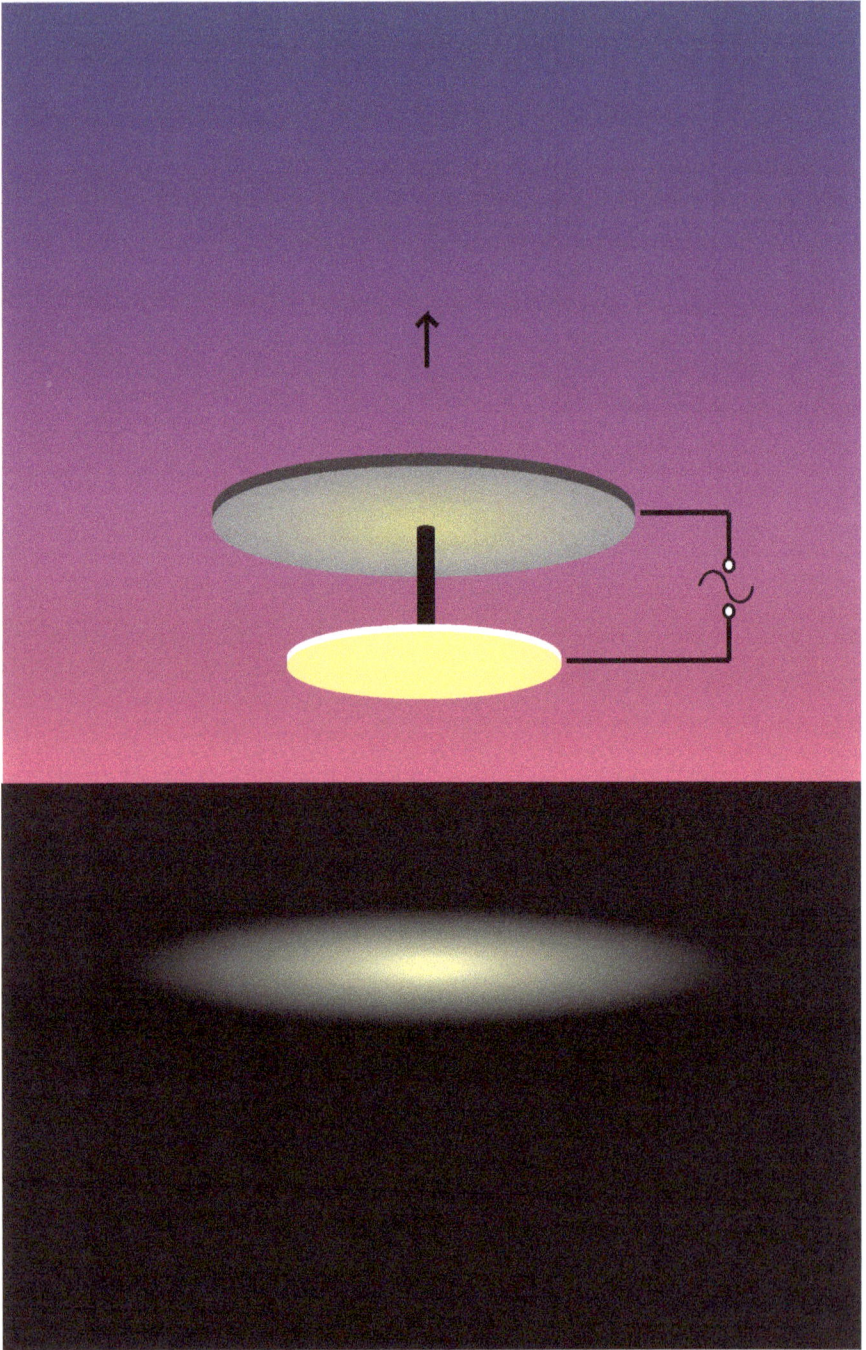

An alternative electrokinetic device from SUNRISE.

APPENDIX F

U.S. Army Experiment on Thomas Townsend Brown's Invention

n September 2002, researchers Thomas B. Bahder and Christian
Fazi from the U.S. Army Research Laboratory (ARL) conducted an
experiment to confirm the movement of Brown's asymmetric
capacitor and to identify the likely explanation for the technology. As
no physics journal had discussed this type of capacitor, the researchers
obtained designs for several asymmetric capacitors from the internet, as
revealed by French experimenter Jean-Louis Naudin, Transdimensional
Technologies and the American Antigravity web sites.

According to their report, dated 27 September 2002, the researchers
built a triangular asymmetric capacitor model, 20cm on each side. The
bottom aluminium electrode had a height of 4cm. Attached to the apex
of the triangular electrode were three thin balsa wood sticks holding a

copper wire to form the top electrode. The total weight of the model was 5 grams.

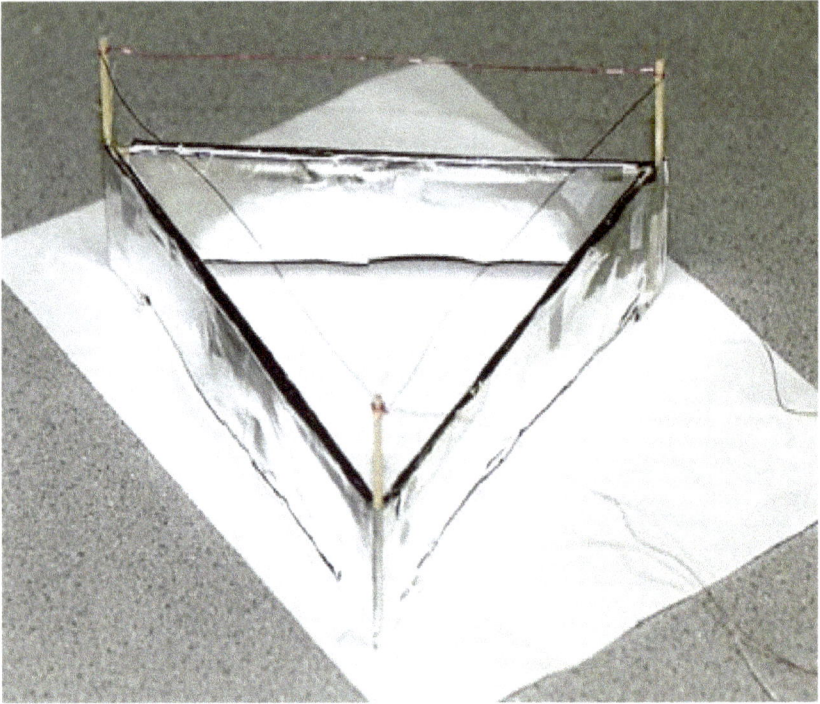

On charging the capacitor with 37kV and a current of about 1.5mA, the authors reported:

> "The capacitor lifted off its resting surface. However, this capacitor was not a vigorous flier, as reported by others on the Internet. One problem that occurred was arcing from the thin wire electrode to the foil. The thin wire electrode was too close to the foil. We have found that arcing reduces the force developed on the capacitor."

The researchers created a second model using a Styrofoam lunch box and plastic drinking straws. The design was very similar to the previous model, except the researchers decided on a rectangular geometry for the bottom electrode of 18cm x 20cm, and the wire for the top electrode could be adjusted to increase the distance from the bottom electrode.

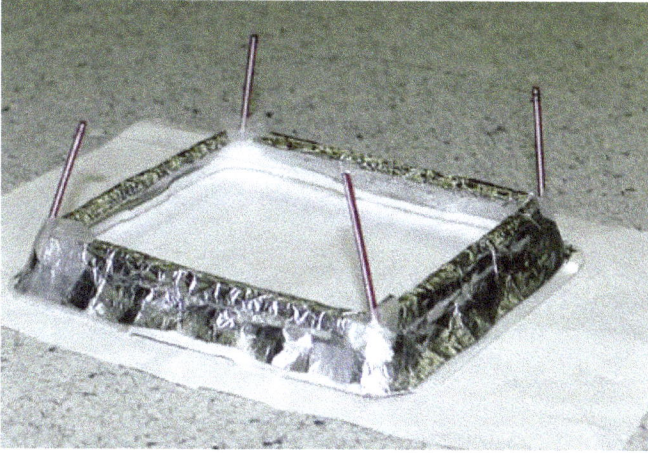

On testing this capacitor, the researchers discovered:

"…that making a 6cm gap [between the electrodes] resulted in little arcing. When 30 kV was applied, the capacitor drew about 1.5 mA, and hovered vigorously above the floor."

Thinking the geometry of the electrodes was somehow affecting the movement, the researchers built a flat, asymmetric capacitor—the bottom electrode was 20cm x 4cm, and they used a wire for the top electrode. The distance between the electrodes was 6.3cm.

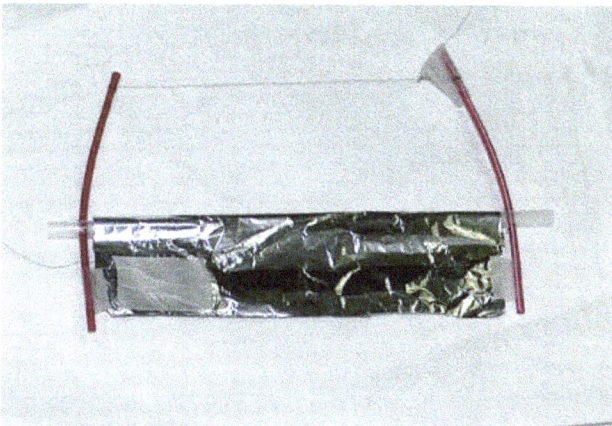

The researchers applied 30kV to the capacitor and noted the following:

"The force on this capacitor greatly exceeded its weight, so much so that it would vigorously fly into the air when the voltage was increased from zero. Therefore, we have concluded that the closed geometry of the electrodes is not a factor in the net force on an asymmetric capacitor."

Furthermore, the researchers realised it was important for the foil in the bottom electrode—the one closest to the wire—to have a rounded (rather than a sharp) edge to minimise the effect of arcing. When the original triangular model was re-built to take into account this rounded edge, the researchers claimed that the capacitor "...showed improved lift when rounded foil was put over the foil electrode closest to the thin wire".

The researchers did not specify a frequency for the oscillating voltage, claiming a direct current voltage was applied to all the asymmetric capacitor designs. Another thing the researchers noted was how the polarity chosen for charging the electrode didn't matter—the movement would be towards the smaller electrode. As the report stated:

"When operated in air, the asymmetric capacitors exhibit a net force toward the smaller conductor, and in all three capacitors, we found that this force is independent of the D.C. voltage polarity. The detailed shape of the capacitor seems immaterial, as long as there is a large asymmetry between the characteristic size of the two electrodes."

In these experiments, the larger electrode was given a negative charge, and the smaller electrode was given a positive charge.

In trying to find an explanation, the researchers looked at the ionic wind theory:

"The first proposed scheme is that there exists an ionic wind in the high field region between the capacitor electrodes, and that this ionic wind causes the electrodes to move as a result of the momentum recoil. This scheme, described in Section A [in the report] below, leads to a force that is incorrect by at least three orders of magnitude compared to what is observed."

When calculating the force of the "ionic wind" on the capacitor, the researchers assumed that either the air molecules were being ionised (i.e., the molecules become positively-charged by the copper wire before receiving their electrons and a little extra from the negatively-charged foil where it gets electrostatically repelled), or heavy ions of copper in the wire were being stripped off.

Using the heaviest ion scenario of copper ions, the "recoiling" force exerted by these ions was still too small by at least three orders of magnitude.

In the case of air molecules, the researchers noted how the smaller electrode (i.e., the wire) generated a stronger electric field (due to its smaller radius) compared to the larger electrode. It was thought that this may be important when it comes to stripping off the electrons of the air molecules. As the positively-charged molecules drift toward the negative electrode, the electrons are suddenly acquired, and with extra electrons added to the molecules, they can be repelled electrostatically by the negatively-charged foil to create a recoiling force. However, the researchers thought the ionised air molecules were too heavy to move fast enough to the opposite electrode. As a result, the molecules tended to stay as plasma around the smaller electrode. So most of the force is likely to be exerted by the electrons hitting the positively-charged wire electrode:

> "…we present the second scheme, which assumes that a drift current exists between the capacitor plates. This scheme is basically a scaling argument, and not a detailed treatment of the force. In this scheme, the order of magnitude of the force on an asymmetric capacitor is correct, however, this scheme is only a scaling theory."

However, the theory that electrons move the capacitor dramatically falls apart at the moment the capacitor's weight is marginally increased. For example, if the capacitor is a mere 5 grams, the researchers could calculate the required force using the ion drift theory. However, if the capacitor is too heavy (even 10 grams or more), the force of electrons is insufficient to move the capacitor. Yet Brown claimed he could move a Coolidge X-ray tube at increasing speeds without any trouble. So too did his three-foot-diameter "airfoils" during his demonstration in the 1950s. If this is true, it is unlikely "ion drift" by electrons or copper ions can adequately explain the movement.

So the researchers took on a thermodynamic analysis of the capacitor focusing on all the possible forces, including electrons in the air and metal atoms being ejected from the charged electrodes, the non-linear dielectric fluid (which is effectively ionised air molecules) and how it moves, and the electromagnetic field generated by the charged electrodes.

When determining the contribution of the electromagnetic field on the capacitor, the researchers stated a constant direct current (D.C.) voltage was applied to the capacitor. Thus, the only field present was the static electric field. Therefore, any momentum being contributed by an electromagnetic field on the capacitor was effectively ruled out from the investigation. An extraordinary decision considering Brown used an oscillating or alternating current (A.C.) voltage to make his devices move, and quite significantly too. Any oscillating voltage would definitely require an analysis of the electromagnetic field emitted by the oscillating charged electrodes to determine the overall force on the capacitor. Unfortunately, the researchers ignored this contribution in their investigation.

The researchers then calculated the likely force due to the fluid motion of the dielectric material between and around the electrodes (e.g., the air). However, the researchers admitted that the force effectively drops to zero in a vacuum. Given that Brown stated the movement was more significant in a vacuum when he tested his capacitor in 1955, it is unlikely the movement is due to the drift motion of ionised air molecules. This leaves us with matter being ejected from the electrodes (i.e., the metal ions). Unfortunately, the researchers were unable to determine definitively if the matter from the electrodes could contribute to this force. So they recommended:

> "...constructing a set of experiments designed to verify if the thermodynamic theory presented here can explain the magnitude and sign of the observed force."

Since the report was written, the explanation for the movement has not been found, and the researchers did not find anything unusual in the devices' rate of acceleration.

In essence, the report helps to make the U.S. Army (and perhaps by implication the rest of the military) look like they have no special interest in Brown's electrokinetic device and is not aware of any significant observations (e.g., the "runaway" acceleration effect), and certainly gives the impression that there was "nothing to hide" regarding any secret USAF work on Brown's invention after the late 1950s. However, the USAF has been using Brown's invention to reduce the weight of its B-2 Stealth Bomber. Furthermore, the Cash-Landrum UFO case where a glowing diamond-shaped UFO was tested by the U.S. military (the USAF would have to be the principal party involved) as black military helicopters surrounded the object and guided it to some unspecific destination, would require the use of Brown's invention to a much more significant level. As the glowing surface is the region where oscillating electromagnetic fields must have been emitted for the object to lift itself off the ground, it is likely the U.S. Army's report on the invention was its own way of indirectly helping to maintain UFO secrecy for the USAF and all its secret experimentation using Brown's invention.

APPENDIX G

Derivation of the Abraham-Lorentz Formula

Presented in this section is the currently-accepted derivation of the controversial "accelerated charge emitting radiation in one direction" scenario. It was prepared by SUNRISE based on available advanced university textbooks on classical electrodynamics that discusses the formula, with appropriate changes to account for increasing frequency of the radiation on the recoiling force on the charge.

While scientists are still debating whether the exponential solution derived from the formula occurs in reality, this section is provided knowing the facts surrounding UFOs and the realisation that the observations are based on this solution. Hopefully this will be seen as an impetus for scientists to carry out the experiment needed to resolve this controversy.

Derivation of the Abraham-Lorentz Formula

1 Introduction

The development of an equation of motion for an accelerating charge began when Max Abraham (1875–1922) published in 1905 a formula under the title "Theorie der Elektrizitat" (Theory of Electricity) in Volume II of *Electromagnetosche Theorie der Strahlung* (Electromagnetic Theory of Radiation)[1]. When the formula is solved, a remarkable runaway (or exponential) accelerating solution is derived. Ever since the publication of the formula and its solution, it has been the subject of intense debate among scientists to this day.

For example, on first observing this runaway solution, Hendrik Antoon Lorentz (1853–1928) thought Abraham's approach to looking at a point charge, such as an electron, might have brought with it certain inherent problems, resulting in some kind of possible violation in the conservation of energy. So he derived from basic principles the formula for the general case of a charged spherical shell of any arbitrary radius. Despite his best efforts, the formula still revealed Abraham's results and the same disturbing solution. Again, Lorentz fiddled around with his equations, including setting more appropriate boundary conditions when integrating his equations. However, another problem arose when it appeared the solution to his formula would indicate pre-acceleration before any force was applied. In his mind, this simply made no sense at all. In the end he had to accept his own original formula was in essential agreement with Abraham's work. The work can be found in his book, *The Theory of Electrons*, first published in 1909.

Other scientists have looked at this conservation law argument and noticed something else. Professor David J. Griffiths and his colleagues at the department of physics at Reed College, Portland, Oregon, wrote in an

---

1. Abraham 1920, p.65. A scanned copy of the original 1905 book is available from https://archive.org/details/theoriederelekt04fpgoog/page/n14. An earlier version for the derivation of this Abraham-Lorentz formula[2] was also published in *Annalen der Physik* by Max Abraham in 1903 titled "Classical Theory of Radiating Electrons", and independently by Hendrik A. Lorentz (1853-1928) in 1904 published in the *Proceedings of the Academy of Sciences* under the title "Electromagnetic phenomena in a system moving with any velocity smaller than that of light".

article titled "Abraham-Lorentz versus Landau-Lifshitz" published in the *American Journal of Physics* in April 2010, that:

> . . . the conservation law argument is manifestly ambiguous because not all of the energy that leaves the particle eventually escapes in the form of radiation—some of it may be temporarily stored in the 'nearby' fields and later reabsorbed by the particle.[3]

In other words, there is the possibility that the charged particle is able to recycle some of its energy. Could this be the clue to how we should interpret the runaway solution?

To this day, scientists have not yet been able to give a proper explanation for this solution. Is this runaway solution a defect of classical electrodynamics? Or can it be put into practice through a real-life technology?

Whatever the truth, let us reveal what is currently the simplest approach to deriving the Abraham-Lorentz formula.

2 The Equation of Motion for a Point Charge

Imagine a single point charge (or tiny spherical object equivalent in size to, say, an electron) of charge Q and mass m resting in the vacuum of space. As there is no motion, there is absolutely no emission of radiation coming off the charged surface. Suddenly the object is made to accelerate in one direction by an initial external force given by the term F_X.

The general non-relativistic[4] classical equation of motion for a single point charge is written as:[5]

$$F_R = m \left(\frac{\mathrm{d}v}{\mathrm{d}t} - \tau \frac{\mathrm{d}^2 v}{\mathrm{d}x^2} \right) \tag{1}$$

where

$$\tau = \frac{Q^2}{6\pi\varepsilon_o c^3 m} \tag{2}$$

---

3. Griffiths et al 2010, p.392.
4. Meaning that the particle is moving much slower than the speed of light. Otherwise the particle's mass would increase sufficiently to affect the equation of motion in a drastic way.
5. Carati & Galgani, p.1; Dixon 2004, p.1; Kwang-Je & Sessler 1998, p.6; O'Connell 2003, p.2.

Breaking down this equation for the readers, we see it is composed of two parts:

$$F_T = F_X - F_R \tag{3}$$

where F_X is the external force designed to set off an initial accelerating motion, F_R is the force of the radiation on the charge, and F_T is the total resultant force on the object. All forces are measured in newtons (N).

The minus (-) sign could quite easily be a plus (+) sign. If it is the former, force of the radiation F_R is counteracting the external force F_X as if it could be an inertial force. However, if the charge is distributed over one end such that F_R reinforces the acceleration, then we have a runaway solution.

Another thing we should mention is that as the object accelerates, it will receive extra radiation at the front end from the surrounding environment. Naturally this would see some of the energy get absorbed. Now in order for the conservation of energy to reign supreme in this situation, the absorbed energy would have to be emitted. Otherwise, the object would get hotter and hotter over time as it accelerated. Therefore, if the object emits this energy, scientists assume this is part of the radiation that is helping with the recoiling force F_R.

However, for the purposes of deriving the Abraham-Lorentz formula in the simplest way possible, we will assume a perfect vacuum containing absolutely no electromagnetic energy. The only energy we need to worry about is what is coming off the charge to make it accelerate.

3 Deriving the Radiation Reaction Force

If it makes things easier to visualise the formula, imagine the object is at rest[6]. Suddenly an external force F_X from, say, your hand pushing on the object, is used to start the acceleration. We are not talking about a lot of force here. Just the slightest touch because you know the object is extremely lightweight enough for you to see the movement. If there is no

---

6. In a perfect vacuum, this is fine. In the real universe, this is not strictly true for an object at rest or moving at constant speed. In the real universe, the radiation in space can collide with the object and cause additional movements to take place at the atomic level for electrons making up the chemical bonds for holding the atoms together. Similarly, if the electrons emit radiation, they will recoil to push against the object. We know there is radiation being emitted by the electrons because this is how we get to see the object. As the collision and emission of radiation occurs right around the object, the object will vibrate ever so slightly in all directions. Thus the object has to be constantly accelerating. However, at a large scale, we cannot see this acceleration. And as we cannot see excess radiation being emitted in one direction to make it accelerate in one direction, we can effectively see the object as either not moving or moving at constant velocity.

charge on the object other than its mass m, then this means the above general equation of motion simplifies to the classical Newtonian equation of motion:

$$\boldsymbol{F}_X = m\boldsymbol{a} \tag{4}$$

Or if the acceleration varies over time in what scientists call non-uniform acceleration, resulting in a force that would potentially vary over time, then we would need to know what this is at any instant in time. Therefore, we could use:

$$\boldsymbol{F}_X = m\frac{\mathrm{d}\boldsymbol{v}}{\mathrm{d}t} \tag{5}$$

or

$$\boldsymbol{F}_X = m\frac{\mathrm{d}^2\boldsymbol{x}}{\mathrm{d}t^2} \tag{6}$$

depending on whether we are more interested in working with velocity or distances when calculating the force. To make life a little easier, the external force F_X on the object's mass will be kept constant in order to create a constant initial acceleration. Therefore, the classical Newtonian equation of motion can be kept to:

$$\boldsymbol{F}_X = m\boldsymbol{a} \tag{7}$$

Also, if we know the object is accelerating along the x-axis, we can drop the vector notation (i.e., bold letters) containing both direction and magnitude information and write only for the magnitude:

$$F_X = m.a \tag{8}$$

Now we are ready to move on to the second part.

For the sake of simplicity in the equations, imagine the external force F_X is suddenly removed while the object is accelerating. There is no need to maintain this external force because if the laws of classical electromagnetism are correct, then according to Alcaine and Llames-Estrada:

> "Accelerated particles carrying electric charge are a source of electromagnetic radiation. The momentum carried away by the radiation field affects the particles' classical motion, imparting a recoil force traditionally known as radiation reaction."[7]

Depending on how lightweight the accelerating object is, turning on the charge Q and emitting a strong enough momentum in the emitted radiation could be enough for the object to continue accelerating. If this is the reality, there is no need to maintain the external force. We will drop F_X.

7. Alcaine & Llames-Estrada 2013, p.033203-1.

3.1 The Problem with the Radiation Reaction Force

But as we will discover when focusing on the radiation reaciton force, solving the differential equation for velocity and acceleration reveals a runaway solution. This is because the energy density of the radiation increases over time with each successive emission as a result of the high acceleration reached by the charge. At first the acceleration will seem slow, but with each passing second, the acceleration apparently gets more dramatic as it continues to emit radiation. There is something about the way the acceleration of the charge creates an electromagnetic field of its own to influence the emitted radiation and so create a positive feedback loop.

Despite this reasonable explanation and basic observations of how electrons behave when accelerating (i.e., radiating energy), it is amazing to find considerable debate in the scientific literature as to how the energy is emitted from the point charge, what the charge is doing to influence the radiated energy and in which direction this energy comes off. For example, some scientists assume the energy is not emitted as a straight line, but rather as a spherical wave and, therefore, it isn't clear how the charge moves. For example, Richard Becker noted this problem in his publication *Electromagnetic Fields and Interactions*, Volume II:

> "...it is not at all clear how the emitted spherical wave influences the electron's motion."

In David J. Griffiths' 1987 publication of *Introduction to Electrodynamics*, he states in chapter 11 on page 461 that the energy is emitted in a torus (or doughnut) shape. As Griffiths said:

> "No power is radiated in the forward or backward direction —rather, it is emitted in a donut about the direction of instantaneous acceleration."

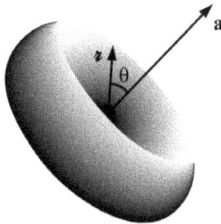

Figure 1: Emission of radiation from a spinning electron

Could these pictures of energy emission be more of a reflection of whether the point charge is rotating in all directions for a spherical emission of energy, or spinning around a particular axis for a doughnut emission of energy?

We will not be concerned about any form of rotation and spinning behaviours of the charged object. We know electrons can spin. However, our picture will be even simpler. For the purposes of deriving the Abraham-Lorentz formula in the simplest way possible, we will assume all energy emissions by radiation and any transfer of energy by an external force to the object to initiate acceleration or vice versa to decelerate it (e.g., resistance provided by the air) is strictly done along the x-axis. In other words, we are getting all the forces to be applied in one direction.

Now if the external force is suddenly removed and a charge is instantly initiated on the object while still in the accelerating phase, and the radiation can maintain the acceleration, our general equation of motion simplifies to:

$$\boldsymbol{F}_T = \boldsymbol{F}_R = m\left(\tau \frac{\mathrm{d}^2\boldsymbol{v}}{\mathrm{d}t^2}\right) \tag{9}$$

This is effectively the Abraham-Lorentz formula for a point charge emitting radiation as it accelerates.

But how did the scientists actually manage to work out this specific part of the radiation reaction force with its charge? Let us take a closer look.

3.2   Derivation of F_R

To derive the Abraham-Lorentz formula for a point charge when F_R is the only force acting on the charge, one must ensure the conservation of energy is always maintained. On looking at the scientific literature, this is achieved using a term called *power*.

Looking at the original general equation of motion, the total power W_T may be written as:

$$\boldsymbol{W}_T = \boldsymbol{W}_X + \boldsymbol{W}_R \tag{10}$$

where W_X is the power transferred by the external force to the object to help raise its kinetic energy, W_R is the power added to the object to raise its kinetic energy if the radiation is colliding with the object to make it accelerate, and W_T is the total power in the object. But since the radiation is coming off the charge to make the object recoil in the opposite direction, the sign must be changed to:

$$\boldsymbol{W}_T = \boldsymbol{W}_X - \boldsymbol{W}_R \tag{11}$$

If the object is uncharged, this results in a total power W_T transferred to the object of:

$$\boldsymbol{W}_T = \boldsymbol{W}_X = \boldsymbol{F}_X.v \tag{12}$$

If the object is charged and remains accelerating and the external force disappears, what we have left is the power W_R of the radiation, which is:

$$\boldsymbol{W}_T = -\boldsymbol{W}_R \tag{13}$$

Thus when an uncharged object at rest receives an external force F_X, or a charged object with mass receives a radiation reaction force F_R, the energy (or power, which is just another way of saying the flow of energy over time) is transferred to the object, thereby increasing its kinetic energy. Kinetic energy is electromagnetic energy that has bound up in the electrons making up the mass of the object. The electrons will vibrate and move around a little more in response to this added energy. Eventually this energy will get transferred again only when something else (e.g., another object) is able to take away some or all of this energy through a collision, or perhaps the energy will increase should the interaction with other objects see their own stored energy transferred and added to the overall kinetic energy of the original object.

As for the minus sign, this is just part of conserving the energy of the system. Of course, the power equation assumes the energy is never returned to the object, or the universal background radiation does not congregate around the charged object and somehow contribute energy in order to maintain charge and hence potentially sustain the process of emitting radiation. All equations described here are considered to be in a perfect vacuum containing absolutely no free electromagnetic energy in space except for the radiation coming off the accelerating charge. In reality, we all know this is untrue. The so-called vacuum of space above the Earth's atmosphere is filled with radiation. We effectively live in a sea of background radiation. This means there is the potential for some of this energy to contribute to maintaining the charge on the object and so help to emit radiation continuously as it accelerates (as occurs for electrons and protons since the experimental value of the charge for both particles does not seem to vary at all when radiation is emitted). However, given the way the equations are written here, we must assume the energy of this background radiation is non-existent. The only energy being transferred is thought to be from the external force to the object to start the acceleration, and what is emitted as radiation by the charged object during acceleration. No other energy sources are being considered. Also, the energy emitted by the accelerating charge in the form of radiation is assumed to disappear into the universe, never to be seen again.

Another thing to keep in mind is that if the radiation reaction force is emitted in a direction that helps to reduce the acceleration of the charged object by taking away some or all of the kinetic energy stored in the object after it has been accelerated by the external force, we would expect W_R to go down over time until the object stops moving. Then W_R should be zero since a stationary charged object does not radiate.

It makes reasonable sense so far.

However, something strange happens when we get the charged object to emit radiation in a direction that increases its power and moves faster and faster. If no air friction (or electromagnetic energy in space at relativistic speeds) can counteract this acceleration, the equation of motion describing this accelerating charge provides a disturbing solution, as we shall see. And it is all because of the fact that the charge Q does not vary during the radiation emission, as we see in electrons and protons.

3.3 Introducing the Larmor Formula

We have now reached a point where scientists have to put in the hard yards of working out a reasonable formula to represent the total power radiated by an accelerating charge. Afterwards, it will be integrated over time for velocity and acceleration to give us the radiation reaction formula.

In electrodynamics, scientists use Poynting's theorem as a statement of the energy conservation for the electromagnetic field. It is a partial differential equation showing the total power of a charge after it has absorbed and emitted electromagnetic energy. If the charge perfectly conserves the energy, total energy going in should equal total energy going out.

However, in the case of the accelerating point charge relevant to this discussion, this will be seen primarily as energy being emitted as radiation into space from the charge itself. The only question is: is this energy lost forever into space or could it be recycled by the charge itself?

In a highly simplistic format, the total power equation (also known as the Larmor Formula) is written as:

$$W_R = \int \mathbf{S}_R . dA \qquad (14)$$

The term $\mathbf{S}_R$ is named the Poynting vector in honour of the British physicist who created it—John Henry Poynting. On its own, $\mathbf{S}_R$ isn't all that enthralling. Fortunately, things do change once we substitute $\mathbf{S}_R$ with slightly more helpful terms. In particular, we can add the electric field E and the magnetic field B of the emitted radiation, like so:

$$W_R = \frac{1}{\mu_o} \int (\mathbf{E} \text{x} \mathbf{B}) . dA \qquad (15)$$

The next step is to work out the formulas for the fields of an accelerating charge. Now, the equation relating an electric field with a non-moving charge is given by:

$$\int E.dA = \frac{Q}{\varepsilon_o} \tag{16}$$

As the electric field emerges from the charge perpendicular to an imaginary surface of a sphere at any radius from the charge and the field strength is the same over this entire surface, E will be constant at a given radius, and the surface area of a sphere is well-known. As the area of a sphere is $4\pi r^2$, the electric field E becomes:

$$E = \frac{Q}{r^2} . \frac{1}{4\pi\varepsilon_o} \tag{17}$$

where r is the distance from the point charge to a position in space where you want to calculate the electric field.

When the charge is moving at a constant velocity v, the equation becomes (we won't bother with its derivation):

$$E = c.B = \frac{Q}{r^2} . \frac{1}{4\pi\varepsilon_o} .v.sin\theta \tag{18}$$

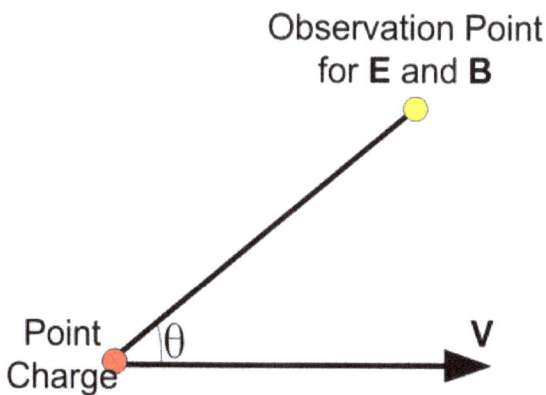

Figure 2: Calculating E and B at a point from a moving point charge

Note the presence of a magnetic field B. In the vacuum of space, B only differs from the electric field E by a constant, which is the speed of light c. Also, Θ is the angle from the point charge moving in one direction to a point in space where you are calculating the fields E or B.

And if the charge is accelerating, the equation becomes:

$$E = c.B = \frac{Q}{r} \cdot \frac{1}{4\pi\varepsilon_o} \cdot \frac{a.sin\theta}{c^2} \qquad (19)$$

where a is acceleration. It is this last equation that we must use to calculate the total power of the energy radiating from the point charge.

In the vacuum of space, the Poynting flux (or power density, or simply the power per unit area dA per unit time dt measured in erg s^{-1} cm^{-2}) in vector form is:

$$\mathbf{S}_R = \frac{\mathbf{E}\mathbf{x}\mathbf{B}}{\mu_o} \qquad (20)$$

However, a vector (shown here in **bold** letters) has additional information in terms of the direction as well as magnitude. As direction is a slightly more complicated thing, especially since we are only going to look at the radiation moving in one direction, it is not necessary to use vectors. A magnitude look at the above equation is sufficient, which is:

$$S_R = \frac{E.B}{\mu_o} \qquad (21)$$

Keeping in mind that electric field E is related to the magnetic field B in a directly proportional manner through the equation:

$$E = c.B \qquad (22)$$

and in a vacuum, the speed of light is a constant equal to:

$$c = \frac{1}{\sqrt{\mu_o\varepsilon_o}} \qquad (23)$$

we can re-write the power density equation as:

$$S_R = \frac{E^2}{\mu_o.c} = \varepsilon_o c E^2 = \left(\frac{Q^2}{r^2} \cdot \frac{1}{16\pi^2\varepsilon_o} \cdot \frac{a^2.sin^2\theta}{c^3} \right) \qquad (24)$$

Next, we have to integrate over the entire surface in all directions for dA. Using the sphere as the simplest solution, total power $(W_T = W_R)$

emitted becomes:

$$W_R = \int S_R.dA$$

$$= \frac{Q^2}{16\pi^2\epsilon_o c^3} \cdot \left(\frac{d\boldsymbol{v}}{dt}\right)^2 \int_{\phi=0}^{2\pi} \int_{\theta=0}^{\pi} \frac{sin^2\theta}{r^2} r sin\theta d\theta r d\phi$$

$$= \frac{Q^2}{16\pi^2\epsilon_o c^3} \cdot \left(\frac{d\boldsymbol{v}}{dt}\right)^2 \cdot (2\pi) \int_{\theta=0}^{\pi} sin^3\theta d\theta \qquad (25)$$

$$= \frac{Q^2}{8\pi\epsilon_o c^3} \cdot \left(\frac{d\boldsymbol{v}}{dt}\right)^2 \cdot \frac{4}{3}$$

$$= \frac{1}{6\pi\epsilon_o} \cdot \frac{Q^2 a^2}{c^3}$$

What we see here is the total power for an accelerating point charge, such as an electron, of charge Q radiating electromagnetic waves. Please note that c is the speed of light in free space, equal to exactly 299,792,458 metres per second (m/s), and ε_o is the permittivity constant of free space equal to $8.8541878176 \times 10^{-12}$ farads per metre (F/m).

3.4 Adding frequency and amplitude

One thing observed in practically all the available published articles showing the derivation of the Abraham-Lorentz formula is how it leaves out two important terms when calculating the total power: frequency and amplitude of the emitted radiation. The lack of such terms pobably explains why Professor Thomas Erber stated in his article titled "The Classical Theories of Radiation Reaction":

"...we may be missing something dynamically important in averaging."

For example, the flow of electromagnetic energy from a point (or spherical) charge through a unit area of an imaginary spherical surface per unit of time just outside the charge means the energy density over this area will vary over time. As the electromagnetic field oscillates, the energy swings from zero to a maximum amount in a periodic manner. However, as frequency increases, a stronger gravitational field is generated by the radiation according to the Unified Field Theory. As the gravitational field compresses the energy, we should expect to see the the energy density increase, and this in turn affects total power.

Amplitude is mainly suitable for low energy radiation such as radio waves. At high energy (or frequency) of the radiation, the energy is compressed to a small size and tends to be emitted by electrons. As electrons have very small sizes and a fixed charge, quantum theory focuses on the frequency and charge, but not on the amplitude. But if we ignore quantum considerations, amplitude is controlled primarily by the physical size of the charge itself (should the charge vary over time), or the amplitude of the sinusoidal motion of the charge (if the charg itselfe is constant). Once you know amplitude, the energy density of an EM wave is proportional to the square of the amplitude of the electric or magnetic field.

Now, energy density is related to the power density in a directly proportional manner. For radiation travelling in the vacuum of space, the power density has been found to be equal to:

$$S_R = c.u \tag{26}$$

where c is the speed of light, and u is the energy density.

In terms of the electric field E, this is related to energy density u, via:

$$u = \varepsilon_o E^2 \tag{27}$$

where ε_o is the permittivity constant of free space.

Or in terms of just the magnetic field, the formula can also be:

$$u = \frac{B^2}{\mu_o} \tag{28}$$

One way to add frequency and amplitude to the Abraham-Lorentz formula through the power equation is to consider a point charge moving side to side while accelerating linearly through space. The size of the sinusoidal motion gives an indication of amplitude A of the radiation, and how quickly the charge moves side-to-side gives an indication of the frequency f. As the charge moves from side-to-side, it will accelerate and decelerate over time. One can calculate this acceleration at any moment in time using the equation:

$$a(t) = \frac{d^2(x(t))}{dt^2} = -A\omega^2 cos(\omega t) \tag{29}$$

where ω is the frequency, A is the amplitude, and t is time.

Or in terms of distance travelled by the point charge:

$$x(t) = A.cos(\omega t) \tag{30}$$

530

Therefore, the radiated power from the charge associated with this additional accelerated motion can be calculated by using the Larmor formula to come up with the following power equation:

$$W_R(t) = \frac{1}{6\pi\epsilon_o} \cdot \frac{Q^2\left(-A\omega^2\cos(\omega t)\right)^2}{c^3} \tag{31}$$

$$\tag{32}$$

$$= \frac{1}{6\pi\epsilon_o} \cdot \frac{Q^2 A^2 \omega^4}{c^3} \cdot \cos^2(\omega t) \tag{33}$$

And if scientists want to get an average result to simplify the equations, this is done by taking the average of the maximum power reached at a certain time, and zero power at another time. The result is:

$$W_{R,avg} = \frac{1}{2} \cdot \frac{1}{6\pi\epsilon_o} \cdot \frac{Q^2 A^2 \omega^4}{c^3} \tag{34}$$

$$\tag{35}$$

$$= \frac{1}{12\pi\epsilon_o} \cdot \frac{Q^2 A^2 \omega^4}{c^3} \tag{36}$$

At any rate, the above formula shows how the radiated power increases with the frequency and amplitude.

However, a point charge wobbling side-to-side can be seen from a great distance as moving in a straight line. Any radiation emissions can be attributed entirely to the charge itself varying over time. In which case, amplitude can be considered effectively the radius of the charge.

At high frequency (or high energy) of radiation, this tends to be generated by electrons and, as such, amplitude is not considered mainly because the electrons are very small. However, let us imagine the radius is of a known and finite size and is constant. Refering to Figure 3 we can see how energy density varies based on the size of the charged plate. For the same amount of charge injected on the two plates, the smaller plate will have a higher energy density. This in turn generates a higher total power output and, therefore, a stronger recoiling force of the emitted radiation. Thus, if we include amplitude in terms of a radius of a charged plate, then:

$$A = \frac{1}{r} \tag{37}$$

Maximum power reached then becomes:

$$W_{R,avg} = \frac{1}{12\pi\epsilon_o} \cdot \frac{Q^2 \omega^4}{r^2 c^3} \tag{38}$$

531

However, as radius of the charge reduces, or charge Q increases, or frequency ω increases, another factor to affect the energy density and with it the power output is the presence of the gravitational field in radiation. Albert Einstein's Unified Field Theory indicates that as the energy density increases with frequency, the gravitational field generated by the electromagnetic field will increase. This will have the effect of compressing the radiated energy and so raising energy density. At some point, it is possible for the energy to bend back on itself. As this may seem to reduce the radiated power output at a sufficient distance, in terms of recoiling the charge, it could have the opposite effect of increasing the recoiling force (i.e., the power output will appear to increase very close to the surface of the charge). This factor needs to be considered as we approach high energy density of the radiation.

3.5 Calculating the Radiation Reaction Force

If we accept the above total power equation for an accelerating charge using the laws of averaging to create a constant energy density in a moment in time and not including frequency or amplitude, it is time to convert total power to force F_R, so later we can see the velocity and acceleration of the charge when the radiation is emitted.

Just a few more things we need to bear in mind. The way the above equation is written assumes the mass of the object plays no part in the motion of the point charge. Only the charge matters here (and how it affects the charge's acceleration). Either it is presumed there is no mass, or any mass present is not being converted to energy via a nuclear reaction to help contribute to the total power being radiated. We are only looking at the kinetic energy of the object stored in the accelerating motions of the electrons within the chemical bonds, and those that orbit the atoms, and certainly not the energy stored in the nuclei or composing the atomic particles themselves. Also, this equation is only valid when the velocity of the object is non-relativistic. In other words, we shall assume the velocity of the object is much less than the speed of light. Additionally, the power equation assumes no variation over time for acceleration. Of course, we all know the charged object's acceleration does change over time during the emission of radiation. Therefore, the Larmor formula would have to be integrated like so:

$$\int_{t_1}^{t_2} \boldsymbol{F}_R.\boldsymbol{v}dt = -\frac{1}{4\pi\varepsilon_o}.\frac{2Q^2}{3c^3}\int_{t_1}^{t_2} a^2 dt \tag{39}$$

where F_R (and hence F_T), v and a are the terms expected to change over time.

When integrating, Lorentz proposed that we integrate this equation over an integral number of full cycles of the radiation, say, from $t_1=0$ seconds to $t_2=2\pi/\omega$ (where ω is the frequency). Indeed, this suggestion turns out to be useful when dealing with the integration-by-parts section to help finish off the derivation of the formula.

Thus, continuing with the integration:

$$\int_{t_1}^{t_2} \boldsymbol{F}_R.\boldsymbol{v}dt = -\frac{Q^2}{6\pi\varepsilon_o c^3} \int_{t_1}^{t_2} \boldsymbol{a}.\boldsymbol{a}dt \tag{40}$$

$$= -\frac{Q^2}{6\pi\varepsilon_o c^3} \int_{t_1}^{t_2} \left(\frac{\mathrm{d}\boldsymbol{v}}{\mathrm{d}t}\right).\left(\frac{\mathrm{d}\boldsymbol{v}}{\mathrm{d}t}\right)dt$$

and integrating by parts, we get:

$$\int_{t_1}^{t_2} \boldsymbol{F}_R.\boldsymbol{v}dt = -\frac{Q^2}{6\pi\varepsilon_o c^3} \left(\left(\boldsymbol{v}.\frac{d\boldsymbol{v}}{dt}\right)\Big|_{t_1}^{t_2} - \int_{t_1}^{t_2} \left(\frac{\mathrm{d}^2\boldsymbol{v}}{\mathrm{d}t^2}\right).\boldsymbol{v}dt\right) \tag{41}$$

Noting that the velocities and accelerations are the same at t_1 and t_2 after following Lorentz's recommendation, the boundary term drops out, and the equation can be written equivalently as:

$$\int_{t_1}^{t_2} \boldsymbol{F}_R.\boldsymbol{v}dt = -\frac{Q^2}{6\pi\varepsilon_o c^3} \int_{t_1}^{t_2} \left(\frac{\mathrm{d}^2\boldsymbol{v}}{\mathrm{d}t^2}\right).\boldsymbol{v}dt \tag{42}$$

or

$$\int_{t_1}^{t_2} \left(\boldsymbol{F}_R - \frac{Q^2}{6\pi\varepsilon_o c^3}\left(\frac{\mathrm{d}^2\boldsymbol{v}}{\mathrm{d}t^2}\right)\right).\boldsymbol{v}dt = 0 \tag{43}$$

On integrating, we can identify the radiation reaction force as:

$$\boldsymbol{F}_R = \frac{Q^2}{6\pi\varepsilon_o c^3}.\frac{\mathrm{d}^2\boldsymbol{v}}{\mathrm{d}t^2} \tag{44}$$

This is the Abraham-Lorentz formula for the radiation reaction force for a charged object emitting radiation in one direction in a perfect vacuum.

If we include the frequency ω and amplitude of the radiation, where amplitude is effectively the radius r of a circular charged plate emitting the radiation, the formula will look like so:

$$\boldsymbol{F}_R = \frac{Q^2\omega^4}{6\pi\varepsilon_o c^3 r^2}.\frac{\mathrm{d}^2\boldsymbol{v}}{\mathrm{d}t^2} \tag{45}$$

Here we assume an average radiated power is taking place, so no oscillations in the energy density to worry about here, even though some kind of oscillation is required to emit radiation. We will assume the charge achieves this by some means while we observe from a great distance what happens to a charge that emits radiation.

3.6 Same result occurrs when charge is of a fixed radius

Due to the runaway solution obtained after solving this differential equation for velocity and acceleration, Lorentz chose a spherical shell of radius R for the charged object as the starting point. Still, Abraham's formula could be seen lurking in Lorentz's more general formula.

Again, we must stress that the Abraham-Lorentz formula is valid and consistent with the conservation of energy. There appears to be nothing inherently wrong with the formula. The problem is clearly in the runaway solution, how to interpret it, and whether the solution can actually work in reality.

4   Where to Go from Here?

Extensive debate continues as to whether this formula is valid at all times for both non-uniform and uniform acceleration or just non-uniform acceleration[8]. Some scientists are even prepared to deny the existence of a point charge in classical electrodynamics to avoid this unpleasant runaway solution. For example, if the radius of a spherical charged shell approaches zero to become an infinitely small point charge, we effectively have no mass. And charge Q cannot exist without mass (as a case in point, only the electron and proton have a charge, and both just so happen to carry mass). So the formula collapses, and no meaningful solution can be obtained from it. Furthermore, even with a very small radius (but not zero), the field is said to be very strong because the mass is confined to an incredibly tiny volume (i.e., trying to calculate the mass/energy density gets ridiculously large), making it too difficult to analyse what is going on. But even if the radius R is not zero, the runaway solution cannot be avoided no matter what scientists do.

So what about for a flat plane?

In 1997, W. N. Hugrass of the Department of Physics at the University of New England in Armidale, NSW, saw the difficulties of using a point charge. So he decided to view the field of the charge as a "superposition of plane waves. . . which propagate away from the charge at the speed of light". Despite this innovative approach not applied by other scientists in the past, he summarised the results of his work as:

---

8. The presumed uniform acceleration being referred to here probably relate to the moment when the external force has been applied to the charge. Otherwise, the radiation with its periodic motion will not produce a uniform acceleration of the charged object as it exerts a recoiling force on the charge. Whatever the truth, there is no general agreement in the scientific literature about whether a uniformly accelerated charge radiates in classical electrodynamics.

"The force acting on the charge due to this [plane wave] field is the well-known Abraham-Lorentz radiation reaction [formula]."[9]

Ultimately it does not matter what scientists try to do to avoid the runaway solution. Assuming classical electrodynamics is correct in saying an accelerating charge radiates, there is nothing anyone can do but to either accept the formula as a remarkable quirk in the laws of electromagnetism, or somehow find clever mathematical or personal ways to "sweep the formula under the rug", so to speak, as a means of ignoring it.

Or maybe the real question we should be asking is, can this solution occur in reality?

It seems the only way this debate can be resolved is to perform an experiment using an appropriate electrical device that will emit the radiation in one direction, and an electrical generator to maintain charge. For if such a device exists and the charge can be maintained during an emission of a sufficiently high-frequency radiation causing a perceptible accelerating motion, it will permit scientists the opportunity to directly observe whether there is an exponential solution in action or not. Looking at the UFO reports, the former is suggested. In other words, as incredible as the solution may seem, there is inherently nothing wrong with it. It is just that scientists cannot believe what they are seeing.

---

9. Hugrass 1997, p.815. He currently works in the Department of Engineering at the University of Tasmania.

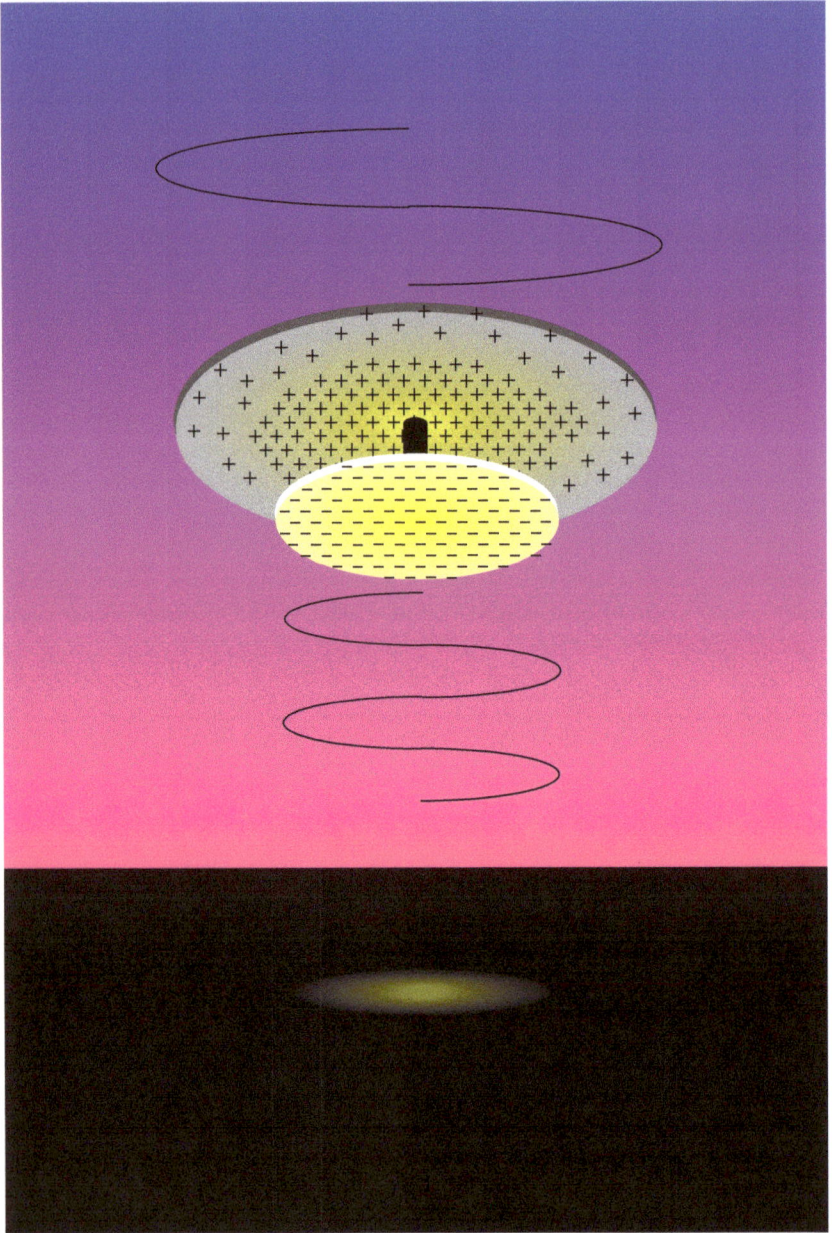

Figure 3: Charge distribution and density for different sized charged plates

APPENDIX H

Letter to Julius Schaefer by General Nathan Twining 17 July 1947

T his letter obtained under the U.S. Freedom of Information (FoI) was allegedly written by General Nathan Farragut Twining (1897–1982), head of the Air Force Matèriel Command based at Wright–Patterson AAF in Dayton, Ohio, USA, in 1947, to a Mr Julius Earl Schaefer (1893–1974) of the Wichita Division of Boeing Airplane Company, on 17 July 1947.

The USAF latter officially explained the visit by Twining (thereby confirming this letter is genuine) as a "routine inspection" and allegedly involved attending a Bomb Commanders' course on 8 July 1947. Apparently, it would be the most important course Twining had ever attended in his life.

BOEING AIRPLANE COMPANY

WICHITA DIVISION

WICHITA, KANSAS

17 July 1947

Dear Earl:

I have received your letter in which you asked us to drop by at Wichita for a brief visit. With deepest regrets we had to cancel our trip to the Boeing factory due to a very important and sudden matter that developed here. All of us were considerably disappointed as Mr. Allen had planned a very fine trip for us; however, we hope to go out at a later time. Will remember your invitation and get out to see you just as soon as we can, as I am very anxious to see the XL-15.

I have been away quite a bit the last couple of weeks so have not had a chance to submit any information to you that you asked for in your round robin letter. I will get on this very shortly.

Best regards,

N. F. TWINING
Lieutenant General, U.S.A.

P.S. Unification looks
like a sure thing now.

Mr. J. E. Schaefer
Boeing Airplane Co.,
Wichita, Kansas

538

BOEING AIRPLANE COMPANY

WICHITA DIVISION
WICHITA, KANSAS

JULY 17, 1947

Dear Earl:

    I have received your letter in which you asked us to drop by at Wichita for a brief visit. With deepest regrets we had to cancel our trip to the Boeing factory due to a very important and sudden matter that developed here. All of us were considerably disappointed as Mr Allen had planned a very fine trip for us; however, we hope to go out at a later time. Will remember your invitation and get out to see you just as soon as we can, as I am very anxious to see the XL-15.

    I have been busy quite a bit the last couple of weeks so have not had a chance to submit any information to you that you asked for in your round robin letter. I will get on this very shortly.

                         Best regards,

                         N. F. TWINING
                         Lieutenant General,
                         U.S.A.

P.S. Unification looks
like a sure thing now.

Mr J. E. Schaefer
Boeing Airplane Col,
Michita, Kansas

APPENDIX I

Memo from
General Nathan Twining
15 July 1947

The secret memo shown in the next three pages (a cleaner version accompanies the original) released under the U.S. Freedom of Information (FoI) was written by General Nathan Farragut Twining (1897–1982). This memo appears to show the real reason for his decision to urgently attend a matter of great importance in New Mexico.

The memo suggests that at least one and possibly two discs were recovered in New Mexico and a preliminary examination of the main wreckage revealing a disc-shaped object reveals various unusual internal components that appear to serve either a previously unknown electromagnetic function or reveals electromagnetic side-effect associated with whatever unknown process the disc operated during flight.

HEADQUARTERS ARMY AIR FORCES

Proving Grounds, New Mexico

Army Air Material Command, Wright Field, Ohio, 15 July 1947

TO: Commanding General, Army Air Forces, Washington 25, D.C.

Re: AIR DEFENSE COMMAND

FOR THE COMMANDING GENERAL

Proving Ground, New Mexico

15 July 1947

TO: Commanding General, Army Air Forces, Washington 25, D.C.,
AIR DEFENSE COMM.

FOR THE COMMANDING GENERAL

543

Proving Ground, New Mexico

Hq. Air Materiel Command, Wright Field, Ohio, 15 July 1947

TO: Commanding General, Army Air Forces, Washington 25, D.C.

Forwarded for your information.

FOR THE COMMANDING GENERAL:

HEADQUARTERS ARMY AIR FORCE

Air Accident Report on Flying Disc Aircraft that
Crashed near White Sands Proving Ground, New Mexico
D333.5   ID   15 Jul 47
Hl., Air Material Command, Wright Field, Ohio,
15 July 1947

To: Commanding General, Army Air Forces,
      Washington 25, D.C.
      HQ. AIR DEFENSE COMMAND
      ATTN: AC/15-2

      Forwarded for your information

      FOR THE COMMANDING GENERAL

                        N. F. TWINING
                        Lieutenant General, U.S.A.
                        Commanding

1. As ordered by Presidential Directive, dated 9 July 1947, a preliminary investigation of a recovered "Flying Disc" and remains of a possible second disc, was conducted by the senior staff of this command. The data furnished in this report was provided by the engineer staff personnel of T-2 and Aircraft Laboratory, Engineering Division T-3. Additional data was supplied by the scientific personnel of the Jet Propulsion Laboratory, CIT and the Army Air Forces Scientific Advisory Group, headed by Dr Theodore von Karman. Further analysis was conducted by personnel from Research and Development.

2. It is the collective view of this investigation body, that the aircraft recovered by the Army and Air Force units near Victorio Peak and Socorro, New Mexico, are not of U.S. manufacture for the following reasons:

    a. The circular, disc-shaped "planform" design does not resemble any design currently under development by this command nor of any Navy project.

    b. The lack of any material propulsion system, power plant, intake, exhaust either for propeller or jet propulsion, warrants this view.

    c. The inability of the German scientists from Fort Bliss and White Sands Proving Ground to make a positive identification of a secret German V weapon out of these discs. Though the possibility that the Russian have managed to develop such aircraft, remains. The lack of any markings, ID numbers or instructions in Cyrillic, has placed serious doubt in the minds of many, that the objects recovered are not of Russian manufacture either.

    d. Upon examination of the interior of the craft, a compartment exhibiting a possible atomic engine was discovered. At least this is the opinion of Dr Oppenheimer and Dr von Karman. A possibility exists that part of the craft itself comprises the propulsion system, thus allowing the reactor to function as a heat exchanger and permitting the storage of energy into a substance for later use. This may allow the converting of mass into energy, unlike the release of energy of our atomic bombs. The description of the power room is as follows:

(1) A doughnut shaped tube approximately thirty-five feet in diameter, made of what appears to be a plastic material, surrounding a central core (see sketch in TAB1). This tube was translucent, approximately ____ inch thick. This tube appeared to be filled with a clear substance, possibly a heavy water. A large rod centered inside the tube, was wrapped in a coil of what appears to be of copper material, run through the circumference of the tube. This may be the reactor control mechanism or a storage battery. There were no moving parts discernible within the power room nor in

_____.

(2) This motivation of a electrical potential is believed to be the primary power to the reactor, though it is only a theory at present. Just how a heavy water reactor functions in this environment is unknown.

(3) Underneath the power plant, was discovered a ball-turret, approximately ten feet in diameter. This turret was encompassed by a series of gears that has a unusual ratio not known by any of our engineers. On the underside of the turret were four circular cavities, coated with some smooth material not identified. These cavities are symmetrical but seem to be movable. Just how is not known. The movement of the turret coincides with the dome-shaped copula compartment above the power room. It is believed that the main propulsion system is a bladeless turbine, similar to current development now underway at AMC and the Mogul Project. A possible theory was devised by Dr August Steinhoff (a Paperclip scientist), Dr Wertner von Braun and Dr Theodore van Karman: as the craft moves through the air, it somehow draws the oxygen from the atmosphere and by a induction process, generates a atomic fusion reaction (see TAB 2). The air outside the craft would be ionized, thus propelling the craft forward. Coupled with the circular air foil for lift, the craft would presumably have an unlimited range and air speed. This may account for the reported

absence of any noise and the apparent blue flame often associated with rapid acceleration.

(4) On the deck of the power room there are what resembles typewriter keys, possibly reactor/powerplant controls. There were no conventional electronics nor wiring to be seen connecting these controls to the propulsion turret.

    e. There is a flight deck located inside the copula section. It is round and domed at the top. The absence of canopy, observation windows/blisters, or any optical projection, lends support to the opinion that this craft is either guided by remote viewing or is remotely controlled.

(1) A semi-circular photo-tube array (possibly television).

(2) Crew compartments were hermetically sealed via a solidification process.

(3) No weld marks, rivets or soldered joints.

(4) Craft components appear to be molded and pressed into a perfect fit.

Origins of the Twining Memo

Not fully endorsed by the USAF, this incomplete three-page document purported to be from General Nathan F. Twining (1897-1982) provides details of a preliminary examination of a disc recovered in New Mexico; the memo was found in amongst a mixture of authentic and forged documents known collectively as the Majestic 12 (MJ-12) documents.[1]

UFO researcher Timothy Cooper, whose father worked at White Sands Missile Range in the 1940s, was the one to receive this document among various others in 1992 when he received one of two packages—the second and last parcel arrived in July 2001—and marked as having been sent from Virginia by an anonymous source. In 2002, Cooper ended his private work as a dedicated UFO researcher and sold the documents to the father and son team of Ryan and Robert Wood, and later entrusted to Stanton Friedman. The Twining document shown here was first published in 1994 by Leonard H. Stringfield (1920–1994) in *UFO Crash Retrievals: Status Report VII*.

Whilst various documents have appeared anonymously since 1984 to different UFO investigators and researchers and all described as MJ-12 documents, this Twining document is considered by many to be genuine. Indeed, neither Twining's son, Nathan Alexander Twining, Jr. (1933–2016) of Baltimore, nor his daughter, Olivia B. Twining[2] (1935–2017) of San Antonio, Texas, have categorically denounced the document as fake.[3] They believed, in fact, that the document supported claims made by their father that something artificial and foreign did crash in the New Mexico desert together with bodies and these were closely examined at Wright-Patterson AFB before being moved to another undisclosed location. The general made this confession only 6 weeks before his death.

---

1   It is believed this tactic was employed by someone within the U.S. government out of concern for information leaks; the logic goes that in order to confound the general public (only those "in the know" could discern the difference), a combination of genuine and fabricated documents were deliberately filed together. This tactic assures that UFO researchers who receive some of these documents and publishes them would be susceptible to criticism and ridicule, and the discovery of a single dubious document could unleash the charge that the entire package of MJ-12 documents was fraudulent.

2   Married as Mrs Haywood S. Hansell III.

3   His second son is retired Air Force Major Richard Grant Twining (1927–2021) of Clearwater Beach, Florida.

Furthermore, an examination of the document's construction showed no obvious signs of forgery using modern means of reproduction; the general consensus is that the typewriter used to create this document is consistent with state of the art technology used by the military in 1947.

Olivia Twining subsequently produced additional documents and memos received or produced by her father during the same time period; these were later declassified by the Library of Congress Manuscript Division and made available to the public via the Freedom of Information Act. Included in the package is a copy of the official letter from Twining in July 1947 to Mr. Julius Earl Schaefer of the Boeing Division claiming that the general was forced on short notice to change his travel plans to accommodate an urgent visit to New Mexico. Stanton Friedman was the first to inspect the material upon its declassification.

APPENDIX J

*Revealing Three UFO
Cases in Detail*

Triangular UFO Abducts U.S. Navy Jets, Puerto Rico

Following a wave of UFO sightings and close encounters that seemed to have begun immediately after an inexplicable 1987 underground explosion in the southwest of the island of Puerto Rico, more than a hundred witnesses in and around Lajas and Cabo Rojo came forward to corroborate evidence for a particular sighting many will never forget.

On the night of 28 December 1988, at approximately 7.45 p.m., witnesses observed a huge triangular-shaped UFO moving silently across the sky in the presence of U.S. Navy military jets flying in the area. Wilson Sosa, one of the witnesses, explained what happened that evening:

"Starting at 6.00 p.m., we saw jet fighters flying over the area. At 7.45 p.m. or so, we heard some other planes that were either from the Puerto Rico National Air Guard or from the U.S. Navy. Even though they were very high you could still clearly hear their engines. I was paying close attention to their fly-over because about a week before, another one of those jets, an F-14 or F-15, chased another UFO—a small one—over the Sierra Bermeja [a small mountain ridge] and the Laguna Cartagena [a lagoon]. I came out to watch them and then saw a big UFO flying over the Sierra Bermeja. It was enormous! It was blinking with many coloured lights. I ran and got my binoculars and could then clearly see that it was triangle-shaped and slightly curved at its rear side.

It made a turn back and then came over lower, appearing much larger. It was then that we noticed two jet fighters right behind it. Then, when the UFO went [to the west], one of the planes tried to intercept it and passed in front of it, at which point the UFO veered to the left and made a turn back, reducing its speed. The jets tried to intercept it three times, and that's when the UFO slowed down and almost stopped in mid-air. It was incredible! How something that big could remain almost motionless was unbelievable. Considering its size, it must have been very heavy.

The second jet remained at the right side of the UFO and the other one positioned itself at the left rear side. Then—I don't know exactly what happened—if the jet entered the UFO by the rear, by its upper side, or what. That was when we all yelled, because we were afraid there would be a collision and maybe an explosion. The jet in the back just disappeared on top or inside of the UFO, because I was observing everything through my binoculars and it didn't come out from the rear, the upper side or the other sides.

The second jet remained very close to the right side of the UFO. It looked very small alongside that huge thing. As the UFO flew a little to the west, the jet disappeared, as well as its engine sound. This was exactly what happened when the first jet seemed to disappear inside the UFO.

That UFO was huge! I tell you that ship was bigger than this community's baseball park. You could observe its grey metallic structure and great central yellow light that was being emitted from a big/bulging luminous circular concave appendage. At the triangle's right wing tip it had brilliant yellow lights, and on the left side it had red ones.

After 'trapping' the jets, the UFO lowered its position and came very close to the ground [over the small Saman Lake]. It stood still in mid-air for a moment then straightened its corners and gave off a big flash of light from the central ball of yellow light. It then divided itself in the middle into two separate and distinct triangular sections. It was just incredible! The triangle to the right was illuminated in yellow, and the other one in red. That's when they both shot away at great speed, one to the southeast and the other one toward the northeast, in the direction of Monte del Estado. You could see red sparks falling from it when it divided itself."[1]

Ivan Cote, a young resident of Lajas, gave a comparable report and further commented that after the jets were abducted, "another jet came, but it flew away, apparently because they [the pilots] saw what happened to the other two jets, and got lost in some clouds while the smaller UFOs with red lights were chasing it. That is all I saw. My grandmother, Josefina Polanco, saw it all too, because I called her out to watch."

Mrs J. Polanco was interviewed and could verify Ivan's account.

Another witness to the UFO scene, Mrs Eduviges Olmeda, recalled that "it was like something out of science fiction. From our balcony we could see everything. Those planes were circling and getting in front of that thing with the big yellow light....It was beautiful....And suddenly it stopped and the planes seemed to disappear inside it."

Her husband, Edwin Olmeda, said:

"That certainly was a UFO, and it was really big. It was glowing with a large yellow light and didn't make a sound as it flew over the area, but the jets did. You know, we recently moved to this place, and this is the second occasion that we have seen UFOs over the area. The first occasion was about three weeks ago... and it was like a flying saucer, but shimmering with light. Something is going on around here."

---

1   All quotes for this UFO case from Good 1990, pp.192–204.

State section director for the Mutual UFO Network and the director of the UFO Information Center in Puerto Rico, Jorge Martin, was immediately informed of the incident by several witnesses. Martin then contacted the Federal Aviation Administration (FAA) office in Isla Verde the following day to gather information about the incident. The supervisor of the FAA, Ed Purcell, said he was unaware of any report of UFOs flying in the area at the stated time, except that "some military movement down in the south-west region in Cabo Rojo and operations manoeuvres were being made by personnel, apparently from the Roosevelt Naval Base in Ceiba".

Inquiries to the Puerto Rico Air National Guard base in Muñiz, Isla Verde, resulted in the same answer as the FAA's, but they added that none of their aircraft were aloft on the night in question.

Representatives of FURA (in English this stands for United Forces of Fast Action), a radar unit set up by the police force to stamp out drug smuggling in the region, said they knew nothing about such an incident but detected a large number of planes overflying at low altitudes in the western zone in regions not normally covered by ordinary domestic or commercial aircraft. This information was later corroborated by an unnamed high-ranking officer of the Criminal Investigations Corps of the Puerto Rico Police Department.

With only the Roosevelt Roads U.S. naval base left to answer his inquires, Jorge Martin hoped that the personnel there would be able to shed some light on the incident. However, Martin was surprised to hear from the supervisor of Air Operations that "we did not have any personnel in that area yesterday or last night. Whoever said that is mistaken".

The FAA was contacted once again but assured Martin that "there were air exercises in that area in Cabo Rojo last night. Wednesdays are the official days for exercises in that sector, and the Administration is officially notified as such". The FAA could not give an explanation as to why the Roosevelt Roads base denied any involvement of jet activity in the area concerned.

Jorge Martin again contacted the FAA at the end of the day. This time another officer, who wished not to disclose his name, admitted, "I don't know anything about the incident, but even so, whenever there is a UFO report we are not allowed to investigate it, since investigation is done by a special division of the FAA, based in Washington, D.C."

There are unmistakable inconsistencies in the statements made by these authorities, as if they're not being informed or are not telling the truth. Whatever the case, there seem to be attempts to whitewash the whole incident.

Without any knowledge of the incident, Carlos Rocafort, supervisor of Air Operations at the El Mani Airport in Mayagüez, near Cabo Rojo, told Martin and others who accompanied him the previous night at 8.30 p.m. (a mere 45 minutes after the UFO incident), four individuals arrived at the airport in a Cessna military aircraft, and it was rumoured that they were going to investigate "something important that had happened in the area".

Apparently, just ten minutes earlier, at around 8.20 p.m., the area where the incident took place was under the scrutiny of black military helicopters. This search operation was reported by Aristides Medina, a retired U.S. Army veteran. He asserts that:

> "At about 8.20 p.m., a bunch of black helicopters arrived, and for hours flew over the Sierra Bermeja and Laguna Cartagena areas without lights, until about midnight. They seemed to be searching for something in the area. Apparently they did not want to be seen and were flying low. Maybe they were looking for the planes that the UFO took away, or for some kind of trace. Apparently they were equipped with infra-red detectors, and this might explain why they were not using any lights in the search."

At daybreak, on 29 December 1988, five U.S. Navy ships and one aircraft carrier were stationed some 15 miles off the coast of Cayo Margarita, Puerto Rico. The area had been the scene of many UFOs emerging and re-entering the ocean, according to fishermen as well as commercial and private pilots.

Diego Segarra, a fisherman with twenty years of experience, told investigators:

> "Shortly before this happened, that aircraft carrier was stationed there for about a month and the area was restricted. The reason was that four mysterious huge metallic tubes apparently protruding from the bottom of the sea appeared in that area, where they had not been before, and just where those

flying saucers are seen entering the water. They can tell us they are just manoeuvres, but to those of us who have seen all that is going on here, the truth is that they are searching and following up the UFO sightings there.

On many nights, when we are fishing late at night, we can see the military overfly the area with huge aircraft equipped with searchlights, looking for something under water. On other occasions, a big AWACS plane has been seen flying over this area as well as the Laguna Cartagena and Sierra Bermeja, at low altitude, escorted by fighters. They are looking for something down there, and we think that 'something' is probably a UFO base. After all we have seen here, that is the most plausible explanation for what is happening."

Aristides Medina, the U.S. Army veteran, said:

"Something abnormal is going on down here. Anyone in his right mind can see what all these events indicate. For some reason, the authorities don't want the people to know what is going on; the mysterious underground explosions that are felt here, and the UFOs that are constantly seen. I have seen them myself. On one occasion, I was fishing late at night near Cayo Margarita, and two of them passed under my boat, radiating a blue light. On other occasions, I have seen them when they emerge from the water and fly away at great speed, and I have also seen them plunge into the water—always in the same area where the Navy ships are now."

Jorge Martin managed to get in touch with a navy officer in Puerto Rico a week after the incident. The officer—name withheld—provided some interesting information. "There are radar tapes that show what happened," the officer claimed, "and they were classified at once and sent to Washington, D.C. to be analysed. We were able to see what happened on the radar systems of the ships that were stationed nearby. We saw when the smaller targets on the radar, which represented the jets, merged with a bigger one [the UFO]. After that, the big target seemed to split, and shot off at great speed. A lid has been placed on the whole incident. Many things like this have been happening, but we are not allowed to comment on anything we see. Many strange things are happening in the waters of Puerto Rico that should be known."

The local Puerto Rican government has since "leased" the area at Sierra Bermeja and Laguna Cartagena to the U.S. federal government. Why? Authorities claim they are organising a Voice of America radio station in the Sierra Bermeja region. But many investigators and residents of the area are pretty convinced that the U.S. military is there to watch over the presence and activities of UFOs.

While the officialdom may deny the existence of the UFO to the public, one thing is certain: the names of the missing pilots have been kept secret to this day.

McMinnville UFO, United States[2]

On 11 May 1950, about 16 kilometres southwest of McMinnville, Oregon, Mrs Paul Trent was feeding her rabbits in the backyard of her farmhouse at about 7.45 p.m. when a bright and silvery disk-shaped object appeared in the overcast sky to the northeast, moving westward. She called her husband, who was inside the house. Mr Trent viewed the unusual craft and decided to run for his camera, which he thought he had left in his car, but Mrs Trent remembered that he had left it in the house and hurried to fetch it. The film of the camera had about five frames left.

The noiseless craft tipped to one side as it approached, as if to slow down or change direction. A dark underside and a slight "breeze" were noted by both witnesses. After receiving the camera from his wife, Mr Trent took a picture of the tilted craft and wound on ready for the next frame. He kept on looking at the object through the viewfinder and moved his position a couple of metres to the right. During this time, the object stabilised in a horizontal position facing side-on to Mr Trent, who then took a second picture some 30 seconds after the first. It glided off westward and "dimly vanished" without making any sound. Mr Trent estimated the flying saucer to be six to nine metres across.

Strangely, they did not consider their sighting as important enough to have their film developed immediately—they assumed the UFO was probably a secret military aircraft. Instead, Mr Trent waited a few days to use up the remaining frames and then had the film developed locally.

---

2   Details (including quotes) of the McMinnville and the Antonio Villas Boas UFO cases can be found in Brookesmith 1980–1984 and other sources.

He mentioned the incident to only a few friends, telling them he was not the type to seek publicity and didn't want to be "in trouble with the government". However, a reporter from a local newspaper heard of the incident from two of Mr Trent's friends. He followed it up and found the negatives on the floor of the Trents' house, under a writing desk where the Trent children had been playing with them!

Today, the two photographs are famous, for no amount of careful analysis of the pictures has revealed any signs of trickery on the part of the witnesses or anyone else. They have been scientifically examined using computer-enhanced and colour-coding techniques, to name just a few, and they are considered by scientists and by those who participated in the analysis to be genuine. Examinations of the Trent photos have confirmed that the object was about 1,300 metres from the camera and had a diameter of about 20 to 30 metres.

Dr William K. Hartmann said, in the conclusion of his detailed analysis of the McMinnville photographs:

"This is one of the few UFO reports in which all factors investigated, geometric, psychological, and physical appear to be consistent with the assertion that an extraordinary flying object, silvery, metallic, disk-shaped, tens of metres in diameter and evidently artificial…flew within sight of two witnesses."

However, further examination of the photos has also revealed evidence that seems to contradict the witnesses' account of some important details. Shadows caused by the overhanging roof edge on the east wall of the Trents' garage suggest the photographs were taken at about 7.30 in the morning, not 7.45 in the evening as the Trents claimed. This expected time was confirmed from local weather records for that day stating that while the sky had been absolutely clear of mist and clouds in the evening, it had been smoky in the morning, as is clearly evident in the photographs.

Apart from this discrepancy, the evidence presently available indicates that there was no sign of deception (intentionally or otherwise) by either of the witnesses—it seemed as if the Trents had not realised the importance of their sighting. Furthermore, additional evidence has been found giving credence to the McMinnville UFO.

In March 1954, a French military pilot photographed a UFO from an aircraft near Rouen, France. The pictures have been analysed and have been described as "one of the few [photographs] which seem authentic". The striking similarity with respect to the shape of these UFOs in both cases is remarkable!

Abduction Case of Antonio Villas Boas, Brazil

On 5 October 1957, near the town of São Francisco de Salles in the Brazilian state of Minas Gerais, a twenty-three-year-old student/farmer named Antonio Villas Boas (1934–1992) and his brother arrived home from a party at about 11.00 p.m., but as they were preparing to sleep that night, they were surprised to see a light through their bedroom window. It moved up from the farmyard below to the roof of their house, and together they watched it shine through the gaps in the roof tiles and window shutters before it left the scene.

On 14 October at about 9.30 p.m., the Villas Boas brothers were confronted once again by a brilliant light, "big and round", as they were ploughing their fields with a tractor. It hovered 90 metres above one end of the field. Antonio asked his brother to help him find out what it was, but his brother refused to accompany him, so Antonio went over for a closer look for himself. However, as if toying with him, the light moved swiftly to the other end of the field. He and the light repeated the same movements for at least a couple of times before the young farmer stopped in exhaustion. Then the light abruptly departed.

The following cold, starlit night, Antonio was working alone, ploughing by the light of the tractor's headlamps just 50 metres from the Grande River and on the same spot he and his brother had been working, when at about 1.00 a.m., he noticed a "large red star" descending from the sky and heading toward the outer fringe of his field. He was standing a short distance from the tractor when he noticed the light. The light got larger and larger as it approached him, and he soon realised that it was a large luminous elongated egg-shaped object about 35 feet long and 23 feet wide. The craft flew overhead, some 45 metres above the tractor, and stopped momentarily. Unexpectedly, the object cast a huge light over the entire field. Antonio was now in the aliens' spotlight.

As Antonio recalled, all this happened "so quickly that it was on top of me before I could make up my mind what to do about it".

Then the object moved slowly away with the three legs extending beneath it. Antonio continued to watch in fascination and fear as the

UFO hovered close to the ground 15 metres in front of him. A large collection of purple lights surrounded the rim of the rounded object, and a red rotating cupola on top of the craft changed to green as its speed diminished. The powerful light source was produced by a big round headlamp on the side facing Antonio. Mesmerised by the scene, he saw three metal legs make contact with the ground, and at this, the terrified farmer started to run to his tractor and attempted to drive off and head home. He didn't travel far, as the engine stopped, in spite of the fact that the tractor had been working smoothly before. Unable to restart it after a few attempts, he leaped from his tractor and started to run away. However, the heavily ploughed field made escape very difficult for him, almost tripping as he went.

After running a few steps, he was forced to stop by someone holding his arm. As he turned, he was startled to see an alien covered from head to toe in a tight-fitting grey suit and wearing a large, broad, metallic helmet. This individual reached only to Antonio's shoulder. At that—and feeling somewhat uncompliant with the wishes of this creature—he knocked his assailant away from him in the hope of renewing his escape. The individual fell to the ground, but he was immediately surrounded and seized by three other humanoids who firmly took hold of him, lifted him off the ground, and carried him toward the machine as he tried to wriggle and shout his way free without success.

> "I noticed that as they were dragging me towards the machine my speech seemed to arouse their surprise or curiosity, for they stopped and peered attentively at my face as I spoke, though without loosening their grip on me. This relieved me a little as to their intentions, but I still did not stop struggling…"

A ladder was lowered from the craft, and his captors tried to carry him in, though not without difficulty. Being extremely uncooperative, Antonio hung on with "life and limb" to a kind of handrail alongside the ladder. Nevertheless, they eventually succeeded in bringing him inside.

The first room Antonio found himself in was square and metallic, brightly lit by recessed square lights in the smooth metallic walls. The door he had entered through closed behind him and appeared suddenly invisible, as if no door had ever existed there before. Then as soon as his feet were lowered to the ground, one of the aliens beckoned the two holding Antonio firmly, like two guards with a prisoner, to an adjoining room. As they were doing so, he noticed that there were five small beings taking part in the abduction operation. The room he was led into was larger and oval-shaped, with a metal column connecting the floor with the ceiling, together with a table and several backless swivel chairs placed over to one side. As Antonio noticed:

> "The table as well as the stools were one-legged, narrowing toward the floor where they were either fixed (such as the table) to it or linked to a moveable ring held fast by three hinges jutting out on each side and riveted to that floor (such as the stools, so that those sitting on them could turn in every direction)."

His captors began to converse with each other:

> "Those sounds were so totally different from anything I had heard until now. They were slow barks and yelps, neither very clear nor very hoarse, some longer, some shorter, at times containing several different sounds all at once, and at other times ending in a quaver. But they were simple sounds, animal barks, and nothing could be distinguished that could be taken as the sound of a syllable or word of a foreign language. Not a thing! To me it sounded alike, so that I am unable to retain a word of it...I still shudder when I think of those sounds, I can't reproduce them...my voice just isn't made for that."

Once the conversation among the aliens ceased, all five set about him, forcibly removing his clothes without harming him. Antonio was "cursing them with loud yells" as he tried to resist their actions, and at times his shouting did arouse their interest—"at which they stopped and stared at me as if trying to make me understand that they were being polite".

As he struggled, Antonio took in the finer details of the aliens and what they wore. He noticed light-coloured eyes through the apertures of the alien's helmet. The helmets looked oversized in an upwards direction above where he thought the heads of the aliens were. He thought, "Probably there was something else hidden under those

562

helmets, placed on top of their heads, but nothing could be seen from the outside."

Each helmet had three tubes (no wider than a common garden hose) protruding from above, the central one running down the middle of the back and entering the clothing; the other two curved away to enter the clothes "at about four inches from the armpits". The tubes looked like a metal, but he couldn't be sure. They could have been made of rubber with metal painted on the surface. Where the tubes entered the clothing, there was nothing to indicate what their purpose was. As Antonio said:

> "I didn't notice anything at all, no hump or lump to show where the tubes were attached, nor any box or contrivance hidden under their clothes."

The aliens' hands were completely covered by thick gloves, and their footwear seemed to be attached directly to the trousers like the gloves to the sleeves; their soles were as thick as five centimetres. A red breastplate or shield "about the size of a slice of pineapple", which reflected light, was set into the uniform of each alien. The plate, positioned over the chest, was attached to a waist belt by a thin, flat strip of laminated metal.

Antonio gave further details of the clothing:

> "No pocket could be seen anywhere, and I don't remember seeing any buttons either. The trousers were also tight-fitting over the buttocks, thighs, and legs, as there was not a wrinkle nor a crease to be seen. There was no visible hem between the trousers and shoes, which were actually a continuation of the former, being part of the self-same garment. The soles of their shoes were different from ours: they were thick, about two or three inches thick, and a little turned up (or arched up) in front, so that the tips looked like those described in the fairy tales of old, though the general appearance was that of common tennis shoes. From what I saw later, they must have fitted loosely, for they were larger than the feet they covered. In spite of this the men's gait was free and easy, and their movements were swift indeed. Perhaps the closed siren-suit they wore did interfere

slightly with their movements because they kept walking very stiffly."

In terms of the height of the aliens, Antonio noticed a difference. He said:

"They [the aliens] were all about my height (1.64 meters tall, in shoes), perhaps a little shorter because of those helmets, except for one of them, the one who had caught hold of me out there—this one did not even reach my chin."

Antonio, now naked, stood there trembling in fear and in the environment of the craft, which had a temperature much the same as the coldish night outside. Then one of the aliens came forward with what seemed to be a wet sponge soaked in a "liquid…as clear as water, but quite thick, and without smell". He said to investigators that "I thought it was some sort of oil, but was wrong, for my skin did not become greasy or oily", but most likely it was a disinfectant as it was rubbed all over his skin.

He was now directed to a door. The door itself had a red inscription over it, generated by light. These hieroglyphic-type characters appeared to levitate some five centimetres away from the metallic door. Antonio tried to memorise this inscription, although it meant nothing to him.

He entered the room behind the door. Two of the aliens followed him in, carrying an instrument consisting mostly of an empty container and two flexible tubes, with one end of the tube shaped like a suction cup. This tube was then secured to his chin, while the other tube was pumped up and down. He was then surprised to watch the small container fill with what appeared to be his own blood. The creatures then walked out of the room, leaving Antonio unaccompanied. He probably thought about making a run for it, but the door closed as soon as the last alien passed through.

As Antonio quietly sat on a soft couch—the only item of furniture in a very much empty room—for about half an hour, he suddenly perceived an unusual odour. He went to investigate the source of the smell and found small metallic tubes just beneath the ceiling. They were releasing greyish smoke through tiny holes in the tubes that spread and dissolved in the air. It made him feel very sick, so much so that he rushed to a comer of the room and vomited.

Approximately half an hour after his ordeal with the greyish smoke, feeling more relaxed and a little less frightened, Antonio heard a noise at the door. The door opened, and much to his astonishment, he saw a woman standing there without a uniform. In fact, she was as naked as he was! The woman began to walk toward him as if she wanted something from him.

"She came in slowly, unhurriedly, perhaps a little amused at the amazement she saw written on my face," Antonio recalled. "I stared, open-mouthed...she was beautiful, though of a different type of beauty compared with that of the women I have known."

According to Antonio, her smooth hair was almost white in colour, parted in the centre, and reached halfway down her neck, curling inward at the ends. She had large blue eyes that "slanted outwards, like those pencil-drawn girls made to look like Arabian princesses...except that they were natural; there was no makeup". Her nose was small and straight; she had thin lips around a small mouth, looking almost like a slit. The woman's wide face with prominent cheekbones (he later noticed they were soft and fleshy to the touch, suggesting that they were not made of bone) converged to a markedly pointed chin, while her ears looked normal but small in size; her feet appeared to be small, and she had long, narrow hands. Antonio also said:

> "Her body was much more beautiful than any I have ever seen before. It was slim, and her breasts stood up high and well-separated. Her waistline was thin, her belly flat, her hips well-developed, and her thighs were large..."[3]

Although Antonio was about 1.6 metres tall, the woman's head reached only to his shoulders. He saw too that the hair underneath her arms and her pubic hair was notably a deep red. While her skin was white, "she was full of freckles on her arms".

When the door closed, she approached the farmer in silence, "looking at me all the while as if she wanted something from me". He smelled no perfume on her "apart from the feminine odour".

Then Antonio explained what happened next:

---

3   Antonio would change his perception of the woman being beautiful to one of being ugly as he feared the experience would be interpreted by others as a fanciful sexual fantasy of his.

"Suddenly she hugged me and began to rub her head against my face [presumably by standing on tiptoe] from side to side. At the same time I also felt her body glued to mine and it also was moving...This sounds quite incredible considering the circumstances.

Alone with this woman, who clearly gave me to understand what she wanted, I became very excited. I forgot everything, seized the woman and responded to her caresses. It was a normal act and she behaved like any other woman, even after repeated embraces. But she did not know how to kiss, unless her playful bites on my chin had the same meaning."

Although she never spoke, she occasionally grunted and growled, and that "nearly spoiled everything, giving the disagreeable impression that I was with an animal".

Antonio added:

"Then we had some petting, followed by another act, but by now she had begun to deny herself to me, to end the matter....Finally she became tired and was breathing heavily."

At this point, she refused further advances.

Antonio also recalled that after their second sexual intercourse, the woman did use a small container to collect a sample of Antonio's sex fluids.

"Shortly after we had separated," Antonio explained, "the door opened. One of the men appeared on the threshold and called the woman. Then she went out. But, before going out, she turned to me, pointed at her belly and then pointed towards me and with a smile (or something like it), she finally pointed towards the sky—I think it was in the direction of the south."

Antonio was naturally a little frightened by this gesture, interpreting the pointing at him and to the sky by the female as suggesting these aliens might permanently abduct the young farmer by taking him away to wherever they came from. In reality, the female was probably doing nothing more than indicating that she would bear their child on some alien world.[4]

---

4   The direction in the sky given by the alien could be significant. The only sun-like star within 12 light years of our sun that is truly south of Earth—and to take into account the number of aliens involved, the somewhat complicated design of the UFO with its various internal rooms

Antonio was told to get dressed. The only thing he could not find in his pocket was his Homero lighter. He was then escorted on a tour of the craft, which he described in great detail, including the giant "dish-shaped cupola" revolving overhead, which whistled like the "sound of air being drawn in by a vacuum cleaner". The walls were smooth, hard and made of a metal. No windows could be seen anywhere on the walls. During this time he made an attempt to embezzle an instrument[5] as a souvenir, only to be angrily denied by one of the crew members. Eventually, he was encouraged to leave the craft of his own accord via the ladder. He did so, and was then signalled by one of the aliens to step away to a safe distance. From there, Antonio turned around and looked at the UFO. He watched the ladder and the metal legs be retracted (leaving no sign of the openings they came from) and the lights began to glow. Antonio said:

> "The craft rose into the air with its cupola turning at great speed and its landing legs were withdrawn into its base. When the craft had risen to about 35 metres above the ground [and making a buzzing noise], the object again increased its brightness with lights around the rim flashing and changing colour [and] finally settling on a bright red. Then, listing slightly to one side [and the noise suddenly increasing], that strange machine shot off like a bullet towards the south…"

It was by then 5.30 a.m., and the abductee's adventure had lasted four hours and fifteen minutes.

At first, the young farmer told this extraordinary tale to no one for fear of ridicule, except his mother. However, in the days that followed, his health deteriorated quickly. Indeed, nausea and headaches kept him indoors, his eyes burned, and he could not sleep. Then small red circles appeared all over his body as a result of unusually infected wounds drying up.

---

and fixed protrusions (i.e., extra mass) and, therefore, the need for a short distance to travel at just the right speed to make the journey times short for the participants—is Alpha Centauri A and B. These two stars are part of a triple star system lying at a distance of 4.3 light-years—and are the ones that resemble our sun most out of any of the 70 nearest stars we know of. As for Tau Ceti and Epsilon Eridani—the only other two potential regions for extraterrestrial life and perhaps intelligence at this distance of 12 light-years—they are more easily observed in the northern sky than our nearest stellar neighbour.

5   Looking like a small box with a glass top. Looking through it showed what appears to be a hand of a clock and several marks corresponding to the 3, 6, 9 and 12 position as if it could have been a clock.

Local doctors, including a specialist from Rio de Janeiro who examined him, strongly believe the farmer had been exposed to radiation. The dark scars found on his chin were evidence of bleeding.

Antonio died in 1992. Despite receiving ridicule from the Brazilian media about his story, especially after he added one extra detail in a television interview regarding the small container he saw the female alien use to collect his sperm—since then, Antonio had made the decision to describe the female alien as "ugly" to avoid people thinking he was describing a wild sexual fantasy—he never confessed his encounter was a hoax. In fact, his telling of the story remained as vivid as when he first reported it.

Glossary

Aluminium: The correct scientific term for the lightest metal in the periodic table. In the US, this is called *aluminum*.

Atom: A microscopic object consisting of electrons, protons and neutrons. The atom is structured such that the electrons move at high speeds in discrete orbits around a fast-spinning nucleus containing protons and neutrons.

Atomic: On the scale of the atom.

Auroras: A spectacular electrical display of yellow-green, red and violet lights in the sky found mainly over the magnetic pole regions. The lights we see are due to ionised nitrogen and oxygen atoms (formed by the collision of the solar wind with the upper atmosphere) as they procure electrons from the atmosphere.

Conductor: A material that allows the rapid movement of electrical, heat or sound energy through the material.

Cosmic rays: High-speed atomic and subatomic particles of similar composition to the solar wind, but which originate from objects outside of the solar system.

Current, electric: The flow of electric charge.

Dielectric: An electrical insulator.

Doppler effect: An apparent variation in the frequency of a wave (light or sound), owing to the motion of the wave source relative to the listener/observer.

Electric field: A region of space where invisible electric forces are present. Affects only charged objects.

Electrokinetic: Pertaining to the physical movement of something through the direct application of electrical energy.

Electromagnetic field: Electric and magnetic fields. Static electromagnetic fields are believed to affect only charged objects. Oscillating electromagnetic fields (also called radiation) affect both charged and uncharged objects.

Electromagnetic waves: Regions of intense electromagnetic energy called photons which move at the speed of light. Also called radiation.

Electromagnetism: The study of electric charge, and the forces that exist between moving and non-moving bodies of varying or non-varying charge. The particle mediator for all the forces of electromagnetism is the photon.

Energy: That which is carried by electromagnetic waves by way of electromagnetic fields in order to create matter and all the forces of nature.

Extraterrestrial: Pertaining to something that does not originate from Earth.

Extraterrestrial hypothesis: The theory that some UFOs are space vehicles piloted by extraterrestrial beings from one or more extraterrestrial civilisations.

Frequency: The number of oscillations (or cycles) in a wave in one second, measured in hertz (Hz).

General Theory of Relativity: A theory devised by the late Dr Albert Einstein. The theory is an extension of the Special Theory of Relativity to take into account the accelerating motion of matter and, at the same time, show how acceleration and gravity are linked.

Gravitational field: A region of space where invisible gravitational forces are present. According to Einstein's Unified Field Theory published in 1929, this is believed to be a fictitious entity and is nothing more than electromagnetic radiation (or photons) moving the charged particles making up so-called uncharged matter (or, more correctly, charged matter). Previously, it was believed that the radiation (or the gravitational field) only moved the mass of the particles and, hence, only uncharged matter. Under the Unified Field Theory, this concept will be challenged.

Inertia: The pressure exerted by electromagnetic radiation on an accelerating body to oppose its tendency to move.

Insulator: A material that inhibits the passage of electrical, heat, or sound energy through it.

Light bending: Predicted by Albert Einstein's General Theory of Relativity and experimentally verified by astronomers, notably Professor Arthur Stanley Eddington during a solar eclipse in 1919. It is caused by light passing through a strong gravitational field, which naturally implies that light must have a gravitational field of its own in order to interact with the gravitational field of another object. This was one of the observations that led Albert Einstein to develop his Unified Field Theory.

Light-year: The distance travelled by a ray of light in one year, or 9.454255×10^{15} metres.

Magnetic field: A region of space where invisible magnetic forces are present. Affects only charged objects.

Mass: A region of concentrated electromagnetic and gravitational energy (or essentially just electromagnetic energy, according to Einstein's Unified Field Theory).

Momentum: A measure of the force or energy of a particle, conserved in all collisions with other particles in a closed system.

Newton's Laws of Motion: Famous English mathematician, physicist and astronomer Sir Isaac Newton (1642–1727) established three laws of classical science to describe the observed behaviour of bodies with mass. The first law states that all bodies have the observed tendency to remain at rest or to continue in motion in a straight line at constant speed until acted upon by a force. The second law states that a body subjected to the actions of a force will accelerate (or decelerate), according to the differential equation:

$$a = \frac{dv}{dt} = \frac{F}{m}$$

where a is acceleration, F is force, m is mass, and dv/dt is the instantaneous velocity v at any specified time t. The acceleration of the body is in a direction governed by the force. In the third law, any body under the action of a force will exert an equal but opposite force, owing to the inertial tendency of the body to remain at rest or move at constant speed.

NiTi: The nickel-titanium alloy (also called nitinol) is the world's most powerful shape-memory alloy and the lightest titanium-based shape-memory alloy, making it ideal for aerospace use. The shape-memory effect is activated by heat and electricity. This distinctively dark-grey alloy began the official scientific investigations into all shape-memory alloys after 1958 when the Naval Ordnance Laboratory (NOL) subjected a highly pure sample of NiTI to a cigarette lighter. However, it is now understood, after the release of several formerly secret USAF/Battelle reports after 1947 that the USAF were the first in the world to express an interest in NiTi at Wright-Patterson AFB and only began following the retrieval of the Roswell materials. It should be noted that the Roswell foil—which was sent to Wright-Patterson AFB for analysis—was described by witnesses as having a strong shape-memory effect, was a metal of some sort (and definitely not a plastic), and it was dark-grey in colour. However, the purity issue had been the stumbling block for the USAF in 1947. To address this shortfall in scientific knowledge, the USAF requested assistance from the Battelle Memorial Institute in 1948 to make highly pure experimental samples of the alloy, among other titanium-based alloys. Unfortunately, such a move would raise serious questions as to how the USAF was able to make so much of the Roswell foil by early July 1947.

Nitinol: The acronym for Nickel-Titanium (NiTi) Naval Ordnance Laboratory (NOL) in remembrance of the people who first discovered the world's most powerful shape-memory effect of the NiTi alloy system around 1958 and, thereby, commencing the official civilian scientific interest into this class of metals.

Nuclear fusion: The fusing together of two or more atomic nuclei to form a new nucleus and, in the process, emit vast amounts of energy to the surroundings.

Oscillation: The variation of the value of a physical quantity over time in a periodic way.

Photon: A region of concentrated electromagnetic energy. A stream of photons propagating through space is described mathematically by scientists as an oscillating electric and magnetic field at right angles to each other and to the direction of motion, also called an electromagnetic wave. In classical physics, a photon is just another form of ordinary matter, albeit invisible.

Radioactive: The spontaneous emission of radiation, protons, electrons, or atomic nuclei from the nucleus of an atom.

Relativistic: Of, or pertaining to, high speeds approaching that of light and that follow the laws of the Special and General Theory of Relativity and of the Unified Field Theory as proposed by the late Dr Albert Einstein.

Science: A body of useful and practical knowledge and a method of obtaining it. The knowledge epitomises the experiences of many people and consists of a set of beliefs or theories that help represent and explain the fundamental and reproducible laws of nature under standard conditions.

Shape-Memory Alloy (SMA): An alloy that can return to its original shape when heated or an electric current is applied after it was physically deformed in the cool temperature or "non-flow of electrical current" state.

Solar wind: A continuous yet sometimes sporadic flow of energetic atomic and subatomic particles from the sun, such as electrons and protons, travelling 300 to 500 kilometres per second.

Special Theory of Relativity: A theory devised by the late Dr Albert Einstein. The theory discusses what happens to time, mass and length of an object as perceived (and measured) by an outside observer when the object travels at very high speeds approaching that of light relative to the observer. The result is that time dilates, mass increases and length contracts for the object as it approaches the speed of light.

Speed of light: Exactly 299,792,458 metres per second.

Stars: Large luminous bodies of hot compressed gases composed mostly of hydrogen and helium, with some heavier elements. Stars form in space like raindrops do on Earth.

Subatomic: On the scale of fundamental particles, such as the electron and proton.

Temperature: A measure of how hot or cold something is. This sensation of hotness or coldness is the result of the quantity and frequency of electromagnetic radiation present in a substance.

TiZr: The zirconium-titanium alloy system used by the USAF in 1948 to compare with the NiTi alloy system. TiZr is a shape-memory alloy, but not as pronounced as NiTi. It is a silver-coloured alloy with a much higher temperature of activation (around 200°C) as compared to NiTi, which is around room temperature (or less with small additions of cobalt).

UFOs: Unidentified Flying Objects. After compiling the world's most common UFO observations, it is now understood that there is a new electromagnetic technology hidden in the genuine UFO reports. It is based on the poorly studied Abraham-Lorentz formula in classical electrodynamics relating to the emission of radiation from the surface of a charged object. The curvature of the heated glowing metal hull of UFOs is the primary means of directing a higher energy density of the radiation emission in one direction and, with it, a higher recoiling force on the UFOs.

Unified Field Theory: A theory devised by the late Dr Albert Einstein and published in 1929. The theory is an extension of the General Theory of Relativity to take into account the motion of the electric charge and the presence of the electromagnetic field and how it affects the strength of the gravitational field. More specifically, it is the oscillating electromagnetic field that generates a gravitational field. Just like the Earth or a tennis ball have a gravitational field of their own, so too does radiation have a gravitational field. And, indeed, radiation should be treated like any other ordinary matter. Therefore, what actually controls the strength of the gravitational field is the energy density of the electromagnetic field and how this varies over time. Thus, factors such as the amount of electric charge and the frequency of the oscillating charge can affect the energy density of the electromagnetic field and, with it, the strength of the gravitational field. The implications of this concept is that it is now feasible to control the effect of gravity on the body and other objects by considering the idea of reducing the energy density of the electromagnetic field. A classic way to achieve this is via Faraday's cage effect. As the laws of electromagnetism tells us, a perfectly symmetrical metal box reduces the radiation levels to zero through destructive interference of the out-of-phase radiation being reflected off the interior metal walls. So if the radiation drops to zero, so will the strength of the gravitational field. This means that the gravitational effects of inertial forces on the body, when it is accelerated, can be eliminated. Another implication of the Unified Field Theory is whether the concept of a gravitational field should be maintained, or if we should get rid of the field altogether and imagine photons (or electromagnetic radiation) doing all the work of the gravitational field. If this is true, then we can say that we are living in a purely electromagnetic universe.

USAF: United States Air Force.

Wavelength: The distance of one repeating unit in a wave. The repeating unit may begin and end at the crests or troughs or any adjacent point of equal phase in a wave.

Weight: A measure of the force exerted by a gravitational field on an object. According to the Unified Field Theory, this is merely radiation pressure. In other words, the universal background radiation is pushing objects together to give the impression of being attracted to one another (or being pulled together by a mysterious force that scientists since the 16th century have called the "gravitational force") and is caused by the radiation shielding effect of the objects to reduce the energy density of the radiation between the objects. Since the energy density of the radiation on the outer surfaces of the objects facing into space is unchanged, the imbalance in forces exerted by radiation on the inner surfaces of the objects facing each other compared to the outer surfaces is what leads to the movement of the objects towards each other.

Bibliography

Abraham, M. "Theorie der Elektrizitat": *Elektromagnetishe Theorie der Strahlung*. Volume II (2 volumes), 1905 and republished in 1920. Leipzig, Berlin: B. G. Teubner.

Abraham, Max. "Classical Theory of Radiating Electrons": *Annalen der Physik*. Volume 10 (1903), Issue 105.

"AIAA Committee Looks at UFO Problem" *Astronautics and Aeronautics (AIAA Journal)*, Volume 6, Number 12, 8 December 1968.

Alcaines, Guillermo Garcia & Llames-Estrada, Felipe J. "Radiation reaction on a classical charged particle: A modified form of the equation of motion": *Physical Review*, Volume E 88, 2013, pp.033203-1 to 15.

Amos, Jonathan. "Ball lightning 'may explain UFOs'": BBC News. 1 December 2010. Online article available from: http://www.bbc.co.uk/news/science-environment-11877842

Ardley, Neil. 1991, *Science for Kids: Light*. Sydney, Australia: Allen & Unwin Australia Pty Ltd.

Arnold, Kenneth & Palmer, Ray. 1952, *The Coming of the Saucers: A Documentary Report on Sky Objects that have Mystified the World*. Boise, Idaho and Amherst, Wisconsin: Palmer Publishing.

Bahder, Thomas B. & Fazi, Christian. "Force on an Asymmetric Capacitor": Army Research Laboratory (Maryland). 27 September 2002 and updated in March 2003.

Basterfield, Keith. 1981, *Close Encounters of an Australian Kind*. Sydney: A.H. & A.W. Reed Pty Ltd.

Basterfield, Keith. 1997, *UFOs: A Report on Australian Encounters*. Kew, Victoria: Reed Books Australia.

Basterfield, Keith. "Revelations - The 'Disclosure Australia' Project": *Australasian Ufologist*. Volume 9, Number 2, 2005, pp.47-53.

Beacom, John F. & Bell, Nicole F. "Do Solar Neutrinos Decay?": *Physical Review D*. April 2002. Volume 65, Issue 11, pp.1-10.

Beckley, Timothy Green. 1989, *MJ-12 and the Riddle of Hangar 18*. New Brunswick, N.J.: Inner Light Publications.

Berlitz, Charles & Moore, William. 1982, *The Roswell Incident*. London: Granada Publishing Ltd.

Berlitz, Charles & Moore, William. 1988, *The Roswell Incident*. New York: Berkley Publishing Group.

Berlitz, Charles & Moore, William. 1991, *The Philadelphia Experiment: Project Invisibility*. New York: Fawcett Crest.

Berry, Adrian. 1995, *The Next 500 Years: Life in the Coming Millenium*. London: Headline Book Publishing. (Chapter 16: Starship).

Blum, Ralph & Judy. 1974, *Beyond Earth: Man's Contact with UFOs*. New York: Bantam Books, Inc.

Blundell, Nigel & Boar, Roger. 1984, *The World's Greatest UFO Mysteries*. London: Octopus Books Ltd.

Boar, Roger. 1984, *The World's Greatest UFO Mysteries*. London: Octopus Books Ltd.

Bord, Janet & Colin. 1980, *Are We Being Watched?* Sydney: Angus & Robertson.

Boyce, Chris. 1979, *Extraterrestrial Encounter: A Personal Perspective*. Wellington, New Zealand: A.H. & A.W. Reed Pty Ltd.

Broderick, Damien. "What If....Enjoy these Two Intriguing UFO Explanations": *Omega Science Digest*. November/December 1981, pp.58-63 & 123.

Brookesmith, Peter (editor). 1980-84, "Appearances and Disappearances", "The UFO Casebook", "The Alien World", "UFOs: Where Do They Come From?" *The Unexplained*. London: Orbis Publishing Ltd.

Brookesmith, Peter (editor). 1984, *The Age of the UFO*. London: Orbis Publishing Ltd.

Brown, Thomas Townsend. U.S. Patent Numbers 2,949,550 and 3,187,206: *Electrokinetic Apparatus*.

Brown, Thomas Townsend. British Patent Number 1,274,875: *A System for Imparting Movement to an Ionizable Dielectric Fluid Medium*.

Cahn, John P. "Flying Saucer Swindlers": *True* Magazine, August 1956. pp.36-37 & 69-71.

Cahn, John P. "The Flying Saucers and the Mysterious Little Men": *True* Magazine, September 1952. pp.17-19 & 102-112.

Calkins, Carroll C (Project Editor). 1982, *Mysteries of the Unexplained: How Ordinary Men and Women have Experienced the Strange, the Uncanny, and the Incredible*. New York: The Reader's Digest Association, Inc.

Carati, Andrea and Galgani, Luigi. *Recent Progress on the Abraham-Lorentz-Dirac Equation*. University of Milan (downloadable from http://www.mat.unimi.it/users/carati/pdf/caserta.pdf).

Chalker, Bill. "The UFO Connection": *Omega Science Digest*. March/April 1985, pp.56-59 & 118-120.

Chalker, Bill. "The Bizarre UFO of Rendlesham Forest": *Omega Science Digest*. July/August 1985, pp.88-91.

Chalker, Bill. "UFOs: Yes, There is a Cover-Up": *Omega Science Digest*. November/December 1983, pp.22-25.

Chalker, Bill. UFOs: "Australian Secret Documents Revealed": *Omega Science Digest*. September/October 1982, pp.22-25, 117 & 125.

Chamberlin, Jo. "The Foo Fighter Mystery": *The American Legion Magazine*. December 1945. Volume 39, Number 6, pp.9, 43-44 & 47.

Chant, Christopher & Hogg, Ian. 1983, *The Nuclear War File*. London: Ebury.

Cheney, Margaret. 1981, *Tesla: Man Out of Time*. New York: Dell Publishing.

Childress, D. Hatcher (compiler). 1990, *The Anti-Gravity Handbook*. Stelle, Illinois, USA: Adventures Unlimited Press.

Childress, D. Hatcher (compiler). 1990*, *Anti-Gravity and the Unified Field*. Stelle, Illinois, USA: Adventures Unlimited Press.

Clark, Jerome. 1993, *Encyclopedia of Strange and Unexplained Physical Phenomena*. Detroit, USA: Gale Research Inc.

Clifford, Katrina. "Love @ first byte": *My Business*. November 2002, pp.20-23.

Cocconi, Giuseppe and Morrison, Philip. "Searching for Interstellar Communications": *Nature*. 1959. Volume 184, Number 4690, pp.844-846.

Collyns, Robin. 1974, *Did Spacemen Colonise the Earth?* England: Granada Publishing Ltd.

Condon, Edward Uhler et al. 1969, *Scientific Study of Unidentified Flying Objects*. USA: Dutton (full text available online).

Cooray, Gerald and Vernon. "Could Some Ball Lightning Observations be Optical Hallucinations Caused by Epileptic Seizures?": *The Open Atmospheric Science Journal*, 2008, 2, pp.101-105.

Cwiklik, Robert. 1987, *Albert Einstein and the Theory of Relativity*. New York: Barron's Educational Series, Inc.

Davies, P.C.W. 1987, *Quantum Mechanics*. London: Routledge & Kegan Paul Ltd.

Deyo, Stan. 1993 (revised), *The Cosmic Conspiracy*. Kalamunda, Australia: West Australian Texas Trading.

Dione, R.L. 1973, *God Drives a Flying Saucer*. London: Corgi Books.

Drukey, D. L. "Radiation from a Uniformly Accelerated Charge": *Physical Review.* Volume 76, Issue 4, pp.543-544.

Durrach, H. B. Jr. & Ginna, Robert. "Have We Visitors from Space?": *LIFE Magazine.* 7 April 1952, pp.80, 90-92, 94 & 96.

Eisen, Jonathan. 1999, *Suppressed Inventions & Other Discoveries: True Stories of Suppression, Scientific Cover-Ups, Misinformation, and Brilliant Breakthroughs.* Garden City Park, New York: Avery Publishing Group, p.285.

Encyclopedia Britannica Inc. (Chicago, USA), 1969, 1980, 1986 and 1988, 1995 editions.

Evans, Hilary & Spencer, John (editors). 1988, *Phenomenon: Forty Years of Flying Saucers.* New York: Avon Books Printing.

Evans, Hilary & Stacey, Dennis. 1997, *A World History of UFOs.* Potts Point, NSW, Australia: Red Sparrow.

Fawcett, Lawrence & Greenwood, Barry J. 1984, *The UFO Cover-Up.* New York: Prentice Hall Press.

Flammonde, Paris. 1976, *UFO Exist!* New York: Ballantine Books.

Fogarty, Quentin. "For Experts & Sceptics, the Message is the Same: The UFO Phenomenon just does not go away": *Omega Science Digest.* September/October 1981, pp.18-21 & 116-117.

Fowler, Raymond E. 1974, *UFOs: Interplanetary Visitors.* New York: Bantam Books, Inc.

Fulton, Thomas. & Rohrlich, Fritz. "Classical radiation from a uniformly accelerated charge": *Annals of Physics.* April 1960. Volume 9, Issue 4, pp.499-517.

Gelman, Rita Golden & Seligson, Marcia. 1978, *UFO Encounters.* New York: Scholastic Book Services.

George, Professor Frank. 1978, *Science Fact.* New York: Sterling Publishing Company, Inc.

Goldberg, Dr Joshua M. 1992, "U.S. Air Force Support of General Relativity: 1956-1972": *Studies in the History of General Relativity (J. Eisenstaedt and A. J. Kox, editors).* Volume 3. Boston, Massachusetts: Center for Einstein Studies (published by Birkhäuser), pp.89-102.

Good, Timothy. 1988, *Above Top Secret: The Worldwide UFO Cover-Up*. First U.S. Edition. London: Sidgwick & Jackson Ltd.

Good, Timothy (editor). 1990, *The UFO Report 1991*. London: Sidgwick & Jackson Ltd.

Good, Timothy (editor). 1991, *The UFO Report 1992*. London: Sidgwick & Jackson Ltd.

Good, Timothy. 1991, *Alien Liaison*. London: Arrow Books Ltd.

Good, Timothy. 1997, *Beyond Top Secret: The Worldwide UFO Security Threat*. London: Pan Books/McMillan Publishers Ltd.

Graham, Ian. 1993, *The Big Book of Flight*. London: Reed International Books Ltd.

Grant, John. 1990, *The Great Unsolved Mysteries of Science*. London/New York: Quintet Publishing Limited/Chartwell Books.

Greenfield, Susan A. (General Editor). 1996, *The Human Mind Explained: An Owner's Guide to the Mysteries of the Mind*. Sydney, Australia: Reader's Digest (Australia) Pty Ltd.

Greenwood, Barry J. & Fawcett, Lawrence. 1984, *The UFO Cover-Up*. New York: Prentice Hall Press.

Griffiths, David J. 1989, *Introduction to Electrodynamics*. 2nd Edition. New Jersey, USA: Prentice Hall Press.

Griffiths, David J., Proctor, Thomas C., Schroeter, Darrell F. "Abraham-Lorentz versus Landau-Lifshitz": *American Journal of Physics*. April 2010. Volume 78 Number 4, pp.391-402.

Gunston, Bill. "Military Power - Engines for military aircraft have to meet a set of criteria different from those of commercial jets and turboprops": *AIR International*. Volume 58, Number 1, January 2000, pp.37-41.

Haish, Bernhard, Rueda, Alfonso & Puthoff, H.E. "Beyond E=mc^2 - A First Glimpse of a Universe Without Mass": *The Sciences*. November/December 1994, Volume 34, Number 6, pp.26-31.

Hall, Richard. 1988, *Uninvited Guests*. Santa Fe, USA: Aurora Press.

Hobana, Ion & Weverbergh, Julian. 1971, *UFOs from Behind the Iron Curtain*. London: Corgi Books.

Holledge, James. 1965, *Flying Saucers over Australia*. Melbourne, Australia: Horwitz Publications Inc., Pty Ltd.

Holmes U.S.N., Captain David C. 1966, *The Search For Life on Other Worlds*. New York: Bantam Books, Inc.

Hough, Peter & Randles, Jenny. 1994, *The Complete Book of UFOs: An Investigation into Alien Contacts and Encounters*. London: Judy Piatkus (Publishers) Ltd.

Hough, Peter & Randles, Jenny. 1991, *Looking for the Aliens: A Psychological, Scientific and Imaginative Investigation*. London: Blandford Press.

Hugrass, W. N. "A New Derivation for the Radiation reaction Force": *Australian Journal of Physics*, 1997, Volume 50, pp.815-826.

Huyghe, Patrick. "Scientists Who Have Seen UFOs": *Omega Science Digest*. March/April 1982, pp.94-99 & 120-121.

Hynek, J. Allen. 1972, *The UFO Experience: A Scientific Inquiry*. London: Abelard-Schuman Ltd.

Hynek, J. Allen & Vallee, Jacques. 1975, *The Edge of Reality - A Progress Report on Unidentified Flying Objects*. Chicago, Illinois: Henry Regnery Company.

Interavia, "Towards Flight without Stress or Strain...or Weight", 23 March 1956, Volume XI, Number 5, pp.373-374.

Jackson, Robert. 1992, *UFOs—The Sightings of Alien People and Spacecraft*. London: Quintet Publishing Ltd.

James, Peter and Thorpe, Nick. 1994, *Ancient Inventions* New York: Ballantine Books, pp.146-157.

Jastrow, Dr Robert. "The UFO Debate Goes On": *Omega Science Digest*. March/April 1981, pp.110-111 & 115.

Jeans, Sir James. 1951, *The Mathematical Theory of Electricity and Magnetism*. Fifth Edition. Cambridge University Press.

Keel, John A. 1970, *Operation Trojan Horse*. New York: G.P. Putnam's Sons & Toronto, Canada: Longman's Canada Ltd.

Kennefick, Daniel J. 1997, *Controversies in the History of the Radiation Reaction problem in General Relativity.*. Cornell University.

Keyhoe, Major Donald E. 1973, *Aliens from Space: The Real Story of Unidentified Flying Objects*. England: Panther Books Ltd.

Keyser, Paul T, "The Purpose of the Parthian Galvanic Cells: A First Century A.D. Electric Battery used for Analgesia": *Journal of Near Eastern Studies*, Volume 52, Number 2 (April 1993), pp.81-83.

Kuettner, J. P. et al. "UFO - An Appraisal of the Problem: UFO – An Appraisal of the Problem: A Statement by the UFO Subcommittee of the AIAA": *Journal of Aeronautics and Astronautics*. November 1970, Volume 8, Number 11, pp.49-51.

Kettelkamp, Larry. 1996, *ETs and UFOs: Are They Real?* New York: William Morrow and Company, Inc.

Keyhoe, Major Donald E. 1975, *Aliens from Space*. England: Panther Books Ltd.

Landsburg, Alan & Sally. 1977, *In Search of Extraterrestrials*. London: Corgi Books.

LaViolette, Paul A. 2008, *Secrets of Antigravity Propulsion: Tesla, UFOs and Classified Aerospace Technology*. Vermont, USA: Bear & Company.

Lorentz, H.A. 1909, *The Theory of Electrons*. First Edition. Leipzig, Berlin: B. G. Teubner.

Lorentz, Hendrik A. "Electromagnetic phenomena in a system moving with any velocity smaller than that of light": *Proceedings of the Royal Netherlands Academy of Arts and Sciences*. Volume 6 (1904), pp.809-831.

Los Angeles Times. "Flying Saucers Explained by Men of New Research University Here". 3 April 1952, Section B, p.1.

Maney, Professor C.A. & Hall, Richard. 1961, *The Challenge of Unidentified Flying Objects*. Washington D.C., USA: National Investigations Committee on Aerial Phenomena.

Marsh, Carole. 1996, *Unidentified Flying Objects and Extraterrestrial Life*. New York: Henry Holt & Company, Inc.

McCampbell, James. 1976, *UFOLOGY: A Major Breakthrough in the Scientific Understanding of UFOs*. California: Celestial Arts.

McGraw-Hill Encyclopedia of Science and Technology. New York: McGraw-Hill Inc., 1987, 1992.

Mead Jr., Franklin B. (editor). June 1972, *Advanced Propulsion Concepts – Project Outgrowth.* Air Force Rocket Propulsion Laboratory (Edwards AFB). Declassified Technical Report AFRPL-TR-72-31.

Milonni, Peter W. 1994, *The Quantum Vacuum: An Introduction to Quantum Electrodynamics.* Academic Press, Inc., Section 3.3.

Missler, Chuck & Eastman, Mark. 1997, *Alien Encounters The Secret Behind the UFO Phenomenon.* USA: Koinonia House.

Montagu, Ashley. "The Unknown Einstein: Never-before-told Conversations": *Omega Science Digest.* January/February 1986, pp.56-59 & 119.

Moore, William L. "T. Townsend Brown: The Wizard of Electro-gravity": *Farout* (Fall 1992). Los Angeles, pp.20-27, 65 & 68.

Mysteries of the Unknown: The UFO Phenomenon, 1987. Richmond, Virginia, USA: Time-Life Books Inc.

Omega Science Digest: "The Wizard who was Tesla". July/August 1983, p.9.

Panofsky, Wolfganf K. H. and Phillips, Melba. 1962, *Classical Electricity and Magnetism.* London: Addison-Wesley Publishing Company, Inc.

Pauli, Wolfgang. 1958, *Theory of Relativity.* London: Pergamon Press.

Pauwels, Louis & Bergier, Jacques. 1974, *Impossible Possibilities.* Frogmore, St Albans, Herts: Mayflower Books Ltd.

Peebles, Curtis. 1994, *Watch the Skies! A Chronicle of the Flying Saucer Myth.* Washington: Smithsonian Institution Press.

Phillips, Colin A. "A Politico-Sociological Evaluation of a Close Encounter of the Third Kind": *UFO Conference Report.* 1981.

Phillips, Sir Raymond. World Patent Number 88/00769: *Method of and Apparatus for Converting Radio Frequency Energy to Direct Current.*

Product Engineering. "Electrogravitics: Science or Daydream?" 30 December 1957, Volume 28, Number 26, p.12.

Prytz, John. "UFOs: Pushing Us Into Progress? An Interest in UFOs Could be the Stimulus for Future Discoveries in Science": *Omega Science Digest*. May/June 1984, p.10.

Randles, Jenny. 1983, *UFO Reality: A Critical Look at the Physical Evidence*. London: R. Hale.

Randles, Jenny. 1987, *The UFO Conspiracy: The First Forty Years*. London: Blandford Press.

Randles, Jenny. 1993, *The UFO Conspiracy: The First Forty Years*. New York: Barnes & Noble Books.

Randles, Jenny. 1992, *UFOs and How To See Them*. London: Anaya Publishers Ltd.

Rassam, Clive. "A Tale of Two Cultures": *New Scientist*. 26 June 1993, Number 1879, pp.30-33.

Rehn, K. Gosta. 1974, *UFOs: Here and Now!* London: Abelard-Schuman Ltd.

Robbins, Anthony. 1992, *Awaken the Giant Within*. New York: Simon & Schuster.

Rose, Colin. 1985, *Accelerated Learning*. New York: Dell Publishing.

Rose, Mason. *The Flying Saucer: The Application of the Biefeld-Brown Effect to the Solution of the Problems of Space Navigation*. University for Social Research, April 8, 1952.

Ruppelt, Edward J. 1956, *The Report on Unidentified Flying Objects*. Garden City, New York: Doubleday & Company, Inc.

Sachs, Margaret. 1980, *The UFO Encyclopedia*. London: Corgi Books/Transworld Publishers Ltd.

Sagan, Carl. 1996, *The Demon-haunted world: Science as a candle in the dark*. London, UK: Headline Book Publishing.

Sagan, Carl & Page, Thornton (editors). 1996, *UFOs: A Scientific Debate*. New York: Barnes & Noble, Inc.

Saulson, Peter R. "Josh Goldberg and the physical reality of gravitational waves": *General Relativity and Gravitation*. Volume 43, Issue 12, December 2011, pp.3289-3299.

Saunders, Keith. "Miracle Metal - There is Absolutely Nothing Like it": *Omega Science Digest*. 1982. New York: The Hearst Corporation, pp.11-13 & 119.

Schuch, Christoph (Director). *Big Bang in Tunguska*. 2008. Halbtotal Film / ZDF.

Scott, W. B. "Black World engineers, scientists encourage using highly classified technology for civil applications.": *Aviation Week & Space Technology*, 9 March 1992, pp. 66–67.

Simon, Seymour. 1980, *Strange Mysterious from Around the World*. New York: Four Winds Press.

Simon, Seymour. 1997, *Strange Mysterious from Around the World*. New York: Morrow Junior Books.

Snyder, Gerald S. 1981, *Are there Alien Beings? - The Story of UFOs*. Second Printing. New York : Julian Messner/Simon & Schuster.

Spencer, John & Evans, Hilary (editors). 1988, *Phenomenon: Forty Years of Flying Saucers*. New York: Avon Books Printing./

Spencer, John. 1991, *UFOs: The Definitive Casebook*. London: Hamlyn Publishing Group Ltd.

Spencer, John (editor). 1991, *The UFO Encyclopedia*. London: Headline Book Publishing PLC.

Stemman, Roy. 1991, *Mysteries of the Universe*. USA: J.G. Ferguson Publishing Company.

Story, Ronald D (editor). 1980, *The Encyclopedia of UFOs*. Garden City, New York, USA: Dolphin Books, Doubleday & Company, Inc.

Stothers, Richard. "UFOs in Classical Antiquity": *The Classical Journal*, Volume 103 Issue 1 (2007), pp.79-92.

Sturrock, Peter A. "An Analysis of the Condon Report on the Colorado UFO Project": *Journal of Scientific Exploration*, Volume 1, No. 1, pp.75-100 (1987).

Swords, Michael. 1997, *Donald E. Keyhoe and the Pentagon: A World History of UFOs* (Edited by Hilary Evans & Dennis Stacy). Potts Point, NSW, Australia: Red Sparrow.

Tambling, Richard. 1978, *Flying Saucers: Where do they come from?* Melbourne, Australia: Horwitz Publications.

Tralli, Nunzio. 1963, *Classical Electromagnetic Theory.* New York: McGraw-Hill Book Company, Inc.

Trench, Brinsley Le Poer. 1974, *The UFO Story: Mysterious Visitors.* London: Pan Books Ltd.

Trench, Brinsley Le Poer. 1977, *The Flying Saucer Story.* London: Universal-Tandem Publishing Co. Ltd.

The Encyclopedia of Space Travel and Astronomy 1979. London: Octopus Books Ltd.

The New Book of Popular Science 1990. New York: Grolier International, Inc.

Vallee, Jacques. 1965, *UFOs in Space: Anatomy of a Phenomenon.* New York: Ballantine Books.

Vallee, Jacques & Janine. 1966, *The UFO Enigma.* New York: Ballantine Books.

Vallee, Jacques. 1988, *Dimensions: Casebook on Alien Contact.* Chicago, USA: Contemporary Books, Inc.

Vallee, Jacques & Janine. 1990, *Challenge to Science: The UFO Enigma.* Sixth Printing. New York: Ballantine Books

Vernon, Jan (editor). *Electronics Today International Circuits* No. 1, 2 & 3. Sydney, Australia: ETI Publications.

White, Paul (Producer). 1992, *The UFO Mystery.* Network 23 Production: Cosmic Capers Pty Ltd.

Williams, Richard (editor) et al. 1992, *Reader's Digest Quest for the Unknown: UFO The Continuing Enigma.* Sydney: The Reader's Digest Association, Inc.

Yale Scientific Magazine (Yale University). Volume XXXVII, Number 7, April 1963.

Yenne, Bill. 1997, *UFO: Evaluating the Evidence.* Rowayton, CT (USA): Saraband Inc.

Zemansky, Mark W., Sears, Francis W. & Young, Hugh D. 1982, *University Physics*. Sixth Edition. Addison-Wesley Publishing Company.

PICTURE CREDITS

Chapter 1

Mansfield UFO observed in 1973: Drawing by Captain Lawrence Coyne; Disposition of Mansfield UFO Incident: U.S. FoIA with original currently held by Computer UFO Network (CUFON); Artist impression of Levelland UFO: Orbis Publishing Ltd and Marshall Cavendish (UK), *The Unexplained* (1980-84); Photograph of Major-General John A. Samford: USAF; Dr Carl Sagan: Planetary Society (NASA).

Chapter 2

Dr Edward U. Condon: National Institute of Standards and Technology (courtesy U.S. government); Photograph of Hermann Oberth: NASA.

Chapter 3

Ian Ridpath: Photograph by Max Alexander; Ultraviolet image of Venus as seen by the Pioneer Venus Orbiter (February 5, 1979), NASA: http://nssdc.gsfc.nasa.gov/photo_gallery/photogallery-venus.html; IFOs (glass reflections) by Shell R. Alpert: U.S. Department of Homeland Security and U.S. Coast Guard (http://www.uscg.mil/hq/g-cp/history/Chronology_Aug.html) from *The Daily Chronology of Coast Guard History* web page; Nevada IFO (Chemical smudge on negative film) 1965: Project Blue Book Files (courtesy U.S. government); Fake UFO (Car wheel hub caps): Photographed by Paul Villa; McMinnville UFO photographs: Taken by Paul and Evelyn Trent and is in the public domain; UFO drawing from witnesses (cleaned up versions): Antonio Villas Boas and Alfred Burtoo.

Chapter 4

3D UFO and close approximation of landscape from the Alfred Burtoo UFO case: SUNRISE; Photograph of Bertrand Russell (taken in 1916): Unknown photographer (now in the public domain). Artist impression of UFO observed in Texas by Betty Cash and Vickie Landrum in 1980: Illustrated by Sergio Drummond; Artist impression of UFO and occupants in the Inacio de Souza UFO case: Orbis Publishing Ltd and Marshall Cavendish (UK), *The Unexplained* (1980-84). Artist impression of UFO from Finland in 1970: Illustrated by Edu Torres.

Chapter 5

Artist impression of UFOs: Orbis Publishing Ltd and Marshall Cavendish (UK), *The Unexplained* (1980-84); Witness drawing of UFO by Antonio Villas Boas: Flying Saucer Review of West Mailing, Maidstone, Kent, England; Triangular UFO photograph in Petit-Rechain, Belgium, on 7 April 1990 by an anonymous photographer: MUFON-CES; Paul Trent UFO and UFO near Rouen: NICAP; "Dazzling UFOs Caused Mysterious Power Blackouts" newspaper article: Unknown; UFO photograph by Stephen Pratt in March 1966: Fortean Picture Library; Triangular UFO drawing from Puerto Rico in December 1988 by Carlos Mercado: *Enigma*; Betty and Barney Hill Artist Impression: Mary Evans Picture Library. Betty and Barney Hill UFO drawings: Barney Hill; All other UFO drawings by anonymous witnesses; *Roswell Daily Record* front page spread for July 9, 1947: Public domain image.

Chapter 6

Ball Lightning: "Globe of Fire Descending into a Room" in *The Aerial World* by Dr G. Hartwig, London, 1886. p.267. Library Call Number QC863.4 H33 1886. Image ID: libr0524, Treasures of the NOAA Library Collection.

Chapter 7

UFOs over Nuremberg, Germany, in 1561: Woodcut by Glaser Hans and displayed in the Zentrabibliothek Museum in Zurich (copy obtained and converted to EPS by SUNRISE); UFOs over Basel, Switzerland, in 1566: Woodcut by Samuel Coccius (copy obtained and converted to EPS by SUNRISE); Foo-fighters photographs: Anonymous U.S. military pilots; Major Jesse Marcel: U.S. government; Colonel William H. Blanchard: USAF http://www.af.mil/information/bios/bio.asp?bioID=4708; Newspaper article of Fraud Committed by Newton and GeBauer, 14 October 1952: Denver Post; Photograph of Donald Keyhoe: Fortean Picture Library; *The Flying Saucers are Real* by Donald Keyhoe book cover: SUNRISE; Vice Admiral Roscoe H. Hillenkoetter in 1957: Wikipedia (https://commons.wikimedia.org/wiki/File:Roscoe_H._Hillenkoetter_(1957).jpg); General Hoyt S. Vandenberg taken prior to 1954: Wikipedia (https://commons.wikimedia.org/wiki/File:Hoyt_S_Vandenberg.jpg); Harry S. Truman: Unknown photographer formerly held by U.S. Department of Defense (public domain image available from Wikipedia); Edward Ruppelt and Project Blue Book team: Fortean Picture Library; Photograph of fake alien (the monkey held by two individuals with two women looking at it): Fortean Picture Library; Donald E. Keyhoe: Fortean Picture Library. Thomas Townsend Brown in his younger years: Courtesy Linda Brown; Marilyn Monroe: Publicity photo in the public domain, taken in 1953; John F. Kennedy: Photograph taken on 20 February 1961 by the White House Press Office (WHPO).

Chapter 8
Light bulb: Mike Bonitz, 18 July 2011; Albert Einstein in his office at the University of Berlin in 1920: Unknown photographer published in *Scientific Monthly*; Max Abraham (circa 1905): Niedersächsische Staats und Universitätsbibliothek, Göttingen; Photograph of standard capacitor: SUNRISE; Photograph of Thomas T. Brown holding an electrokinetic device: Brown's family archive; Charge distribution diagram: SUNRISE; Artist impression of UFO seen by policeman Lonnie Zamora: Top image from unknown artist, and bottom image by Kari Rajanen.

Chapter 9
Cartoons of SETI scientist being visited by aliens and waiting too long for the evidence: Concept by SUNRISE and illustrated by Sergio Drummond.

Appendix A
Photographs of IFOs
Aurora Borealis: L. E. Daniel Larsson taken at Lofoten, Norway, on 27 January 2014; Sun dog: Siri Spjelkavik taken on 10 March 2018; Comet Lulin: Cooleewinds; Lenticular clouds: Marcus Ward taken on 7 March 2015; Birds (Cranes): USFWS Mountain-Prairie; Light flare: Matt Baily taken on 10 November 2015; Fireworks: Brian Uhreen taken on 1 July 2011; Weather balloon: IBM Research taken on 16 August 2016; Kite: Michael Swan taken on 7 May 2017; Drone: Thierry Uhrmann taken on 9 October 2013; Paraglider: Photographed by Sangeet Rao.

Appendix B
Albert Einstein: Photograph by Oren Jack Turner courtesy of the U.S. Library of Congress.

Appendix C
Enrico Fermi: Photograph held by U.S. Department of Energy; Rocket diagram: SUNRISE; Space Shuttle: Anonymous photographer available from Pexels.com; Cosmos 1: Illustration by John Ballentine; Paul Davies: Self-portrait; Nikola Tesla: A photograph of Nikola Tesla (1856-1943) at age 40 taken by an anonymous photographer (in the public domain and available at Wikipedia); RF-to-DC Converter Circuit: Sir Raymond Phillips, Texas, from World Patent Number WO88/00769.

Appendix E
Thomas T. Brown's electrokinetic devices: British and U.S. Patent Offices; Colour diagram of alternative electrokinetic device: SUNRISE.

Appendix F
Photographs from U.S. Army Research Laboratory (ARL).

Appendix H and I
Documents obtained under the U.S. FoI.

Appendix J
Antonio Villas Boas UFO and closest 3D representation of farm landscape where the encounter took place: SUNRISE.

Although every effort has been made to trace all copyright holders, we apologise in advance for any unintended omissions and would be pleased to insert the appropriate acknowledgements in any subsequent edition of this book.

ABOUT THE AUTHOR

SUNRISE Information Services (SUNRISE) is a private research centre aimed at producing original, stable, interesting and easy-to-read educational and research information for the global community, while uncovering new and original knowledge.